Lecture Notes in Mathematics 1850

Editors:
J.–M. Morel, Cachan
F. Takens, Groningen
B. Teissier, Paris

V. D. Milman   G. Schechtman

Geometric Aspects of Functional Analysis

Israel Seminar 2002–2003

2002–2003

 Springer

Editors

Vitali D. Milman

Department of Mathematics
Tel Aviv University
Ramat Aviv
69978 Tel Aviv
Israel

e-mail: milman@post.tau.ac.il
http://www.math.tau.ac.il/~milman

Gideon Schechtman

Department of Mathematics
The Weizmann Institute of Science
P.O. Box 26
76100 Rehovot
Israel

e-mail: gideon.schechtman@weizmann.ac.il
http://www.wisdom.weizmann.ac.il/~gideon

Library of Congress Control Number: 2004108636

Mathematics Subject Classification (2000): 46-06, 46B07, 52-06 60-06

ISSN 0075-8434
ISBN 3-540-22360-6 Springer Berlin Heidelberg New York
DOI: 10.1007/b98686

Springer is a part of Springer Science+Business Media
springeronline.com
© Springer-Verlag Berlin Heidelberg 2004
Printed in Germany

Typesetting: Camera-ready TEX output by the author

Printed on acid-free paper
41/3142/du - 5 4 3 2 1 0

Preface

During the last two decades the following volumes containing the proceedings of the Israel Seminar in Geometric Aspects of Functional Analysis appeared:

1983-1984   Published privately by Tel Aviv University
1985-1986   Springer Lecture Notes, Vol. 1267
1986-1987   Springer Lecture Notes, Vol. 1317
1987-1988   Springer Lecture Notes, Vol. 1376
1989-1990   Springer Lecture Notes, Vol. 1469
1992-1994   Operator Theory: Advances and Applications, Vol. 77, Birkhauser
1994-1996   MSRI Publications, Vol. 34, Cambridge University Press
1996-2000   Springer Lecture Notes, Vol. 1745
2001-2002   Springer Lecture Notes, Vol. 1807.

Of these, the first six were edited by Lindenstrauss and Milman, the seventh by Ball and Milman and the last two by the two of us.

As in the previous volumes, the current one reflects general trends of the Theory. Most of the papers deal with different aspects of Asymptotic Geometric Analysis, which includes classical "Local Theory of Normed Spaces" type questions, but now with the main emphasis on geometric problems. In addition, the volume contains papers on related aspects of Probability, classical Convexity and also Partial Differential Equations and Banach Algebras. Last but not least, the volume contains two expository articles on two subjects which seem at first sight quite far removed from the main topic the Seminar deals with but which proved to be very much connected. One is Statistical Learning Theory and the other is Models of Statistical Physics. We are grateful to the two authors for agreeing to contribute these surveys to our volume, and we very much hope this inclusion will substantially contribute to the development of the connection of these fields with Asymptotic Geometric Analysis.

All the papers here are original research papers (or invited expository papers) and were subject to the usual standards of refereeing.

As in previous proceedings of the GAFA Seminar, we also list here all the talks given in the seminar as well as talks in related workshops and conferences. We believe this gives a sense of the main directions of research in our area.

We are grateful to Ms. Diana Yellin for taking excellent care of the typesetting and some text editing aspects of this volume.

Vitali Milman
Gideon Schechtman

The Start of GAFA Seminar Notes:
Some Memories After 20 Years of Activity

In 1983/1984 Tel Aviv University published the first volume of what became the well-known series of GAFA – Geometric Aspects of Functional Analysis – Israel Seminar Notes. Seven years later the same acronym GAFA was adopted for a new journal, Geometric And Functional Analysis, which was founded by M. Gromov and V. Milman. Now, 20 years on, I would like to relate how and why the GAFA seminar started, and how the journal GAFA came to the same name.

Gilles Pisier once said to me while holding a copy of the 1983/84 volume of GAFA, published privately in Tel Aviv: "It looks exactly like our 'Séminaire D'Analyse Fonctionnelle'; you 'stole' it!". Yes and no. We didn't "steal" it because we purposely copied it (which is not to say that we liked its appearance or not)! The volumes of Seminar Notes of Ecole Polytechnique which were first produced under the title "Séminaire Maurey-Schwartz" and then, when Pisier took over from Maurey, under the title "Séminaire D'Analyse Fonctionnelle", were privately published by the Centre de Mathématiques, Ecole Polytechnique. These volumes played an enormously important role in collecting and spreading the most up-to-date knowledge and information on Geometric Functional Analysis, contributing to important developments in this theory in the 1970's and early 1980's. So, when G. Pisier declared that he was "tired" of producing the Seminar Notes and that the 1980/81 volume had been his last, I was most upset and decided to act. Actually, since the 1980/1981 academic year, Joram Lindenstrauss and I had been holding regular Israeli Seminars in Tel Aviv, the geographical center of Israel. The stream of foreign visitors brought important scientific news, and so the research center for Geometric Functional Analysis shifted to Israel. So, in 1983, Joram Lindenstrauss and I agreed to collect all the talks we had that year and to continue in the French tradition by copying the appearance of "Séminaire D'Analyse Fonctionnelle". I was lucky; an excellent student by the name of Haim Wolfson, was finishing his Ph.D. that year under my supervision. He already had a couple of results, known today to every expert in the field (unfortunately for us, he changed his direction to Computational Biology, and became a top expert in that field). So, Haim Wolfson was put in charge of the actual production (and later the distribution) of the volume. It was his idea to create the name "GAFA". He felt that the name should somehow be connected to Tel Aviv, and GAFA was the closest we could come to Jaffa, the oldest and most historical part of Tel Aviv. I accepted the idea immediately, but realized only later how strong and appealing this short expression is. The production of the first issue was too costly for us. Having no way of funding the next issue, Joram Lindenstrauss and I agreed to turn to Springer Lecture Notes. Today, it seems strange that we were reluctant to do so, and were actually forced into it.

As for the logo of GAFA, it was created by one of the most remarkable and distinguished Israeli painters, Jan Rauchwerger. Though a great painter, he was not a graphic artist, but being my closest friend, it was difficult for him to refuse my request for help. Years later, he also controlled the design of the cover of the journal GAFA, as well as any slides I used, or posters of conferences I organized. Today, his taste can be seen in many places in Mathematics, and some people think it is my taste.

Now, continuing to the journal GAFA, in 1987, a Birkhäuser representative approached me with the suggestion of producing a new journal. It was also around that time, in 1988/89, that Gromov and I decided that it would be worthwhile to create a forum to discuss the results related to what we called "Geometric Analysis". I must emphasize here that our understanding of this combination of words, which we first made up in the mid 1980's, was not the mechanical use of Analysis in Geometry or vice versa. It was the influence of analytical ideas in Geometry on a conceptual level and the asymptotic study of Geometry (asymptotics and estimates are the standard way of thinking in Analysis). The only meaning for this is the asymptotics by dimension, namely, the study of geometric problems in high dimensional spaces, with special emphasis on understanding the asymptotics by dimension. The Concentration Phenomenon and the immense progress in the asymptotic study of normed spaces during the mid 1980's showed that the theory already existed and required attention. To our dismay another group decided to launch a journal with the same name "Geometric Analysis" just six months before our journal was scheduled to start. We tried negotiating but to no avail, and we had to urgently find a new name. I was in Zurich at that time for a 2-month visit and half that time I devoted to trying to resolve the matter. Then "GAFA", a name I already liked very much, sprung to mind and Gromov instantly agreed. The dilemma was resolved and GAFA was born. Of course, this time instead of "Geometric Aspects of Functional Analysis", GAFA became the acronym for "Geometric And Functional Analysis".

Vitali Milman

Contents

A Topological Obstruction to Existence of Quaternionic Plücker Map

S. Alesker

School of Mathematical Sciences, Tel Aviv University, Tel Aviv 69978, Israel
semyon@post.tau.ac.il

Summary. It is shown that there is no continuous map from the quaternionic Grassmannian $^{\mathbb{H}}Gr_{k,n}(k \geq 2, \; n \geq k+2)$ to the quaternionic projective space $\mathbb{H}P^{\infty}$ such that the pullback of the first Pontryagin class of the tautological bundle over $\mathbb{H}P^{\infty}$ is equal to the first Pontryagin class of the tautological bundle over $^{\mathbb{H}}Gr_{k,n}$. In fact some more precise statement is proved.

1 Introduction

This note is a bi-product of an attempt to understand linear algebra over the (noncommutative) field of quaternions $\mathbb{H}$. For the basic material on linear algebra over noncommutative fields we refer to [Art], [GGRW]. For quaternionic linear algebra, see [As], [Al1], [GRW], and for further applications of it to quaternionic analysis, see [Al1], [Al2]. The main results of this note are Theorem 1 and its refinement Theorem 2 below. But first let us discuss what motivated them.

Roughly speaking, it is shown that there is a topological obstruction to fill in the last row in the last column of the following table (compare with the table in [Ar]).

$\mathbb{R}$	$\mathbb{C}$	$\mathbb{H}$
w_1	c_1	p_1
$\mathbb{R}P^{\infty} = K(\mathbb{Z}/2\mathbb{Z}, 1)$	$\mathbb{C}P^{\infty} = K(\mathbb{Z}, 2)$	rat. equiv. $\mathbb{H}P^{\infty} \quad \longrightarrow \quad K(\mathbb{Z}, 4)$
$w_1(V) = w_1(\bigwedge^{top} V)$	$c_1(V) = c_1(\bigwedge^{top} V)$	$p_1(V) = p_1(\,?\,)$

In this table, V denotes a vector bundle (respectively real, complex, or quaternionic) over some base, and $K(\pi, n)$ denotes the Eilenberg–MacLane space.

Consider the category of (say) right $\mathbb{H}$-modules. It is not clear how to define in this category the usual notions of linear algebra-like tensor products, symmetric products, and exterior powers. (However, it was discovered by D. Joyce [Jo] that one can define these notions in the category of right $\mathbb{H}$-modules with some additional structure, namely, with some fixed *real* subspace generating the whole space as $\mathbb{H}$-module. See also [Qu] for further discussions.) We show that in a sense there is a *topological* obstruction to the existence of the maximal exterior power of finite dimensional $\mathbb{H}$-modules.

We will use the following notation. For a field $K = \mathbb{R}, \mathbb{C}, \mathbb{H}$ we will denote by ${}^K Gr_{k,n}$ the Grassmannian of k-dimensional subspaces in K^n, and by ${}^K Gr_{k,\infty}$ the inductive limit $\varinjlim_N {}^K Gr_{k,N}$. The homotopy classes of maps from a topological space X to Y will be denoted by $[X, Y]$. For the field $K = \mathbb{R}$ or $\mathbb{C}$ we will consider the Plücker map

$$\eta : {}^K Gr_{k,\infty} \longrightarrow K P^\infty$$

given by $E \mapsto \bigwedge^k E$.

For $K = \mathbb{R}, \mathbb{C}, \mathbb{H}$ the isomorphism classes of K-vector bundles of rank k over X correspond bijectively to the homotopy classes of maps $[X, {}^K Gr_{k,\infty}]$. In particular, the isomorphism classes of K-line bundles over X are in bijection with the homotopy classes of maps to projective space $[X, K P^\infty]$.

Recall that in the case of commutative field K the k-th exterior power provides a canonical way to produce from a k-dimensional vector space a 1-dimensional one. In particular over the field $K = \mathbb{R}$ or $\mathbb{C}$ this gives a functorial construction of a line bundle from a vector bundle of rank k. Hence we get a map $\xi : [X, {}^K Gr_{k,\infty}] \to [X, K P^\infty]$ which is just the composition of a map from X to ${}^K Gr_{k,\infty}$ with the Plücker map η.

However, the spaces $\mathbb{R}P^\infty$ and $\mathbb{C}P^\infty$ are Eilenberg–MacLane spaces $K(\mathbb{Z}/2\mathbb{Z}, 1)$ and $K(\mathbb{Z}, 2)$, respectively. Hence $[X, \mathbb{R}P^\infty] = H^1(X, \mathbb{Z}/2\mathbb{Z})$, and $[X, \mathbb{C}P^\infty] = H^2(X, \mathbb{Z})$. If $f \in [X, {}^\mathbb{R} Gr_{k,\infty}]$ corresponds to a real vector bundle V over X then $\xi(f)$ corresponds to its first Stiefel–Whitney class $w_1(V) \in H^1(X, \mathbb{Z}/2\mathbb{Z})$. (Similarly if $f \in [X, {}^\mathbb{C} Gr_{k,\infty}]$ corresponds to V then $\xi(f)$ corresponds to its first Chern class $c_1(V) \in H^2(X, \mathbb{Z})$.) Thus the Plücker map $\eta : {}^\mathbb{R} Gr_{k,\infty} \to K(\mathbb{Z}/2\mathbb{Z}, 1) = \mathbb{R}P^\infty$ corresponds to the first Stiefel–Whitney class of the tautological bundle over ${}^\mathbb{R} Gr_{k,\infty}$. Analogously the Plücker map $\eta : {}^\mathbb{C} Gr_{k,\infty} \to K(\mathbb{Z}, 2) = \mathbb{C}P^\infty$ corresponds to the first Chern class of the tautological bundle over ${}^\mathbb{C} Gr_{k,\infty}$. Thus *the homotopy class of the Plücker map over $\mathbb{R}$ (resp. $\mathbb{C}$) is characterized uniquely by the property that the pull-back of the first Stiefel–Whitney (resp. Chern) class of the tautological bundle over $\mathbb{R}P^\infty$ (resp. $\mathbb{C}P^\infty$) is equal to the first Stiefel–Whitney (resp. Chern) class of the tautological bundle over ${}^\mathbb{R} Gr_{k,\infty}$ (resp. ${}^\mathbb{C} Gr_{k,\infty}$).*

One can try to obtain a quaternionic analogue of the last statement. Note first of all that the quaternionic projective space $\mathbb{H}P^\infty$ has the rational homotopy type of the Eilenberg–MacLane space $K(\mathbb{Z}, 4)$, but is not homotopically equivalent to it. More precisely, $\pi_i(\mathbb{H}P^\infty) = 0$ for $i < 4$, $\pi_4(\mathbb{H}P^\infty) = \mathbb{Z}$, and $\pi_i(\mathbb{H}P^\infty)$ is finite for $i > 4$. (This can be seen from the quaternionic Hopf fibration $S^3 \to S^\infty \to \mathbb{H}P^\infty$ and the fact that $\pi_3(S^3) = \mathbb{Z}$ and $\pi_i(S^3)$ is finite for $i > 3$.) By gluing cells, one can construct an embedding $\tau : \mathbb{H}P^\infty \to K(\mathbb{Z}, 4)$, which induces an isomorphism of homotopy groups up to order 4. This map is unique up to homotopy.

One can easily see that this map corresponds to the first Pontryagin class of the tautological bundle over $\mathbb{H}P^\infty$ (recall that $[X, K(\mathbb{Z}, 4)] = H^4(X, \mathbb{Z})$). On the other hand, the first Pontryagin class of the tautological bundle over the quaternionic Grassmannian ${}^{\mathbb{H}}Gr_{k,\infty}$ defines a map $\omega : {}^{\mathbb{H}}Gr_{k,\infty} \to K(\mathbb{Z}, 4)$ which is unique up to homotopy. Our main result is

Theorem 1. *There is no map $\eta : {}^{\mathbb{H}}Gr_{k,\infty} \to \mathbb{H}P^\infty$, $k \geq 2$, which would make the following diagram commutative:*

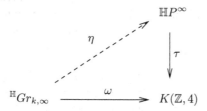

In other words, there is no map $\eta : {}^{\mathbb{H}}Gr_{k,\infty} \to \mathbb{H}P^\infty$ such that the pull-back under η of the first Pontryagin class of the tautological bundle over $\mathbb{H}P^\infty$ is equal to the first Pontryagin class of the tautological bundle over ${}^{\mathbb{H}}Gr_{k,\infty}$.

Remark. It will be clear from the proof of Theorem 1 that there is no map $\eta : {}^{\mathbb{H}}Gr_{k,n} \to \mathbb{H}P^\infty$ for $k \geq 2$, $n \geq k+2$ with the same property (here ${}^{\mathbb{H}}Gr_{k,n}$ denotes the quaternionic Grassmannian of k-subspaces in $\mathbb{H}^n$).

In fact we will prove a slightly more precise statement. Consider the Postnikov tower for $X = \mathbb{H}P^\infty$:

$$K(\mathbb{Z}, 4) = X_4 \leftarrow X_5 \leftarrow X_6 \leftarrow X_7 \leftarrow \cdots .$$

Theorem 2. *Let $k \geq 2$, $\infty \geq n \geq k+2$. The map $\omega : {}^{\mathbb{H}}Gr_{k,n} \to K(\mathbb{Z}, 4)$ corresponding to the first Pontryagin class of the tautological bundle over ${}^{\mathbb{H}}Gr_{k,n}$ can be factorized through X_6, but cannot be factorized through X_7.*

Clearly Theorem 2 implies Theorem 1.

2 Proof of Theorem 2

(1) Let us show that ω cannot be factorized through X_7. We first reduce the statement to the case $k = 2$, $n = 4$.

Let $k \geq 2$, $n \geq k + 2 \geq 4$. Fix a decomposition $\mathbb{H}^n = \mathbb{H}^4 \oplus \mathbb{H}^{n-4}$. Since $n - 4 \geq k - 2$ we can fix a subspace $\mathbb{H}^{k-2} \subset \mathbb{H}^{n-4}$. Consider the embedding i: $^{\mathbb{H}}Gr_{2,4} \hookrightarrow {}^{\mathbb{H}}Gr_{k,n}$ as follows: for every $E \subset \mathbb{H}^4$ let $i(E) = E \oplus \mathbb{H}^{k-2}$. Then the restriction of the topological bundle over $^{\mathbb{H}}Gr_{k,n}$ to $^{\mathbb{H}}Gr_{2,4}$ is equal to the sum of the topological bundle over $^{\mathbb{H}}Gr_{2,4}$ and the trivial bundle $\mathbb{H}^{k-2}$. Hence the Pontryagin classes of these two bundles over $^{\mathbb{H}}Gr_{2,4}$ are equal. Thus it is sufficient to prove the statement for $^{\mathbb{H}}Gr_{2,4}$.

Claim 3. Consider the embedding $i : {}^{\mathbb{H}}Gr_{2,4} \hookrightarrow {}^{\mathbb{H}}Gr_{2,\infty}$. It induces an isomorphism of homotopy groups of order ≤ 7.

Let us prove Claim 3. It is well known (see e.g. [P-R]) that the cell decomposition of quaternionic Grassmannians is given by the quaternionic Schubert cells of real dimensions $0,4,8,12,\ldots$ Moreover i induces an isomorphism of 8-skeletons of both spaces. Hence Claim 3 follows.

Let us denote by Sp_n the group of symplectic matrices of size n, namely, the group of quaternionic matrices of size n satisfying $A^*A = Id$ where $(A^*)_{ij} = \overline{A_{ji}}$ and the bar denotes the quaternionic conjugation.

Claim 4. $\pi_7(^{\mathbb{H}}Gr_{2,4}) = 0$.

By Claim 3 it is sufficient to show that $\pi_7(^{\mathbb{H}}Gr_{2,\infty}) = 0$. Consider the bundle $Sp_2 \to {}^{\mathbb{H}}St_{2,\infty} \to {}^{\mathbb{H}}Gr_{2,\infty}$, where $^{\mathbb{H}}St_{2,\infty}$ is the quaternionic Stiefel manifold, i.e. pairs of orthogonal unit vectors in $\mathbb{H}^\infty$. Since $^{\mathbb{H}}St_{2,\infty}$ is contractible $\pi_7(^{\mathbb{H}}Gr_{2,\infty}) = \pi_6(Sp_2)$. But the last group vanishes by [M-T]. Claim 4 is proved.

Claim 5. $\pi_7(\mathbb{H}P^\infty) \neq 0$.

Indeed using the quaternionic Hopf bundle $S^3 \to S^\infty \to \mathbb{H}P^\infty$ one gets $\pi_7(\mathbb{H}P^\infty) = \pi_6(S^3) = \mathbb{Z}/12\mathbb{Z}$ (see [Hu] Ch. XI.16 for the last equality).

Now let us assume the opposite to our statement, namely, assume that there exists a map $\eta' : {}^{\mathbb{H}}Gr_{2,4} \to X_7$ which makes the following diagram commutative:

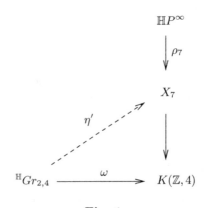

Fig. 1

where ρ_7 is the standard map for the Postnikov system. Let $S \subset {}^{\mathbb{H}}Gr_{2,4}$ denote the 4-skeleton of ${}^{\mathbb{H}}Gr_{2,4}$ consisting of Schubert cells. Then S can be described as follows. Fix a pair $\mathbb{H}^1 \subset \mathbb{H}^3$ inside $\mathbb{H}^4$. Then $S = \{E \in {}^{\mathbb{H}}Gr_{2,4} \mid \mathbb{H}^2 \subset E \subset \mathbb{H}^3\}$. Clearly, $S \simeq \mathbb{H}P^1 \simeq S^4$. Our requirement on $\omega : {}^{\mathbb{H}}Gr_{2,4} \to K(\mathbb{Z}, 4)$ that the pull-back under ω of the fundamental class $\kappa \in H^4(K(\mathbb{Z}, 4), 4)$ is equal to the first Pontryagin class of the topological bundle over ${}^{\mathbb{H}}Gr_{2,4}$, is equivalent to saying that $(\omega \circ j)^*(\kappa) \in H^4(S, \mathbb{Z})$ is the canonical generator of $H^4(S, \mathbb{Z}) = H^4(\mathbb{H}P^1, \mathbb{Z}) = \mathbb{Z}$ (here $j : S \hookrightarrow {}^{\mathbb{H}}Gr_{2,4}$ denotes the identity embedding). This is also equivalent to the fact that $\omega \circ j : S \to K(\mathbb{Z}, 4)$ can be lifted to $h : S \to \mathbb{H}P^\infty$ so that the diagram

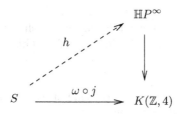

is commutative, and moreover h is homotopic to the composition $S \xrightarrow{id} \mathbb{H}P^1 \hookrightarrow \mathbb{H}P^\infty$, where $\mathbb{H}P^1 \hookrightarrow \mathbb{H}P^\infty$ is the standard embedding. But the embedding $\mathbb{H}P^1 \hookrightarrow \mathbb{H}P^\infty$ induces the *surjective* map $\pi_7(\mathbb{H}P^1) \to \pi_7(\mathbb{H}P^\infty)$ since $\mathbb{H}P^1$ is the 7-skeleton of $\mathbb{H}P^\infty$ under the subdivision to Schubert cells. Let us choose any element $\varphi \in \pi_7(S)$ that is under the composition $S \xrightarrow{id} \mathbb{H}P^1 \hookrightarrow \mathbb{H}P^\infty$ which is mapped to a nonzero element. (Recall that by Claim 4, $\pi_7(\mathbb{H}P^\infty) \neq 0$.) Consider the space $Z := S \cup_\varphi D_8$, where D_8 is the 8-dimensional disk with the boundary $\partial D_8 = S^7$. Since (by Claim 4) $\pi_7({}^{\mathbb{H}}Gr_{2,4}) = 0$ there is a map $f : Z \to {}^{\mathbb{H}}Gr_{2,4}$ such that $f|_S$ is just the identity embedding $S \hookrightarrow {}^{\mathbb{H}}Gr_{2,4}$.

We have assumed that there exists a factorization as in Fig. 1. Consider $\eta' \circ f : Z \to X_7$. The restriction of this map to $S \subset Z$ maps $\varphi \in \pi_7(S)$ to an element of $\pi_7(X_7) = \pi_7(\mathbb{H}P^\infty)$, which is not zero by our choice. But this element must vanish by construction of Z. We get a contradiction.

(2) It remains to prove that there is a factorization of $\omega : {}^{\mathbb{H}}Gr_{k,n} \to K(\mathbb{Z}, 4)$ through X_6. First let us show that ω factorizes through X_5. We have the diagram

$$
\begin{array}{ccc}
& X_5 & \\
& \downarrow & \\
{}^{\mathbb{H}}Gr_{k,n} \xrightarrow{\ \omega\ } & X_4 = K(\mathbb{Z}, 4) \xrightarrow{\ k_4(X)\ } & K(\pi_5, 6)\ ,
\end{array}
$$

where X denotes for brevity $\mathbb{H}P^\infty$, $\pi_5 = \pi_5(X)$, $k_4(X)$ is the 4-th k-invariant of X. The composition $k_4(X) \circ \omega : {}^{\mathbb{H}}Gr_{k,n} \to K(\pi_5, 6)$ is homotopic to the constant map. Indeed $H^6({}^{\mathbb{H}}Gr_{k,n}, \pi_5) = 0$ since all Schubert cells have dimensions divisible by 4. Hence the map ω can be lifted to a map $g : {}^{\mathbb{H}}Gr_{k,n} \to X_5$ such that the following diagram is commutative (see [Mo-Ta], Ch.13):

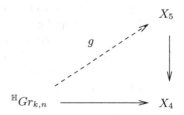

Since $H^7({}^{\mathbb{H}}Gr_{n,k}, \pi_6) = 0$ a similar argument shows that the constructed map g can be lifted to a map $h : {}^{\mathbb{H}}Gr_{k,n} \to X_6$. This map h gives the necessary factorization. $\qquad\square$

Acknowledgement. We would like to thank J. Bernstein, M. Farber, D. Kazhdan, and H. Miller for useful discussions.

References

[Al1]   Alesker, S.: Non-commutative linear algebra and plurisubharmonic functions of quaternionic variables. Bull. Sci. Math., **127**, no. 1, 1–35 (2003). Also: math.CV/0104209

[Al2]   Alesker, S.: Quaternionic Monge-Ampère equations. J. Geom. Anal., **13**, no. 2, 183–216 (2003). Also: math.CV/0208005

[Ar]   Arnold, V.I.: Polymathematics: Is mathematics a single science or a set of arts? In: Arnold, V., Atiyah, M., Lax, P., Mazur, B. (eds) Mathematics: Frontiers and Perspectives. American Mathematical Society, Providence, RI (2000)

[Art]   Artin, E.: Geometric Algebra (reprint of the 1957 original). John Wiley and Sons, Inc., New York (1988)

[As]   Aslaksen, H.: Quaternionic determinants. Math. Intelligencer, **18**, no. 3, 57–65 (1996)

[GGRW]   Gelfand, I., Gelfand, S., Retakh, V., Wilson, R.L.: Quasideterminants. math.QA/0208146

[GRW]   Gelfand, I., Retakh, V., Wilson, R.L.: Quaternionic quasideterminants and determinants. math.QA/0206211

[Hu]   Hu, S.T.: Homotopy theory. Pure and Applied Mathematics, vol. VIII. Academic Press, New York–London (1959)

[Jo]   Joyce, D.: Hypercomplex algebraic geometry. Quart. J. Math. Oxford Ser. (2) **49**, no. 194, 129–162 (1998)

[M-T]   Mimura, M., Toda, H.: Homotopy groups of SU(3), SU(4) and Sp(2). J. Math. Kyoto Univ., **3**, 217–250 (1963/1964)

[Mo-Ta]   Mosher, R.E., Tangora, M.C.: Cohomology Operations and Applications in Homotopy Theory. Harper and Row, Publishers, New York–London (1968)

[P-R]   Pragacz, P., Ratajski, J.: Formulas for Lagrangian and orthogonal degeneracy loci; $\tilde{Q}$-polynomial approach. Compositio Math., **107**, no. 1, 11–87 (1997)

[Qu]   Quillen, D.: Quaternionic algebra and sheaves on the Riemann sphere. Quart. J. Math. Oxford Ser. (2) **49**, no. 194, 163–198 (1998)

Hard Lefschetz Theorem for Valuations and Related Questions of Integral Geometry

S. Alesker

School of Mathematical Sciences, Tel Aviv University, Tel Aviv 69978, Israel
semyon@post.tau.ac.il

Summary. We continue studying the properties of the multiplicative structure on valuations. We prove a new version of the hard Lefschetz theorem for even translation invariant continuous valuations, and discuss the problems of integral geometry that lie behind these properties. Then we formulate a conjectural analogue of this result for odd valuations.

0 Introduction

In this article we continue studying properties of the multiplicative structure on valuations introduced in [A3]. In [A4] we have proven a certain version of the hard Lefschetz theorem for translation invariant even continuous valuations. Its statement is recalled in Section 4 of this article (Theorem 4.1). The main result of this article is Theorem 2.1 where we prove yet another version of it for even valuations which is more closely related to the multiplicative structure. As a consequence, we obtain a version of it for valuations invariant under a compact group acting transitively on the unit sphere and containing the operator $-Id$ (Corollary 2.6). Then in Section 3 we state a conjecture which is an analogue of the hard Lefschetz theorem for odd valuations.

It turns out that behind the hard Lefschetz theorem for even valuations, are results on the Radon and the cosine transforms on the Grassmannians (see the proof of Theorem 2.1 in this article, and also the proof of Theorem 1.1.1 in [A4] which is also a version of the hard Lefschetz theorem). One passes from even valuations to functions on Grassmannians using the Klain–Schneider imbedding (see Section 1). The case of odd valuations turns out to be related to integral geometry on partial flags which seems to be not well understood. On the other hand, various integral geometric transformations on such spaces can be interpreted sometimes as intertwining operators (or their compositions) for certain representations of $GL_n(\mathbb{R})$, and thus can be reduced to a question of representation theory. This point of view was partly used in [AB] in the study of the cosine transform on the Grassmannians, and more explicitly in [A5] in the study of the generalized cosine transform. It would be of interest to understand these problems for odd valuations.

Note also that the connection between the hard Lefschetz theorem for valuations and integral geometry turns out to be useful in both directions. Thus in [A4] it was applied to obtain an explicit classification of unitarily invariant translation invariant continuous valuations.

For a general background on convexity we refer to the book by Schneider [S1]. For the classical theory of valuations we refer to the surveys by McMullen and Schneider [McS] and McMullen [Mc3].

1 Background

In this section we recall some definitions and known results. Let V be a real vector space of finite dimension n. Let $\mathcal{K}(V)$ denote the family of convex compact subsets of V.

Definition 1.1.   1. *A function $\phi : \mathcal{K}(V) \to \mathbb{C}$ is called a valuation if for any K_1, $K_2 \in \mathcal{K}(V)$ such that their union is also convex one has*

$$\phi(K_1 \cup K_2) = \phi(K_1) + \phi(K_2) - \phi(K_1 \cap K_2).$$

   2. *A valuation ϕ is called continuous if it is continuous with respect the Hausdorff metric on $\mathcal{K}(V)$.*
   3. *A valuation ϕ is called translation invariant if $\phi(K + x) = \phi(K)$ for every $x \in V$ and every $K \in \mathcal{K}(V)$.*
   4. *A valuation ϕ is called even if $\phi(-K) = \phi(K)$ for every $K \in \mathcal{K}(V)$.*
   5. *A valuation ϕ is called homogeneous of degree k (or k-homogeneous) if for every $K \in \mathcal{K}(V)$ and every scalar $\lambda \geq 0$, we have $\phi(\lambda \cdot K) = \lambda^k \phi(K)$.*

We will denote by $\mathrm{Val}(V)$ the space of translation invariant continuous valuations on V. Equipped with the topology of uniform convergence on compact subsets of $\mathcal{K}(V)$ it becomes a Fréchet space. We will also denote by $\mathrm{Val}_k(V)$ the subspace of k-homogeneous valuations from $\mathrm{Val}(V)$. We will need the following result due to P. McMullen [Mc1].

Theorem 1.2. ([Mc1]) *The space $\mathrm{Val}(V)$ decomposes as follows:*

$$\mathrm{Val}(V) = \bigoplus_{k=0}^{n} \mathrm{Val}_k(V)$$

where $n = \dim V$.

In particular note that the degree of homogeneity is an integer between 0 and $n = \dim V$. It is known that $\mathrm{Val}_0(V)$ is one-dimensional and is spanned by the Euler characteristic χ, and $\mathrm{Val}_n(V)$ is also one-dimensional and is spanned by a Lebesgue measure [H]. One has a further decomposition with respect to parity:

$$\mathrm{Val}_k(V) = \mathrm{Val}_k^{ev}(V) \oplus \mathrm{Val}_k^{odd}(V),$$

where $\mathrm{Val}_k^{ev}(V)$ is the subspace of even k-homogeneous valuations, and $\mathrm{Val}_k^{odd}(V)$ is the subspace of odd k-homogeneous valuations.

Let us recall the imbedding of the space of valuations into the spaces of functions on partial flags essentially due to D. Klain [K1], [K2] and R. Schneider [S2]. It will be used in Section 2 to reduce the hard Lefschetz theorem for even valuations to integral geometry of Grassmannians.

Let us denote by $Gr_i(V)$ the Grassmannian of real linear i-dimensional subspaces in V. For a manifold X we will denote by $C(X)$ (resp. $C^\infty(X)$) the space of continuous (resp. infinitely smooth) functions on X. Assume now that V is a Euclidean space. Let us describe the imbedding of $\mathrm{Val}_k^{ev}(V)$ into the space of continuous functions $C(Gr_k(V))$ which we call Klain's imbedding. For any valuation $\phi \in \mathrm{Val}_k^{ev}(V)$ let us consider the function on $Gr_k(V)$ given by $L \mapsto \phi(D_L)$ where D_L denotes the unit Euclidean ball inside L. Thus we get a map $\mathrm{Val}_k^{ev}(V) \to C(Gr_k(V))$. The nontrivial fact due to D. Klain [K2] (and heavily based on [K1]) is that this map is injective.

Now we would like to recall the Schneider imbedding of $\mathrm{Val}_k^{odd}(V)$ into the space of functions on a partial flag manifold. Let us denote by $\tilde{F}(V)$ the manifold of pairs (ω, M) where $M \in Gr_{k+1}(V)$, and $\omega \in M$ is a vector of unit length. Let us denote by $C^-(\tilde{F}(V))$ the space of continuous functions on $\tilde{F}(V)$ which change their sign when one replaces ω by $-\omega$. Let us describe the imbedding of $\mathrm{Val}_k^{odd}(V)$ into $C^-(\tilde{F}(V))$ (following [A2]) which we call Schneider's imbedding since its injectivity is an easy consequence of a nontrivial result due to R. Schneider [S2] about characterization of odd translation invariant continuous valuations. Fix a valuation $\phi \in \mathrm{Val}_k^{odd}(V)$. Fix any subspace $M \in Gr_{k+1}(V)$. Consider the restriction of ϕ to M. By a result of P. McMullen [Mc2] any k-homogeneous translation invariant continuous valuation ψ on $(k+1)$-dimensional space M has the following form. There exists a function $f \in C(S(M))$ (here $S(M)$ denotes the unit sphere in M) such that for any subset $K \in \mathcal{K}(M)$

$$\psi(K) = \int_{S(M)} f(\omega) dS_k(K, \omega).$$

Moreover the function f can be chosen to be orthogonal to any linear functional (with respect to the Haar measure on the sphere $S(M)$), and after this choice it is defined uniquely. We will always make such a choice of f. If the valuation ψ is odd then the function f is also odd. Thus applying this construction to $\phi|_M$ for any $M \in Gr_{k+1}(V)$ we get a map $\mathrm{Val}_{k+1}^{odd}(V) \to C^-(\tilde{F}(V))$ defined by $\phi \mapsto f$. This map turns out to be continuous and injective (see [A2] Proposition 2.6, where the injectivity is heavily based on [S2]).

Let us recall the definition of the Radon transform on the Grassmannians. The orthogonal group acts transitively on $Gr_i(V)$, and there exists a unique $O(n)$-invariant probability measure (the Haar measure).

The Radon transform $R_{j,i} : Gr_i(V) \to Gr_j(V)$ for $j < i$ is defined by $(R_{j,i}f)(H) = \int_{F \supset H} f(F) \cdot dF$. Similarly for $j > i$, it is defined by $(R_{j,i}f)(H) = \int_{F \subset H} f(F) \cdot dF$. In both cases the integration is with respect to the invariant probability measure on all subspaces containing (or contained in) the given one. The Radon transform on real Grassmannians was studied in [GGR], [Gri].

Recall now the definition of the cosine and sine of the angle between two subspaces. Let $E \in Gr_i(V)$, $F \in Gr_j(V)$. Assume that $i \leq j$. Let us call *cosine of the angle* between E and F the following number:

$$| \cos(E, F) | := \frac{\mathrm{vol}_i(Pr_F(A))}{\mathrm{vol}_i(A)},$$

where A is any subset of E of non-zero volume, Pr_F denotes the orthogonal projection onto F, and vol_i is the i-dimensional measure induced by the Euclidean metric. (Note that this definition does not depend on the choice of a subset $A \subset E$.) In the case $i \geq j$ we define the cosine of the angle between them as cosine of the angle between their orthogonal complements:

$$| \cos(E, F) | := | \cos(E^\perp, F^\perp) |.$$

(It is easy to see that if $i = j$ both definitions are equivalent.)

Let us call *sine of the angle* between E and F the cosine between E and the orthogonal complement of F:

$$| \sin(E, F) | := | \cos(E, F^\perp) |.$$

The following properties are well known (and rather trivial):

$$| \cos(E, F) | = | \cos(F, E) | = | \cos(E^\perp, F^\perp) |,$$

$$| \sin(E, F) | = | \sin(F, E) | = | \sin(E^\perp, F^\perp) |,$$

$$0 \leq | \cos(E, F) |, \; | \sin(E, F) | \leq 1.$$

For any $1 \leq i, j \leq n - 1$ one defines the cosine transform

$$T_{j,i} : C\big(Gr_i(V)\big) \to C\big(Gr_j(V)\big)$$

as follows:

$$(T_{j,i}f)(E) := \int_{Gr_i(V)} | \cos(E, F) | f(F) dF,$$

where the integration is with respect to the Haar measure on the Grassmannian. The cosine transform was studied in [M1], [M2], [GoH1], [GoH2], [GoHR], [AB].

Let us state some facts on the multiplicative structure on valuations. Let us briefly recall a construction of multiplication from [A3], Section 1. A measure μ on a linear space V is called polynomial if it is absolutely continuous with respect to a Lebesgue measure, and the density is a polynomial. Let μ and

ν be two polynomial measures on V (the case of Lebesgue measures would be sufficient for the purposes of this article). Let A, $B \in \mathcal{K}(V)$. Consider the valuations $\phi(K) = \mu(K + A)$, $\psi(K) = \nu(K + B)$ where "+" denotes the Minkowski sum of convex sets. Let $\Delta : V \hookrightarrow V \times V$ denote the diagonal imbedding. Then the product of valuations is computed as follows:

$$(\phi \cdot \psi)(K) = (\mu \boxtimes \nu)(\Delta(K) + (A \times B))$$

where $\mu \boxtimes \nu$ denotes the usual product measure. Product of linear combinations of measures of the above form is defined by distributivity. The product turns out to be well defined and it is determined uniquely by the above expressions since the valuations of the above form are dense in the space of polynomial valuations with respect to some natural topology.

We will discuss now only the translation invariant case though in [A3] the product was defined on a wider class of polynomial valuations. Actually in this article we will need only the following result.

Proposition 1.3. *Let $p_1 : V \to W_1$ and $p_2 : V \to W_2$ be surjective linear maps. Let μ_1 and μ_2 be Lebesgue measures on W_1 and W_2, respectively. Consider on V the valuations $\phi_i(K) := \mu_i(p_i(K))$, $i = 1, 2$. Then*

$$(\phi_1 \cdot \phi_2)(K) = (\mu_1 \boxtimes \mu_2)\big((p_1 \oplus p_2)(K)\big)$$

where $p_1 \oplus p_2 : V \to W_1 \oplus W_2$ is given by $(p_1 \oplus p_2)(v) = (p_1(v), p_2(v))$ and $\mu_1 \boxtimes \mu_2$ is the usual product measure on $W_1 \oplus W_2$.

Proof. First recall the following well-known formula (which can be easily checked). Let I be a segment of unit length in the Euclidean space V orthogonal to a hyperplane H. Then one has

$$\frac{\partial}{\partial \lambda}\Big|_0 \mathrm{vol}(K + \lambda I) = \mathrm{vol}_{n-1}(Pr_H K)$$

where Pr_H denotes the orthogonal projection onto H. Now let $I_1, \ldots, I_k$ be pairwise orthogonal unit segments in V. Let L be the orthogonal complement to their span. Thus $\dim L = n - k$. By the inductive application of the above formula one obtains

$$\frac{\partial^k}{\partial \lambda_1 \ldots \partial \lambda_k}\Big|_0 \mathrm{vol}_n\Big(K + \sum_{j=1}^{k} \lambda_j I_j \Big) = \mathrm{vol}_{n-k}(Pr_L K).$$

Let us return to the situation of our proposition. We may assume that W_i, $i = 1, 2$, are subspaces of V and p_i are orthogonal projections onto them. We may also assume that the measures μ_i coincide with the volume forms induced by the Euclidean metric. Let us fix $I_1^{(i)}, \ldots, I_{m_i}^{(i)}$, $i = 1, 2$, pairwise orthogonal unit segments in the orthogonal complement to W_i. Then

$$\phi_i(K) = \frac{\partial^{m_i}}{\partial\lambda_1\ldots\partial\lambda_{m_i}}\bigg|_0 \mathrm{vol}_n\left(K + \sum_{j=1}^{m_i}\lambda_j I_j^{(i)}\right).$$

Hence using the construction of the product described above we get

$$(\phi_1\cdot\phi_2)(K) = \frac{\partial^{m_1}}{\partial\lambda_1\ldots\partial\lambda_{m_1}}\frac{\partial^{m_2}}{\partial\mu_1\ldots\partial\mu_{m_2}}\bigg|_0$$

$$\mathrm{vol}_{2n}\left(\Delta(K) + \sum_{j=1}^{m_1}\lambda_j(I_j^{(1)}\times 0) + \sum_{l=1}^{m_2}\mu_l(0\times I_l^{(2)})\right)$$

$$= \mathrm{vol}_{2n-m_1-m_2}\left(Pr_{W_1\oplus W_2}(\Delta(K))\right)$$

$$= \mathrm{vol}_{2n-m_1-m_2}\left((p_1\oplus p_2)(K)\right). \qquad\qquad \square$$

2 Hard Lefschetz Theorem for Even Valuations

To formulate our main result let us recall a definition from the representation theory. Let G be a Lie group. Let ρ be a continuous representation of G in a Fréchet space F. A vector $v \in F$ is called *G-smooth* if the map $G \longrightarrow F$ defined by $g \longmapsto g(v)$ is infinitely differentiable. It is well known (and easy to prove) that smooth vectors form a linear G-invariant subspace which is dense in F. We will denote it by F^{sm}. It is well known (see e.g. [Wa]) that F^{sm} has a natural structure of a Fréchet space, and the representation of G in F^{sm} is continuous with respect to this topology. Moreover $(F^{sm})^{sm} = F^{sm}$. In our situation the Fréchet space is $F = \mathrm{Val}(V)$ with the topology of uniform convergence on compact subsets of $\mathcal{K}(V)$, and $G = GL(V)$. The action of $GL(V)$ on $\mathrm{Val}(V)$ is the natural one, namely, for any $g \in GL(V)$, $\phi \in \mathrm{Val}(V)$ one has $(g(\phi))(K) = \phi(g^{-1}K)$.

Theorem 2.1. *Let $0 \leq i < n/2$. Then the multiplication by $(V_1)^{n-2i}$ induces an isomorphism $\mathrm{Val}_i^{ev}(V)^{sm}\tilde{\rightarrow}\mathrm{Val}_{n-i}^{ev}(V)^{sm}$. In particular for $p \leq n - 2i$ the multiplication by $(V_1)^p$ is an injection from $\mathrm{Val}_i^{ev}(V)^{sm} \hookrightarrow \mathrm{Val}_{i+p}^{ev}(V)^{sm}$.*

Remark 2.2. We would like to explain the use of the name "hard Lefschetz theorem". The classical hard Lefschetz theorem is as follows (see e.g. [GrH]). Let M be a compact Kähler manifold of complex dimension n with Kähler form ω. Let $[\omega] \in H^2(M,\mathbb{R})$ be the corresponding cohomology class. Then for $0 \leq i < n$ the multiplication by $[\omega]^{n-i}$ induces an isomorphism $H^i(M,\mathbb{R})\tilde{\rightarrow}H^{2n-i}(M,\mathbb{R})$.

First recall that in [A3], Theorem 2.6, we have shown that $(V_1)^j$ is proportional to V_j with a non-zero constant of proportionality. Recall also the Cauchy–Kubota formula (see e.g. [SW]):

$$V_k(K) = c\int_{E\in Gr_k(V)} \mathrm{vol}_k(Pr_E K)dE.$$

Lemma 2.3. *Let $F \in Gr_i(V)$. Let $\phi(K) = \mathrm{vol}_i(Pr_F(K))$. Then the image of $V_k \cdot \phi$ in $C(Gr_{k+i})$ under the Klain imbedding is given by the function*

$$g(L) = c \int_{Gr_{k+i}(V) \ni R \supset F} |\cos(L, R)| dR$$

where dR is the (unique) $O(i) \times O(n-i)$-invariant probability measure on the Grassmannian of $(k + i)$-subspaces in V containing F, and c is a non-zero normalizing constant.

Proof. Using the Cauchy–Kubota formula and Proposition 1.3 we have

$$(V_k \cdot \phi)(K) = c \int_{E \in Gr_k(V)} \mathrm{vol}_{k+i}\big((Pr_F \oplus Pr_E)(K)\big) dE.$$

Let $K = D_L$ be the unit ball in a subspace $L \in Gr_{k+i}(V)$. Then the image of $V_k \cdot \phi$ in $Gr_{k+i}(V)$ under the Klain imbedding is

$$g(L) = (V_k \cdot \phi)(D_L) = c \int_{E \in Gr_k(V)} \mathrm{vol}_{k+i}\big((Pr_E \oplus Pr_F)(D_L)\big) dE.$$

Claim 2.4.

$$\mathrm{vol}_{k+i}\big((Pr_E \oplus Pr_F)(D_L)\big) = \kappa \big|\cos\big(L, (E + F)\big)\big| \cdot |\sin(E, F)|$$

where κ is a non-zero constant.

Let us postpone the proof of this claim and let us finish the proof of Lemma 2.3. We get

$$g(L) = c' \int_{E \in Gr_k(V)} \big|\cos\big(L, (E + F)\big)\big| \cdot |\sin(E, F)| dE.$$

For a fixed subspace $F \in Gr_i(V)$ let us denote by U_F the open dense subset of $Gr_k(V)$ consisting of subspaces intersecting F trivially. Clearly the complement to U_F has smaller dimension. We have the map $T : U_F \to Gr_k(V/F)$ given by $T(E) := (E+F)/F$. Clearly T commutes with the action of the stabilizer of F in the orthogonal group $O(n)$ (which is isomorphic to $O(i) \times O(n-i)$). Let $m : Gr_k(V/F) \to Gr_{k+i}(V)$ be the map sending a subspace in V/F to its preimage in V under the canonical projection $V \to V/F$. Then in this notation we have

$$g(L) = c' \int_{E \in U_F} \big|\cos\big(L, (m \circ T)(E)\big)\big| \cdot |\sin(E, F)| dE.$$

Let us consider the following submanifold $M \subset Gr_k(V/F) \times Gr_k(V)$ given by

$$M = \{(R, E)| E \subset m(R)\}.$$

Then U_F is isomorphic to M via the map $T_1 : U_F \to M$ defined by $T_1(E) = (T(E), E)$. Clearly T_1 commutes with the natural action of the group $O(i) \times O(n - i)$ which is the stabilizer of F in $O(n)$. We have also the projection $t : M \to Gr_k(V/F)$ given by $t(R, E) = R$. Then

$$g(L) = c' \int_{R \in Gr_k(V/F)} dR \big| \cos\big(L, m(R)\big) \big| \left[\int_{E \in Gr_k\left(t^{-1}(R)\right)} |\sin(E, F)| d\mu_R(E) \right]$$

where μ_R is a measure on $Gr_k(t^{-1}(R))$. Note that the integral in square brackets in the last expression is positive and does not depend on R since the map $T = t \circ T_1$ commutes with the action of $O(i) \times O(n - i)$. Hence $g(L) = c'' \int_{R \in Gr_k(V/F)} |\cos(L, m(R))| dR$. Lemma 2.3 is proved.   □

Proof of Claim 2.4. We may assume that $E \cap F = 0$. Set for brevity $p := Pr_E \oplus Pr_F$. Then p factorizes as $p = q \circ Pr_{E+F}$ where $q : E + F \to E \oplus F$ is the restriction of $Pr_E \oplus Pr_F$ to the subspace $E + F$. Then we have

$$\text{vol}_{k+i}\big((Pr_E \oplus Pr_F)(D_L)\big)$$
$$= \text{vol}_{k+i}\big(Pr_{E+F}(D_L)\big) \cdot \frac{\text{vol}_{k+i}(q(D_{E+F}))}{\text{vol}_{k+i}(D_{E+F})}$$
$$= \text{vol}_{k+i}(D_L) \cdot \big| \cos\big(L, (E + F)\big) \big| \cdot \frac{\text{vol}_{k+i}(q(D_{E+F}))}{\text{vol}_{k+i}(D_{E+F})}.$$

Let us compute the last term in the above expression. Note that the dual map $q^* : E \oplus F \to E + F$ is given by $q^*((x, y)) = x + y$. Clearly

$$\frac{\text{vol}(q^*(D_{E \oplus F}))}{\text{vol} D_{E \oplus F}} = |\sin(E, F)|.$$

But the left hand side in the last expression is equal to $\frac{\text{vol}(q(D_{E+F}))}{\text{vol}(D_{E+F})}$. Thus Claim 2.4 is proved.   □

We will need one more lemma.

Lemma 2.5. *Let $\phi \in \text{Val}_i^{ev}(V)$ be a valuation given by*

$$\phi(K) = \int_{F \in Gr_i(V)} f(F) \text{vol}_i(Pr_F K) dF$$

with $f \in C(Gr_i(V))$. Then the image of $V_k \cdot \phi$ in $C(Gr_{i+k}(V))$ under the Klain imbedding is given by

$$g(L) = c T_{k+i, k+i} \circ R_{k+i, i}(f)$$

where c is a non-zero normalizing constant.

Proof. By Lemma 2.3 one has

$$g(L) = c \int_{F \in Gr_i(V)} dF f(F) \int_{Gr_{k+i}(V) \ni R \supset F} |\cos(L, R)| dR$$

$$= c \int_{R \in Gr_{k+i}(V)} dR |\cos(L, R)| \int_{F \in Gr_i(R)} f(F) dF$$

$$= c T_{k+i, k+i} \circ R_{k+i, i}(f).$$ □

Proof of Theorem 2.1. Let us consider the Klain imbedding $\mathrm{Val}_i^{ev}(V)^{sm} \hookrightarrow C^\infty(Gr_l(V))$. In [AB] it was shown that the image of $\mathrm{Val}_l^{ev}(V)^{sm}$ is a closed subspace which coincides with the image of the cosine transform $T_{l,l} : C^\infty(Gr_l(V)) \to C^\infty(Gr_l(V))$. By Lemma 2.5, $V_{n-2i} \cdot T_{i,i}(f) = T_{n-i, n-i} \circ R_{n-i, i}(f)$. It was shown in [GGR] that $R_{n-i, i} : C^\infty(Gr_i(V)) \to C^\infty(Gr_{n-i}(V))$ is an isomorphism. Hence the image under the Klain imbedding of $V_{n-2i} \cdot \mathrm{Val}_i^{ev}(V)^{sm}$ coincides with the image of $T_{n-i, n-i} : C^\infty(Gr_{n-i}) \to C^\infty(Gr_{n-i})$ which is equal to $\mathrm{Val}_{n-i}^{ev}(V)^{sm}$. Hence the multiplication by $V_1^{n-2i} : \mathrm{Val}_i^{ev}(V)^{sm} \to \mathrm{Val}_{n-i}^{ev}(V)^{sm}$ is onto. Let us check that the kernel is trivial. Indeed this operator commutes with the action of the orthogonal group $O(n)$. The spaces $\mathrm{Val}_i^{ev}(V)^{sm}$ and $\mathrm{Val}_{n-i}^{ev}(V)^{sm}$ are isomorphic as representations of $O(n)$ (see [A4], Theorem 1.2.2). Since every irreducible representation of $O(n)$ enters with finite multiplicity (in fact, at most 1) then any surjective map must be injective.   □

Now let us discuss an application to valuations invariant under a group. Let G be a compact subgroup of the orthogonal group $O(n)$. Let us denote by $\mathrm{Val}^G(V)$ the space of G-invariant translation invariant continuous valuations. Assume that G acts transitively on the unit sphere. Then it was shown in [A1] that $\mathrm{Val}^G(V)$ is finite dimensional. Also it was shown in [A4] (see the proof of Corollary 1.1.3) that $\mathrm{Val}^G(V) \subset \mathrm{Val}^{sm}(V)$. We have also McMullen's decomposition with respect to the degree of homogeneity:

$$\mathrm{Val}^G(V) = \bigoplus_{i=0}^n \mathrm{Val}_i^G(V)$$

where $\mathrm{Val}_i^G(V)$ denotes the subspace of i-homogeneous G-invariant valuations. Then it was shown in [A3], Theorem 0.9 that $\mathrm{Val}^G(V)$ is a finite dimensional graded algebra (with grading given by the degree of homogeneity) satisfying the Poincaré duality (i.e. it is a so-called Frobenius algebra). A version of the hard Lefschetz theorem was given in [A4]. Now we would like to state another version of it.

Corollary 2.6. *Let G be a compact subgroup of the orthogonal group $O(n)$ acting transitively on the unit sphere. Assume that $-Id \in G$. Let $0 \le i < n/2$ where $n = \dim V$. Then the multiplication by $(V_1)^{n-2i}$ induces an isomorphism $\mathrm{Val}_i^G(V) \tilde{\to} \mathrm{Val}_{n-i}^G(V)$. In particular for $p \le n - 2i$ the multiplication by $(V_1)^p$ induces an injection $\mathrm{Val}_i^G(V) \hookrightarrow \mathrm{Val}_{i+p}^G(V)$.*

This result follows immediately from Theorem 2.1.

3 The Case of Odd Valuations

In this section we state the conjecture about an analogue of the hard Lefschetz theorem for odd valuations and discuss it briefly.

Conjecture. Let $0 \leq i < n/2$. Then the multiplication by $(V_1)^{n-2i}$ is a map $\mathrm{Val}_i^{odd}(V)^{sm} \to \mathrm{Val}_{n-i}^{odd}(V)^{sm}$ with trivial kernel and dense image. In particular for $p \leq n - 2i$ the multiplication by $(V_1)^p$ is an injection from $\mathrm{Val}_i^{odd}(V)^{sm} \hookrightarrow \mathrm{Val}_{i+p}^{odd}(V)^{sm}$.

Let us show that it is enough to prove only either injectivity or density of the image of the multiplication by $(V_1)^{n-2i}$. This is a particular case of the following slightly more general statement.

Proposition 3.1. *Let* $L : \mathrm{Val}_i^{odd}(V)^{sm} \to \mathrm{Val}_{n-i}^{odd}(V)^{sm}$ *be a continuous linear operator commuting with the action of the orthogonal group* $O(n)$. *Then it has a dense image if and only if its kernel is trivial.*

Proof. Note that the spaces $\mathrm{Val}_i^{odd}(V)^{sm}$ and $(\mathrm{Val}_{n-i}^{odd}(V)^*)^{sm} \otimes \mathrm{Val}_n(V)$ are isomorphic as representation of the full linear group $GL(V)$ by the Poincaré duality proved in [A3]. Let us replace all spaces by their Harish–Chandra modules. All irreducible representations of $O(n)$ are selfdual. Hence the subspaces of $O(n)$-finite vectors in $\mathrm{Val}_i^{odd}(V)^{sm}$ and $\mathrm{Val}_{n-i}^{odd}(V)^{sm}$ are isomorphic as $O(n)$-modules. Moreover each irreducible representation of $O(n)$ enters into these subspaces with finite multiplicity (since by [A2] both spaces are realized as subquotients in so-called degenerate principal series representations of $GL(V)$, namely, representations induced from a character of a parabolic subgroup). Then it is clear that the operator L between the spaces of $O(n)$-finite vectors commuting with the action of $O(n)$ is surjective if and only if it is injective. $\square$

4 Hard Lefschetz Theorem for Even Valuations from [A4]

In this short section we recall another form of the hard Lefschetz theorem for even valuations as was proven in [A4]. Note that it was also heavily based on the properties of the Radon and cosine transforms on Grassmannians.

Let us fix on V a Euclidean metric, and let D denote the unit Euclidean ball with respect to this metric. Let us define on the space of translation invariant continuous valuations an operation Λ of mixing with the Euclidean ball D, namely,

$$(\Lambda\phi)(K) := \frac{d}{d\varepsilon}\Big|_{\varepsilon=0} \phi(K + \varepsilon D)$$

for any convex compact set K. Note that $\phi(K + \varepsilon D)$ is a polynomial in $\varepsilon \geq 0$ by McMullen's theorem [Mc1]. It is easy to see that the operator Λ preserves parity and decreases the degree of homogeneity by one. In particular we have

$$\Lambda : \mathrm{Val}_k^{ev} \to \mathrm{Val}_{k-1}^{ev}(V).$$

The following result is Theorem 1.1.1 in [A4].

Theorem 4.1. *Let $n \geq k > n/2$. Then $\Lambda^{2k-n} : (\mathrm{Val}_k^{ev})^{sm} \to (\mathrm{Val}_{n-k}^{ev}(V))^{sm}$ is an isomorphism. In particular $\Lambda^i : \mathrm{Val}_k^{ev} \to \mathrm{Val}_{k-i}^{ev}(V)$ is injective for $1 \leq i \leq 2n - k$.*

References

[A1]   Alesker, S.: On P. McMullen's conjecture on translation invariant valuations. Adv. Math., **155**, no. 2, 239–263 (2000)

[A2]   Alesker, S.: Description of translation invariant valuations on convex sets with solution of P. McMullen's conjecture. Geom. Funct. Anal., **11**, no. 2, 244–272 (2001)

[A3]   Alesker, S.: The multiplicative structure on polynomial continuous valuations. Geom. Funct. Anal., to appear. Also: math.MG/0301148

[A4]   Alesker, S.: Hard Lefschetz theorem for valuations, complex integral geometry, and unitarily invariant valuations. J. Differential Geom., to appear. Also: math.MG/0209263

[A5]   Alesker, S.: The α-cosine transform and intertwining integrals on real Grassmannians. Preprint

[AB]   Alesker, S., Bernstein, J.: Range characterization of the cosine transform on higher Grassmannians. Adv. Math., to appear. Also: math.MG/0111031

[GGR]   Gelfand, I.M., Graev, M.I., Roşu, R.: The problem of integral geometry and intertwining operators for a pair of real Grassmannian manifolds. J. Operator Theory, **12**, no. 2, 359–383 (1984)

[GoH1] Goodey, P., Howard, R.: Processes of flats induced by higher-dimensional processes. Adv. Math., **80**, no. 1, 92–109 (1990)

[GoH2] Goodey, P.R., Howard, R.: Processes of flats induced by higher-dimensional processes. II. Integral Geometry and Tomography (Arcata, CA, 1989), 111–119, Contemp. Math., 113. Amer. Math. Soc., Providence, RI (1990)

[GoHR] Goodey, P., Howard, R., Reeder, M.: Processes of flats induced by higher-dimensional processes. III. Geom. Dedicata, **61**, no. 3, 257–269 (1996)

[GrH]   Griffiths, P., Harris, J.: Principles of algebraic geometry. Pure and Applied Mathematics, John Wiley and Sons, New York (1978)

[Gri]   Grinberg, E.L.: Radon transforms on higher Grassmannians. J. Differential Geom., **24**, no. 1, 53–68 (1986)

[H]   Hadwiger, H.: Vorlesungen über Inhalt, Oberfläche und Isoperimetrie (German). Springer-Verlag, Berlin-Göttingen-Heidelberg (1957)

[K1]   Klain, D.: A short proof of Hadwiger's characterization theorem. Mathematika, **42**, no. 2, 329–339 (1995)

[K2]   Klain, D.: Even valuations on convex bodies. Trans. Amer. Math. Soc., **352**, no. 1, 71–93 (2000)

[M1]   Matheron, G.: Un théorème d'unicité pour les hyperplans poissoniens (French). J. Appl. Probability, **11**, 184–189 (1974).

[M2]   Matheron, G.: Random sets and integral geometry. Probability and Mathematical Statistics, John Wiley and Sons, New York-London-Sydney (1975)

[Mc1]   McMullen, P.: Valuations and Euler-type relations on certain classes of convex polytopes. Proc. London Math. Soc. (3) **35**, no. 1, 113–135 (1977)

[Mc2]   McMullen, P.: Continuous translation-invariant valuations on the space of compact convex sets. Arch. Math. (Basel) **34**, no. 4, 377-384 (1980)

[Mc3]   McMullen, P.: Valuations and dissections. Handbook of Convex Geometry, Vol. A, B, 933–988. North-Holland, Amsterdam (1993)

[McS]   McMullen, P., Schneider, R.: Valuations on convex bodies. Convexity and its Applications, 170–247. Birkhäuser, Basel (1983)

[S1]   Schneider, R.: Convex bodies: the Brunn–Minkowski theory. Encyclopedia of Mathematics and its Applications, **44**. Cambridge University Press, Cambridge (1993)

[S2]   Schneider, R.: Simple valuations on convex bodies. Mathematika, **43**, no. 1, 32–39 (1996)

[SW]   Schneider, R., Weil, W.: Integralgeometrie (German) [Integral geometry]. Teubner Skripten zur Mathematischen Stochastik [Teubner Texts on Mathematical Stochastics]. B.G. Teubner, Stuttgart (1992)

[Wa]   Wallach, N.R.: Real reductive groups. I. Pure and Applied Mathematics, 132. Academic Press, Inc., Boston, MA (1988)

$SU(2)$-Invariant Valuations

S. Alesker

School of Mathematical Sciences, Tel Aviv University, Tel Aviv 69978, Israel
semyon@post.tau.ac.il

Summary. We obtain an explicit classification of $SU(2)$-invariant translation invariant continuous valuations on $\mathbb{C}^2 \simeq \mathbb{R}^4$.

In this paper we obtain explicit classification of $SU(2)$-invariant translation invariant continuous valuations on $\mathbb{C}^2 \simeq \mathbb{R}^4$. The main result is Theorem 1 below. Let us recall the relevant definitions. Let V be a finite dimensional real vector space. Let $\mathcal{K}(V)$ denote the class of all convex compact subsets of V.

Definition. a) *A function* $\phi : \mathcal{K}(V) \longrightarrow \mathbb{C}$ *is called a valuation if for any* $K_1, K_2 \in \mathcal{K}(V)$ *such that their union is also convex one has*

$$\phi(K_1 \cup K_2) = \phi(K_1) + \phi(K_2) - \phi(K_1 \cap K_2).$$

b) *A valuation* ϕ *is called continuous if it is continuous with respect to the Hausdorff metric on* $\mathcal{K}(V)$.

Recall that the Hausdorff metric d_H on $\mathcal{K}(V)$ depends on the choice of Euclidean metric on V and is defined as follows: $d_H(A, B) := \inf\{\varepsilon > 0 | A \subset (B)_\varepsilon \text{ and } B \subset (A)_\varepsilon\}$, where $(U)_\varepsilon$ denotes the ε-neighborhood of a set U. Then $\mathcal{K}(V)$ becomes a locally compact space (by the Blaschke selection theorem).

For the classical theory of valuations we refer to the surveys McMullen–Schneider [McMS], McMullen [McM3]. For the classical convexity, particularly for the notion of intrinsic volumes which will be used in this paper, we refer to Schneider's book [S]. The theory of continuous translation invariant valuations has numerous applications to integral geometry, see e.g. Klain–Rota [KR], Alesker [A3].

For applications to integral geometry, classifications of special classes of valuations are of particular importance. The famous classification of $SO(n)$- (or $O(n)$-)invariant translation invariant continuous valuations on an n-dimensional Euclidean space was obtained by Hadwiger [H]. More recently the author [A3] obtained an explicit classification of $U(n)$-invariant translation invariant continuous valuations on an n-dimensional Hermitian space (for $n = 2$ this classification was obtained earlier by the author in [A1]). The paper [A3] also contains applications to complex integral geometry.

Recently, Tasaki [T] obtained integral geometric results on $\mathbb{C}^2(\simeq \mathbb{H}^1 \simeq \mathbb{R}^4)$ corresponding to the group $SU(2)(\simeq Sp(1))$. Our article was partly motivated by his work. Note that the space of $U(2)$-invariant translation invariant continuous valuations is 6-dimensional by [A1]. If one replaces the group $U(2)$ by the group $SU(2)$ then the space of valuations is 10-dimensional as follows from Theorem 1 of this paper.

The proofs of this article are heavily based on the description of K-type structure of the space of even translation invariant continuous valuations in $\mathbb{R}^4$ obtained in [A1] (in higher dimensions it was obtained in [A2]). Note also that, in turn, the above mentioned description of the K-type structure of even valuations used earlier results of Beilinson–Bernstein [BB], Goodey–Weil [GW], Howe–Lee [HoL], Klain [K1], [K2], McMullen [McM1], [McM2].

The description of the K-type structure implies easily the computation of the dimension of the space of $SU(2)$-invariant valuations. So the main point of this paper article is to exhibit an explicit basis in geometric terms. However we do not claim that the basis we describe in Theorem 1 below is distinguished from any point of view. It is quite possible that there exists a better choice of a basis.

To formulate our main result let us fix some notation. Let us identify $\mathbb{C}^2 \simeq \mathbb{R}^4$ with the quaternionic line $\mathbb{H}$. On $\mathbb{H}$ one has a family of complex structures parameterized by 2-dimensional sphere. Namely such complex structures are given by multiplication on the right by quaternions of the form $ai + bj + ck$ with $a^2 + b^2 + c^2 = 1$ (note that the square of each such quaternion is equal to -1). Let us denote by Q the group of norm one quaternions. Then Q acts on $\mathbb{H}$ by left multiplication. This gives a homomorphism $Q \to U(2)$ (where $U(2)$ denotes the group of unitary transformations of $\mathbb{H}$ equipped with any of the above complex structures; namely this group depends on this choice). This homomorphism is an isomorphism of Q and $SU(2)$. We have also homomorphism

$$\Phi : Q \times Q \to SO(4) \tag{1}$$

given by $\Phi((q_1, q_2))(x) = q_1 x q_2^{-1}$. This map is an epimorphism with the kernel $\{(1, 1), (-1, -1)\}$. Thus $SO(4) = (Q \times Q)/\{(1, 1), (-1, -1)\}$.

To formulate the main result of this article let us introduce some notation. For a point u on the sphere S^2 parameterizing complex structures let us denote by I_u the corresponding complex structure. Let us denote by $\mathbb{C}P_u^1$ the complex projective line consisting of complex lines in $\mathbb{H}$ with respect to the complex structure I_u. Also let us introduce the following valuations:

$$Z_u(K) := \int_{\xi \in \mathbb{C}P_u^1} \mathrm{vol}_2\big(Pr_\xi(K)\big) d\xi$$

where Pr_ξ denotes the orthogonal projection onto ξ, and the integration is with respect to the standard $SU(2)$-invariant probability measure on $\mathbb{C}P^1$. It is easy to see that Z_u is an $Sp(1)$-invariant translation invariant 2-homogeneous continuous valuation. Let V_i denote the i-th intrinsic volume (see e.g. [S]). Our main result is as follows:

Theorem 1. *The valuations*

$$V_i, \, 0 \le i \le 4,$$

$$Z_i, \, Z_j, \, Z_{\frac{i+j}{\sqrt{2}}}, \, Z_{\frac{i+k}{\sqrt{2}}}, \, Z_{\frac{j+k}{\sqrt{2}}}$$

form a basis in the space of $Sp(1)$-invariant translation invariant continuous valuations.

Let $Gr_{2,4}^+$ denote the Grassmannian of oriented 2-planes in $\mathbb{R}^4$. We want to describe an explicit identification of it with $S^2 \times S^2$. Let

$$E_0 := \text{span}\{1, i\} \in Gr_{2,4}^+$$

with the orientation induced by the complex structure i. Let us denote by T the circle $\{z \in \mathbb{C} | \, |z| = 1\} \subset Q$.

Proposition 2. (1) *The stabilizer of $E_0 \in Gr_{2,4}^+$ in $Q \times Q$ is equal to $T \times T$.*
(2) *The stabilizer of* $\text{span}\{1, i\}$ *in non-oriented Grassmannian $Gr_{2,4}$ is equal to $T \times T \cup j(T \times T)$.*

Proof. Clearly $T \times T \subset \text{Stab}_{E_0}$. Hence we have a covering

$$(Q \times Q)/(T \times T) \to Gr_{2,4}^+.$$

Since $Gr_{2,4}^+$ is simply connected, this covering is an isomorphism. This proves part (1).

Let us prove part (2). Clearly $(T \times T) \cup j(T \times T)$ is a subgroup contained in the stabilizer of $\text{span}\{1, i\} \in Gr_{2,4}$. Let $A := ((T \times T) \cup (jT \times T))/(T \times T)$. This is a group of two elements acting on $Gr_{2,4}^+$. Then we have a covering $Gr_{2,4}^+/A \to Gr_{2,4}$. Since $\pi_1(Gr_{2,4}) = \mathbb{Z}/2\mathbb{Z}$ the last map is an isomorphism. $\square$

Let $\theta : Q \to Q$ be the involutive automorphism given by $\theta(q) = iqi^{-1}$. The subgroup of fixed points of θ is equal to T.

Let us recall the construction of the Cartan imbedding. Let $\theta : G \to G$ be an involutive automorphism of a group G. Let K be the subgroup of fixed points of θ. The Cartan imbedding of $G/K \hookrightarrow G$ is given as follows:

$$gK \mapsto g(\theta g)^{-1}.$$

It is easy to see that this map is well defined and injective. Let us write down the Cartan imbedding explicitly in our case:

$$f(qT) = q(iqi^{-1})^{-1} = [q, i]. \tag{2}$$

Note that $Q/T = \mathbb{C}P^1 = S^2$. Thus $f : Q/T = S^2 \hookrightarrow Q$.

Proposition 3. *The image of f coincides with*

$$S^2 = \{q \in Q | \, q = t + jy + kz\} \subset \mathbb{R} \oplus \mathbb{C}j.$$

Proof. First let us show that the image of f is contained in the above set. Indeed

$$Re(f(q)i) = Re(qiq^{-1}) = Re(iq^{-1}q) = Re\,i = 0.$$

Thus $Im f$ is a smooth 2-submanifold of S^2. Hence it coincides with S^2.   □

Let us write down the map f explicitly. We will write quaternions as a pair of complex numbers (z, w):

$$q = z + jw.$$

We have

$$\bar{q} = \bar{z} + \bar{w}\bar{j} = \bar{z} - jw$$
$$i\bar{q}i^{-1} = \bar{z} + jw.$$

Hence we get

$$\begin{aligned} qi\bar{q}i^{-1} &= (z + jw)(\bar{z} + jw) \\ &= |z|^2 - |w|^2 + 2jw\bar{z} \\ &= 2|z|^2 - 1 + j(2w\bar{z}). \end{aligned}$$

Thus for $q \in Q$, $q = z + jw$, one has

$$f(z + jw) = (2|z|^2 - 1) + j \cdot (2w\bar{z}). \tag{3}$$

Recall that $E_0 = \mathrm{span}\{1, i\} = \mathrm{span}\{(1, 0), (i, 0)\}$. We will compute explicitly $|\cos(E_0, E)|$ for $E \in Gr_{2,4}$. The expression for it is given by formula (4) below. Let $(q_1, q_2) \in Q \times Q$. Set

$$\begin{aligned} E &:= q_1 E_0 q_2^{-1} \\ &= \mathrm{span}\{q_1 q_2^{-1}, q_1 i q_2^{-1}\} \\ &= \mathrm{span}\{(z_1 + jw_1)(\bar{z}_2 - jw_2), (z_1 + jw_1)i(\bar{z}_2 - jw_2)\} \\ &= \mathrm{span}\{(z_1\bar{z}_2 + \bar{w}_1 w_2) + j \cdot (w_1\bar{z}_2 - \bar{z}_1 w_2), i(z_1\bar{z}_2 - \bar{w}_1 w_2) \\ &\quad + j \cdot (i(w_1\bar{z}_2 + \bar{z}_1 w_2))\} \\ &=: \mathrm{span}\{x_1 + jy_1, x_2 + jy_2\}. \end{aligned}$$

Note also that if we have two 2-planes E_1, E_2 and orthonormal bases in them $\{u_1, v_1\}$, $\{u_2, v_2\}$, respectively, then

$$|\cos(E_1, E_2)| = \left| \det \begin{bmatrix} (u_1, u_2) & (u_1, v_2) \\ (v_1, u_2) & (v_1, v_2) \end{bmatrix} \right|.$$

Hence

$$
\begin{aligned}
|\cos(E_0, E)| &= \left| \det \begin{bmatrix} Re\, x_1 & Re(x_1\bar{i}) \\ Re\, x_2 & Re(x_2\bar{i}) \end{bmatrix} \right| \\
&= \left| \det \begin{bmatrix} Re\, x_1 & Im\, x_1 \\ Re\, x_2 & Im\, x_2 \end{bmatrix} \right| \\
&= |(Re\, x_1)(Im\, x_2) - (Re\, x_2)(Im\, x_1)| \\
&= |Im\,(\bar{x}_1 x_2)| \\
&= |Im\,(\bar{z}_1 z_2 + w_1 \bar{w}_2)i(z_1 \bar{z}_2 - \bar{w}_1 w_2)| \\
&= |-Re\,(\bar{z}_1 z_2 + w_1 \bar{w}_2)(z_1 \bar{z}_2 - \bar{w}_1 w_2)| \\
&= |-Re\,(|z_1|^2|z_2|^2 - |w_1|^2|w_2|^2 - \bar{z}_1 z_2 \bar{w}_1 w_2 + w_1 \bar{w}_2 z_1 \bar{z}_2)| \\
&= |-Re\,(|z_1|^2|z_2|^2 - |w_1|^2|w_2|^2)| \\
&= |-|z_1|^2|z_2|^2 + |w_1|^2|w_2|^2| \\
&= |(1 - |z_1|^2)(1 - |z_2|^2) - |z_1|^2|z_2|^2| \\
&= |1 - |z_1|^2 - |z_2|^2|.
\end{aligned}
$$

Thus

$$|\cos(E_0, E)| = |1 - |z_1|^2 - |z_2|^2|. \tag{4}$$

We have

$$Gr_{2,4}^+ = Q/T \times Q/T \overset{f \times f}{\to} S^2 \times S^2 \tag{5}$$

where each copy of S^2 is identified with the set of quaternions of norm one of the form $t + yj + zk$. Combining (3) and (4) we obtain

Proposition 4. *Let $E \in Gr_{2,4}^+$ correspond under the identification (5) to a pair $(t_1 + y_1 j + z_1 k, t_2 + y_2 j + z_2 k)$. Then*

$$|\cos(E_0, E)| = \frac{1}{2}|t_1 + t_2|.$$

Let us denote the action of $(q_1, q_2) \in Q \times Q$ on $E \in Gr_{2,4}^+(\equiv S^2 \times S^2)$ by $q_1 E q_2^{-1}$. In this notation we have

Corollary 5.

$$\int_Q |\cos(qE, E_0)| dq = \frac{1}{4}(1 + t_2^2)$$

where the integration is with respect to the Haar measure on Q, and the total measure of Q is equal to 1.

Proof. The left-hand side is equal to

$$\frac{1}{2} \int_{S^2} |t_1 + t_2| = \frac{1}{4} \int_{-1}^1 |t_1 + t_2| dt_1 = \frac{1}{4}(1 + t_2^2). \qquad \square$$

Recall now that we have the epimorphism $\Phi : Q \times Q \to SO(4)$ (see (1)) with the kernel $\{(1,1),(-1,-1)\}$. Recall that irreducible representations of $SO(4)$ are parameterized by highest weights corresponding to the pairs of integers (m_1, m_2) satisfying $m_1 \geq |m_2|$. Recall also that the group Q is isomorphic to $SU(2)$, and irreducible representations of $SU(2)$ are parameterized by non-negative integers. Hence irreducible representations of $Q \times Q$ are parameterized by the pairs of arbitrary non-negative integers (m_1, m_2).

Let ρ be an irreducible representation of $SO(4)$ with highest weight (m_1, m_2), $m_1 \geq |m_2|$. Since Φ is an epimorphism the representation $\rho \circ \Phi$ is also irreducible. It is easy to see that the highest weight of $\rho \circ \Phi$ is equal to $(m_1 - m_2, m_1 + m_2)$.

We will need the following result which was proved in [A1], Corollary 6.2.

Lemma 6. *The representation of $SO(4)$ in the space of even 2-homogeneous translation invariant continuous valuations on $\mathbb{R}^4$ is multiplicity-free and consists exactly of irreducible representations with highest weights (m_1, m_2) satisfying*

$$m_1 \geq |m_2|, \ m_1 \text{ is even }, \ m_2 = 0 \text{ or } \pm 2.$$

Generalization of this theorem to higher dimensions and to arbitrary degrees of homogeneity of even valuations was obtained in [A2].

Corollary 7. *The space of $Sp(1)$-invariant 2-homogeneous translation invariant continuous valuations has dimension 6.*

Proof. Note that all $Sp(1)$-invariant valuations are even. By Lemma 6 the representation of $Q \times Q = Sp(1) \times Sp(1)$ in the space of even 2-homogeneous translation invariant continuous valuations on $\mathbb{R}^4$ is multiplicity-free and consists exactly of irreducible representations with highest weights (m_1, m_1) with $m_1 \in 2\mathbb{Z}_+$, or $(m_1 - 2, m_1 + 2)$ with $m_1 \in (2 + 2\mathbb{Z}_+)$, or $(m_1 + 2, m_2 - 2)$ with $m_1 \in (2 + 2\mathbb{Z}_+)$. Note that the irreducible representations of $Sp(1) \times Sp(1)$ containing a vector invariant with respect to the first copy of $Sp(1)$ have precisely highest weights of the form $(0, m)$ with $m \geq 0$. Hence the only representations of such form which appear in the space of even 2-homogeneous translation invariant continuous valuations have highest weights $(0, 0)$ or $(0, 4)$. The first representation is the identity representation (in particular it is one dimensional). The irreducible representation of $Sp(1)$ with the highest weight 4 is 5-dimensional. Since 1+5=6, Corollary 7 is proved. $\square$

Now let us prove the main result.

Proof of Theorem 1. For a finite dimensional real vector space let us denote by Val(V) the space of translation invariant continuous valuations on V. A valuation ϕ is called k-homogeneous (k is a non-negative integer) if $\phi(\lambda K) = \lambda^k \phi(K)$ for any $K \in \mathcal{K}(V)$, $\lambda \geq 0$. Let us denote by Val$_k(V)$ the subspace of Val(V) of k-homogeneous valuations. It is a result due to P. McMullen [McM1] that

$$\mathrm{Val}(V) = \oplus_{k=0}^{n} \mathrm{Val}_k(V)$$

where $n = \dim V$. For a subgroup G of the group of linear invertible transformations let us denote by $\mathrm{Val}^G(V)$ (resp. $\mathrm{Val}_k^G(V)$) the subspace of G-invariant valuations in $\mathrm{Val}(V)$ (resp. $\mathrm{Val}_k(V)$). It follows that

$$\mathrm{Val}^G(V) = \oplus_{k=0}^{n} \mathrm{Val}_k^G(V).$$

In particular we get $\mathrm{Val}^{Sp(1)}(\mathbb{H}) = \oplus_{k=0}^{4} \mathrm{Val}_k^{Sp(1)}(\mathbb{H})$. It is obvious that $\mathrm{Val}_0(V) = \mathbb{C} \cdot V_0$. It was shown by H. Hadwiger [H] that $\mathrm{Val}_n(V) = \mathbb{C} \cdot \mathrm{vol}_n$ where vol_n is a Lebesgue measure. It was shown in [A4] that if a compact group G acts transitively on the unit sphere then $\mathrm{Val}_1^G(V)$ is spanned by the intrinsic volume V_1, and $\mathrm{Val}_{n-1}^G(V)$ is spanned by the intrinsic volume V_{n-1}. Thus in our situation it remains to show that the valuations V_2, Z_i, Z_j, $Z_{\frac{i+j}{\sqrt{2}}}$, $Z_{\frac{i+k}{\sqrt{2}}}$, $Z_{\frac{j+k}{\sqrt{2}}}$ form a basis in $\mathrm{Val}_2^{Sp(1)}(\mathbb{H})$. By Corollary 7 it suffices to show that these valuations are linearly independent.

For $E \in Gr_{2,4}^+$ let D_E denote the unit disk in E. Let u be a complex structure on $\mathbb{H}$ compatible with the quaternionic structure. In the formulas below we will write the action of $(q_1, q_2) \in Q \times Q$ on $E \in Gr_{2,4}^+$ by $q_1 \cdot E \cdot q_2^{-1}$.

Let $E_u := \mathrm{span}\{1, u\}$. Then

$$V_2(D_E) = \mathrm{const}, \ Z_u(D_E) = \mathrm{vol}_2(D_2) \int_{U \in Sp(1)} |\cos(UE, E_u)| dU$$

where D_2 denotes the standard 2-dimensional unit disk. Let us choose $q_u \in Q$ such that $q_u i q_u^{-1} = u$. Then $E_u = q_u \cdot E_0 \cdot q_u^{-1}$. Hence

$$Z_u(D_E) = \mathrm{vol}_2(D_2) \int_{U \in Sp(1)} |\cos(UE \cdot q_u, E_0)| dU.$$

Assume that under the identification (5) an element $E \in Gr_{2,4}^+$ corresponds to $(v_1, v_2) = (t_1 + y_1 j + z_1 k, t_2 + y_2 j + z_2 k)$, and $E \cdot q_u$ corresponds to $(v_1', v_2') = (t_1' + y_1' j + z_1' k, t_2' + y_2' j + z_2' k)$. Then by Corollary 5

$$\int_{U \in Sp(1)} |\cos(UE \cdot q_u, E_0)| dU = \frac{1}{4}(1 + (t_2')^2).$$

Let us compute t_2'. Let $e := (1, 0, 0)$. Note that $t_2 = (v_2, e)$. Then $t_2' = (v_2, q_u(e))$. Moreover since e corresponds to E_0, by (2) $q_u(e) = [q_u \cdot 1, i] = ui^{-1}$. Hence

$$Z_i(D_E) = \frac{\mathrm{vol}_2(D_2)}{4}(1 + t_2^2)$$

$$Z_j(D_E) = \frac{\mathrm{vol}_2(D_2)}{4}(1 + z_2^2)$$

$$Z_{\frac{i+j}{\sqrt{2}}}(D_E) = \frac{\mathrm{vol}_2(D_2)}{4}\left(1 + (t_2 + z_2)^2/2\right)$$

$$Z_{\frac{i+k}{\sqrt{2}}}(D_E) = \frac{\mathrm{vol}_2(D_2)}{4} \left(1 + (t_2 - y_2)^2/2\right)$$

$$Z_{\frac{j+k}{\sqrt{2}}}(D_E) = \frac{\mathrm{vol}_2(D_2)}{4} \left(1 + (z_2 - y_2)^2/2\right).$$

Thus in order to prove Theorem 1 it is enough to check that the functions

$$1, \ 1 + t_2^2, 1 + z_2^2, \ 1 + (t_2 + z_2)^2/2, \ 1 + (t_2 - y_2)^2/2, \ 1 + (z_2 - y_2)^2/2$$

are linearly independent on S^2. This easily follows from the linear independence of the functions

$$1, \ t_2^2, \ z_2^2, \ t_2 z_2, \ t_2 y_2, \ z_2 y_2.$$

But the linear independence of these functions on S^2 is obvious.   □

Acknowledgement. I thank H. Tasaki for informing me about his recent work [T]. I thank J. Bernstein for useful remarks on the first version of this article.

References

[A1]   Alesker, S.: On P. McMullen's conjecture on translation invariant valuations. Adv. Math., **155**, no. 2, 239–263 (2000)

[A2]   Alesker, S.: Description of translation invariant valuations with the solution of P. McMullen's conjecture. Geom. Funct. Anal., **11**, no. 2, 244–272 (2001)

[A3]   Alesker, S.: Hard Lefschetz theorem for valuations, complex integral geometry, and unitarily invariant valuations. J. Differential Geom., to appear. Also: math.MG/0209263

[A4]   Alesker, S.: The multiplicative structure on polynomial continuous valuations. Geom. Funct. Anal., to appear. Also: math.MG/0301148

[BB]   Beilinson, A., Bernstein, J.: Localisation de g-modules (French). C.R. Acad. Sci. Paris Sr. I Math, **292**, no. 1, 15–18 (1981)

[GW]   Goodey, P., Weil, W.: Distributions and valuations. Proc. London Math. Soc., (3) **49**, no. 3, 504–516 (1984)

[H]   Hadwiger, H.: Vorlesungen über Inhalt, Oberfläche und Isoperimetrie (German). Springer-Verlag, Berlin-Göttingen-Heidelberg (1957)

[HoL]   Howe, R., Lee, S.T.: Degenerate principal series representations of $GL_n(\mathbb{C})$ and $GL_n(\mathbb{R})$. J. Funct. Anal., **166**, no. 2, 244–309 (1999)

[K1]   Klain, D.: A short proof of Hadwiger's characterization theorem. Mathematika, **42**, no. 2, 329–339 (1995)

[K2]   Klain, D.: Even valuations on convex bodies. Trans. Amer. Math. Soc., **352**, no. 1, 71–93 (2000)

[KR]   Klain, D., Rota, G.-C.: Introduction to geometric probability. (English. English summary) Lezioni Lincee. [Lincei Lectures] Cambridge University Press, Cambridge (1997)

[McM1]   McMullen, P.: Valuations and Euler-type relations on certain classes of convex polytopes. Proc. London Math. Soc., (3) **35**, no. 1, 113–135 (1977)

[McM2] McMullen, P.: Continuous translation-invariant valuations on the space of compact convex sets. Arch. Math. (Basel), **34**, no. 4, 377–384 (1980)

[McM3] McMullen, P.: Valuations and dissections. Handbook of Convex Geometry, Vol. A, B, 933–988. North-Holland, Amsterdam (1993)

[McMS] McMullen, P., Schneider, R.: Valuations on convex bodies. In: Convexity and Its Applications, 170–247. Birkhäuser, Basel (1983)

[S]   Schneider, R.: Convex bodies: The Brunn–Minkowski theory. Encyclopedia of Mathematics and its Applications, **44**. Cambridge University Press, Cambridge (1993)

[T]   Tasaki, H.: Integral geometry under the action of the first symplectic group. Arch. Math. (Basel), **80**, no. 1, 106–112 (2003)

The Change in the Diameter of a Convex Body under a Random Sign-Projection

S. Artstein[*]

School of Mathematical Sciences, Tel Aviv University, Tel Aviv 69978, Israel
artst@post.tau.ac.il

1 Introduction

In this note we study the behavior of the Euclidean diameter of a convex body under a certain type of linear mapping, called sign-projections. A sign-projection of level k is a linear mapping $S_k : \mathbb{R}^n \to \mathbb{R}^k$ of the form

$$S_k x = \big((x, \xi_1), \ldots, (x, \xi_k) \big),$$

where for $i = 1, \ldots, k$ the vectors ξ_i are vertices of the n-dimensional cube of edge length $2/\sqrt{n}$, that is, $\xi_i = (\xi_{i,1}, \ldots, \xi_{i,n}) \in \{ -\frac{1}{\sqrt{n}}, \frac{1}{\sqrt{n}} \}^n$. For most choices of such $\xi_1, \ldots, \xi_k$ the mapping S_k is indeed of rank k. In the special case when $\xi_1, \ldots, \xi_k$ are orthonormal, S_k is an orthogonal projection. We equip the set of level k sign-projections with a probability measure in the following way: for every $i = 1, \ldots, k$, the vector ξ_i is chosen uniformly in $\{ -\frac{1}{\sqrt{n}}, \frac{1}{\sqrt{n}} \}^n$. Thus, for a given dimension n, we have 2^{nk} different sign-projections available, and each has probability 2^{-nk}.

Not much is known about the behavior of the Euclidean diameter of convex sets under such mappings. Much more is known if, instead, the vectors $\xi_1, \ldots, \xi_k$ are chosen to be independent Gaussian random vectors, see e.g. [DS]. The other well-understood case is when the vectors $\xi_1, \ldots, \xi_k$ are chosen to be a random *orthonormal system* in R^n. Let us explain what is known about the situation then: It is equivalent to studying projections onto random k-dimensional subspaces, i.e. onto elements in the Grassmann manifold $G_{n,k}$ endowed with the normalized Haar measure. In this case it follows from the classical Dvoretzky theorem, as observed by Milman in [M], that for every convex body K in R^n there exists some critical dimension k^* (depending on K) such that, for all integers $k \geq k^*(K)$, the diameter of K shrinks by a

---

[*] Supported in part by the Israel Science Foundation and by the Minkowski Center for Geometry at Tel Aviv University.

factor close to $\sqrt{k/n}$ when K is projected onto a random k-dimensional sub-space, and for $k \leq k^*$ the projection of K onto a random k-dimensional sub-space is already, with high probability, an isomorphic Euclidean ball of radius $\simeq \sqrt{k^*/n}\,\mathrm{diam}(K)$. (Whenever we use the notation $\simeq$ we mean equivalence up to universal constants independent of the body and of the dimension.) After defining the parameter

$$M^*(K) = \int_{S^{n-1}} \sup\{\langle x, u\rangle : x \in K\}du,$$

which is half of the mean width of K (and will play a central role in this note), there is a simple formula for this critical dimension $k^*(K)$:

$$k^* = \left(\frac{M^*(K)}{\mathrm{diam}(K)}\right)^2 n. \tag{1}$$

Thus for dimension $\simeq k^*$ (and hence for all lower dimensions) for a very large measure of subspaces E in $G_{n,k}$, the projection of K onto E is close to a Euclidean ball of radius $M^*(K)$. A precise formulation of the above can be found for example in [MSz], Proposition 3.3.

In this note we show that for $k \geq k^*$, the same decay of diameter oc-curs when considering random level k sign-projections instead of random or-thogonal projections (and k^* is defined in (1)). The question regarding sign-projections of levels smaller than k^* remains open, and we provide some details concerning this question at the end of this note.

Acknowledgement. I wish to thank my supervisor, Prof. V. Milman of Tel Aviv University, and my host during my visit to University College London, Prof. K. Ball, for their help, advice, enlightening discussions and support. I also thank Prof. S. Bobkov for pointing out to me inequality (3) and the fact that it can be used to prove inequality (2).

2 Decay of Norm for a Single Point

The first step in investigating how a random sign-projection affects the diam-eter of a convex body is to see how it affects the Euclidean norm of a single point. Then one can hope to use this for, say, a *net* of points inside the convex body, or, as will be done here, for a *family* of nets. Another way to view this is that one first investigates the diameter decay for the simplest of all convex sets - a line segment.

The first lemma is an easy estimate, yet will serve as our main technical tool throughout the paper.

Lemma 1. *In the above notation, for a given point $x \in R^n$ and any $t > 1$,*

$$P\left[\, |S_k x| > t\sqrt{\frac{k}{n}}|x| \,\right] \le \exp-\left(\frac{t^2 - 1 - 2\log t}{2}k\right) \qquad (2)$$

(where $P(A)$ denotes the probability of the set of level k sign-projections S_k satisfying A).

Proof of Lemma 1. To generate the measure on the S_k's we choose independently nk signs $\varepsilon_{i,j}$ for $i = 1,\ldots,k$ and $j = 1,\ldots,n$, and define the vectors $\xi_i = (\xi_{i,1},\ldots,\xi_{i,n})$ by $\xi_{i,j} = \varepsilon_{i,j}/\sqrt{n}$. The random operator is then defined by

$$S_k(x) = \sum_{i=1}^{k}(x,\xi_i)e_i$$

(where e_i denotes the standard unit vector basis in R^n).

We denote $\varepsilon_i = (\varepsilon_{i,1},\ldots,\varepsilon_{i,n})$. Thus ε_i is a random vector whose n coordinates are ± 1 with probabilities $1/2$, and

$$P\left[\, |S_k(x)| > t\sqrt{\frac{k}{n}}|x| \,\right] = P\left[\sum_{i=1}^{k}(x,\xi_i)^2 > t^2\frac{k}{n}|x|^2\right]$$

$$= P\left[\sum_{i=1}^{k}(x,\varepsilon_i)^2 > t^2 k|x|^2\right].$$

Notice that (denoting $x = (x_1,\ldots,x_n)$)

$$E\big((x,\varepsilon_i)^2\big) = E\left(\sum_{j=1}^{n}x_j\varepsilon_{i,j}\right)^2 = |x|^2$$

(where $E(B)$ is the expectation of the random variable B).

We will make use of the following well-known inequality which was pointed out to us by S. Bobkov: For $\lambda \in [0,\frac{1}{2})$ and $|y| = 1$,

$$E\left(\exp\left(\lambda(y,\varepsilon_i)^2\right)\right) \le \frac{1}{\sqrt{1-2\lambda}}. \qquad (3)$$

For completeness we provide a proof of this inequality at the end of this note. It uses the fact that one can majorize this expression by the same expression for appropriate Gaussians, see e.g. [Ach], [T1].

Using Chebyshev's inequality, the independence of the ε_i's, and the above estimate, we get that for $\lambda \in (0, \frac{1}{2|x|^2})$,

$$P\left[\sum_{i=1}^{k}(x,\varepsilon_i)^2 > t^2 k|x|^2\right] = P\left[\exp\left(\lambda\sum_{i=1}^{k}(x,\varepsilon_i)^2\right) > \exp(\lambda t^2 k|x|^2)\right]$$

$$\le \exp(-\lambda t^2 k|x|^2)E\left(\exp\left(\sum_{i=1}^{k}\lambda(x,\varepsilon_i)^2\right)\right)$$

$$= \exp(-\lambda t^2 k |x|^2) E \left(\exp \left(\lambda (x, \varepsilon_1)^2 \right) \right)^k$$

$$\leq \exp(-\lambda t^2 k |x|^2) \left(\frac{1}{\sqrt{1 - 2\lambda |x|^2}} \right)^k.$$

For $t > 1$ we optimize over λ, choosing $\lambda |x|^2 = \frac{1}{2} - \frac{1}{2t^2}$, and get that

$$P \left[|S_k(x)| > t \sqrt{\frac{k}{n}} |x| \right] \leq \exp - \left(\frac{t^2 - 1 - 2 \log t}{2} k \right),$$

which completes the proof of (2). □

Lemma 1 gives only an upper estimate for the norm. It shows that for a given point $x \in R^n$ (regardless of its position), with large probability, the Euclidean norm of a level k sign-projection of x is not much bigger than $\sqrt{k/n}$ times the Euclidean norm of x. One can also show that actually this is the right decrease, and that with large probability the norm does not decrease more. The following lemma is proved in [Ach]:

Lemma 2. *In the above notation, for a given point $x \in R^n$ and $0 < \varepsilon < 1$,*

$$P \left[|S_k x| < \sqrt{1 - \varepsilon} \sqrt{\frac{k}{n}} |x| \right] \leq \exp - \frac{k}{2} (\varepsilon^2 / 2 - \varepsilon^3 / 3).$$

The main difference between the two lemmas is that in Lemma 1 we can force the probability of the event to be as small as we want by choosing a suitable (large enough) t. In this sense, Lemma 1 can be considered as isomorphic, while Lemma 2 is not.

3 Decay of Diameter of a Convex Body

In this section we prove the main result of this note which is

Theorem 3. *There exist universal constants C_1, C_2 and c_3 such that for every dimension n, for every convex body $K \subset R^n$, for $k \geq C_1 k^*$, the probability of level k sign-projections such that*

$$\mathrm{diam}(S_k K) \leq C_2 \sqrt{\frac{k}{n}} \mathrm{diam}(K)$$

is greater than $1 - 2^{-c_3 k}$, where k^* is defined in (1).

Remark. In fact the theorem we prove is slightly stronger; for *every* choice of $C_1 > 0$ and $c_3 > 0$ there exists C_2 such that the statement of the theorem holds true.

The idea of the proof is to use the information from Lemma 1 on individual points in order to get an estimate for the whole convex body together. The standard way to do so is to use a covering of the body by small Euclidean balls, and operate on each ball individually, hoping that the probabilities are big enough for many "good" events to happen all at once. The standard estimate for covering numbers (how many Euclidean balls are needed to cover the convex body K) is Sudakov's inequality (see e.g. [Pi2], page 68), which, as it turns out, is not strong enough to prove Theorem 3. Instead we use the (stronger) theorem of M. Talagrand, called the Majorizing Measure Theorem. This theorem can be formulated in several different ways, of which the following is most suitable for our needs. We refer the reader to the paper [T2] and to the forthcoming book [T3] for a proof, different formulations, and related topics.

Theorem 4. (Majorizing Measures) *There exists a universal constant C such that for any dimension n, for every convex body $K \subset R^n$, there exist families of points $B_0 \subset \ldots \subset B_{m-1} \subset B_m \subset \ldots \subset K$ with cardinality $|B_m| \leq 2^{2^m}$ such that for every $x \in K$*

$$\sum_{m=0}^{\infty} d(x, B_m)\sqrt{2^m} \leq CM^*(K)\sqrt{n}$$

where $d(x, B_m)$ denotes the distance of the point x to the mth family, i.e., $d(x, B_m) = \inf\{d(x,y) : y \in B_m\}$.

Proof of Theorem 3. Fix a constant $C_1 > 0$ and let $k > C_1 k^*$. We assume without loss of generality that k is a power of 2, and denote by m_0 the number such that $2^{m_0} = k$. For $x \in K$ write

$$x = v_{m_0}(x) + \sum_{m=m_0+1}^{\infty} v_m(x) - v_{m-1}(x)$$

where $v_m(x) \in B_m$ is the member of B_m closest to x. Thus, as $B_m \subset B_{m-1}$, we see that denoting $u_m(x) = v_m(x) - v_{m-1}(x)$ assures $|u_m(x)| \leq 2d(x, B_{m-1})$. We now denote $A_m = \{u_m(x) : x \in K\}$, and have $|A_m| \leq 2^{2^{m+1}}$. Rewrite

$$x = v_{m_0} + \sum_{m=m_0+1}^{\infty} u_m(x)$$

$$= v_{m_0} + \sum_{m=m_0+1}^{\infty} \left(\frac{d(x, B_{m-1})\sqrt{2^{m-1}}}{CM^*(K)\sqrt{n}} \frac{|u_m(x)|}{2d(x, B_{m-1})} \right)$$

$$\cdot \left(\frac{u_m(x)}{|u_m(x)|} \frac{2CM^*(K)\sqrt{n}}{\sqrt{2^{m-1}}} \right).$$

The Majorizing Measure Theorem tells us that the sum of the (left hand) coefficients in the infinite sum is less than 1, and therefore that for any linear operator S it is true that

$$|S(x)| \leq |S(v_{m_0}(x))| + \max_{m_0 < m < \infty} \frac{|S(u_m(x))|}{|u_m(x)|} \frac{2CM^*(K)\sqrt{n}}{\sqrt{2^{m-1}}}.$$

Hence

$$\text{diam}(S(K)) \leq \max_{v \in B_{m_0}} |S(v)| + \sup_{m_0 < m < \infty, u_m \in A_m} \frac{|S(u_m)|}{|u_m|} \frac{2CM^*(K)\sqrt{n}}{\sqrt{2^{m-1}}}.$$

We will use Lemma 1 to show that for a random level k sign-projection the right-hand side is less than $C_2\sqrt{k/n}$ with probability $1 - 2^{-c_3 k}$, for any c_3, and a suitably chosen C_2 depending only on C_1 and c_3. We do this by showing that the following two events occur simultaneously with probability greater than $1 - 2^{-c_3 k}$ (c_3 is any absolute constant):

(a) For every $v \in B_{m_0}$,

$$|S_k(v)| \leq C_2'\sqrt{\frac{k}{n}}|v|.$$

(b) For every $m > m_0$, for every $u_m \in A_m$,

$$|S_k(u_m)| \leq C_2'' \frac{\text{diam}(K)\sqrt{2^{m-1}}}{M^*(K)\sqrt{n}} \sqrt{\frac{k}{n}}|u_m|.$$

Then we can simply take $C_2 = C_2' + 2CC_2''$.

Taking care of (a) is immediate, since the size of B_{m_0} is exponential in k. Thus we may choose in Lemma 1, say, $C_2' = t = 2 + c_3$, so that $\frac{1}{2}(t^2 - 1 - 2\log t) > 2 + 2c_3$ and get that for measure $1 - \exp(-k(2 + 2c_3))$ of the level k sign-projections, (a) holds true. Notice that in this part the interplay between the constants is straightforward: if we allow smaller c_3, we may choose a better (smaller) C_2', and C_1 does not yet enter into the picture at all.

For (b) we need indeed the isomorphic part of Lemma 1. For every m, the probability that (b) happens for this specific m is (denoting $d = \text{diam}(K)$, and $M^* = M^*(K)$) at least

$$1 - 2^{2^{m+1}} P\left[|S_k(x)| > t\sqrt{\frac{k}{n}}|x|\right],$$

for $t = \frac{d}{M^*} C_2'' \frac{\sqrt{2^{m-1}}}{\sqrt{n}}$.

Since we consider $m > m_0$, we have that $2^{m-1} \geq k$, and hence $t \geq C_2'' \frac{d}{M^*}\sqrt{k/n} \geq C_2'' C_1$. Thus for $C_2'' C_1$ big enough we, first of all, have $t > 1$ and we can apply Lemma 1. Secondly, in Lemma 1 for t which is big enough, one sees that the probability of failure is (for example) less than $e^{-t^2 k/4}$. Thus the above expression, for a specific m and for C_2'' big enough (depending only on C_1), is at least

$$1 - 2^{2^{m+1}} e^{-\frac{k}{4} \frac{2^{m-1} C_2''^2}{n} (\frac{d}{M^*})^2}.$$

We are assuming that $kd^2/nM^{*2} \geq C_1^2$, which bounds this probability from below again by

$$1 - 2^{2^{m+1} - 2^{m-3} C_1^2 C_2''^2 / \ln 2}.$$

Making the choice $C_2''^2 = 16(2 + c_3) \ln 2 / C_1^2$, this probability is at least $1 - 2^{-2^{m+1}(1+c_3)}$. Adding up for all m the probability of failure in (b), and adding also the probability of failure in (a) as the first summand (where we have exchanged e by 2, only increasing the probability), we see that

$$\sum_{m=m_0}^{\infty} 2^{-2^{m+1}(c_3+1)} \le 2^{-c_3 k},$$

which completes the proof. $\square$

We remark again that, as is seen in the proof, for every choice of the two constants c_3 and C_1, there exists a choice of $C_2(c_3, C_1)$ such that the conclusion of the theorem holds.

Secondly notice that Lemma 2 shows that the theorem is sharp in the sense that no smaller bound for the diameter shrinking in a random sign-projection can be obtained (up to the constants involved). Indeed, simply take the point with largest Euclidean norm; under a random sign-projection the Euclidean norm of this specific point is close to $\sqrt{k/n}$ times its original norm, and so the diameter is no smaller.

4 Remarks

It seems appropriate to introduce the parameter $r^*(K)$ which is defined as

$$r^*(K) = \text{Ave}_{\varepsilon_i = \pm 1} \sup \left\{ \sum_i \varepsilon_i x_i : x \in K \right\}.$$

That is, we average the dual norm, not on the whole sphere as in the definition of $M^*(K)$, but only on the vertices of the cube. This parameter is known to be rather close to $\sqrt{n} M^*(K)$. Indeed, it is easy to show that there exists a universal constant C such that for every symmetric convex body K one has

$$\frac{1}{C} r^*(K) \le \sqrt{n} M^*(K) \le \sqrt{\log n}\, C r^*(K).$$

That is, up to a factor which is logarithmic in the dimension, the two parameters $r^*(K)$ and $\sqrt{n} M^*(K)$ are equivalent. Moreover, if the space whose unit ball is K° (the polar body of K) has some non-trivial cotype then the two parameters $r^*(K)$ and $\sqrt{n} M^*(K)$ are equivalent up to a universal constant (for definitions and proofs see [Pi1]).

Another point to notice is that we cannot expect the diameter of a random level k sign-projection to ever be smaller than $r^*(K)/\sqrt{n}$, since the diameter of a random level k sign-projection is clearly a decreasing function of k, and for $k = 1$ the average diameter is exactly $r^*(K)/\sqrt{n}$ (and therefore, by a standard concentration argument, this is true with high probability).

Let us discuss briefly the special example where the body K is the unit ball of l_1, that is, $K = \text{conv}\{\pm e_i\}$. In this case, a sign-projection of level k is simply the convex hull in R^k of the points

$$\big\{ \pm (\xi_{1,i}, \xi_{2,i}, \ldots, \xi_{k,i}), \ i = 1 \ldots n \big\}.$$

That is, we pick n random vertices from the 2^k vertices of the k-dimensional cube (of edge length $2/\sqrt{n}$), and take their (absolute) convex hull. The diameter of this set is *exactly* $\sqrt{k/n}$ times the diameter of the original K, and this holds true for *every* k. Thus we have an example where the decay of the diameter continues beyond the level $k^* = n(M^*)^2$ (which is in this case around $\log n$) and goes all the way to dimension $1 = r^{*2}$.

To conclude, we know how the diameter decreases for sign projections of levels greater than k^*. However, opposed to the orthogonal case in which we know that all projections to dimensions smaller than $\simeq k^*$ are Euclidean balls of radius $\simeq M^*(K)$, in the sign-projections investigation we have no such natural reason for the decrease in diameter to stop, though we know that at dimension 1 it must be, with high probability, around $r^*/\sqrt{n}$, which is at least $M^*(K)/(C \log n)$.

5 Appendix

Let us show a proof of the fact that for $t_i = \pm 1$ (standard Bernoulli random variables), if $\sum y_i{}^2 = 1$ and $\lambda \in [0, \frac{1}{2})$, then

$$E\left(\exp \left(\lambda \Big(\sum y_i t_i \Big)^2 \right) \right) \leq \frac{1}{\sqrt{1-2\lambda}} \tag{4}$$

(where E is taken with respect to the t_i's, that is, $E(A) = \text{Ave}_{t_i = \pm 1} A$). First notice that for any fixed t_i, $i = 1, \ldots, n$,

$$\exp \left(\lambda \Big(\sum y_i t_i \Big)^2 \right) = E_g \exp \left(\sqrt{2\lambda} \Big(\sum y_i t_i \Big) g \right)$$

where g is a standard Gaussian and E_g denotes expectation with respect to it. Using the independence we see that

$$E\left(\exp \left(\lambda \Big(\sum y_i t_i \Big)^2 \right) \right) = E E_g \left(\exp \left(\sqrt{2\lambda} \Big(\sum y_i t_i \Big) \right) g \right)$$

$$= E_g \prod E_{t_i} \left(\exp \left(\sqrt{2\lambda} y_i t_i \right) g \right)$$

$$= E_g \prod \cosh \left(\sqrt{2\lambda} y_i g \right)$$

$$\leq E_g \prod \exp \lambda g^2 y_i^2 = E_g \exp \lambda g^2 \sum y_i^2$$

$$= \frac{1}{\sqrt{1-2\lambda}}$$

where we have used the easily verified inequality $\cosh x \leq e^{x^2/2}$. This completes the proof of inequality (4). □

References

[Ach] Achlioptas, D.: Database-friendly random projections. Proceedings of PODS 01, 274–281 (2001)

[DS] Davidson, K.R., Szarek, S.J.: Local operator theory, random matrices and Banach Spaces. Handbook of the Geometry of Banach Spaces, Vol. I, 317–366. North Holland, Amsterdam (2001)

[M] Milman, V.D.: A note on a low M^*-estimate. Geometry of Banach Spaces (Strobl, 1989), London Math. Soc. Lecture Notes Ser., **158**. Cambridge University Press, Cambridge, 219–229 (1990)

[MSz] Milman, V.D., Szarek, S.J.: A geometric lemma and duality of entropy numbers. Geometric Aspects of Functional Analysis (1996–2000), Lecture Notes in Math., **1745**. Springer, Berlin-New York, 191-222 (2000)

[Pi1] Pisier, G.: Probabilistic methods in the geometry of Banach spaces. Probability and Analysis (Varenna, 1985) Lecture Notes in Math., **1206**. Springer, Berlin, 167–241 (1986)

[Pi2] Pisier, G.: The volume of convex bodies and Banach space geometry. Cambridge Tracts in Mathematics, **94**. Cambridge University Press, Cambridge (1989)

[T1] Talagrand, M.: The Sherrington–Kirkpatrick model: a challenge for mathematicians. Probability Theory and Related Fields, **110**, 109–176 (1998)

[T2] Talagrand, M.: Majorizing measures without measures. Ann. Probab., **29** (1), 411–417 (2001)

[T3] Talagrand, M.: The generic chaining. Springer-Verlag Publications, to appear

An Elementary Introduction to Monotone Transportation[*]

K. Ball

Department of Mathematics, University College London, Gower Street, London
WC1E 6BT, UK kmb@math.ucl.ac.uk

Towards the end of the 18th century Gaspard Monge posed a problem concerning the most efficient way to move a pile of sand into a hole of the same volume. In modern language the problem was something like this:

Given two probability measures μ and ν on $\mathbf{R}^2$ (or more generally $\mathbf{R}^n$), find a map $T : \mathbf{R}^2 \to \mathbf{R}^2$ which transports μ to ν and which minimises the total cost

$$\int \|x - Tx\| \, d\mu(x).$$

(Thus the cost of moving a unit mass from x to Tx is just the distance moved.)

The statement that T transports μ to ν means that for each measurable set A,

$$\mu\left(T^{-1}(A)\right) = \nu\left(A\right) \tag{1}$$

or equivalently, that for any bounded continuous real-valued function f

$$\int f\left(T(x)\right) d\mu(x) = \int f(x) \, d\nu(x). \tag{2}$$

Monge gave a number of examples to illustrate the difficulties of the problem. Plainly if arbitrary measures are allowed it may not be possible to transport at all. If μ is a single point mass and ν consists of two point masses, each with weight $1/2$, then no map T can "split μ in half". Where transportation maps do exist, there may be many different optimal ones, even for measures on the line. If μ consists of two equal masses at (say) 0 and 1, and ν consists of two equal masses at 2 and 3 then it does not matter whether we choose

$$
\begin{array}{ccc}
0 \to 2 & & 0 \to 3 \\
& \text{or} & \\
1 \to 3 & & 1 \to 2
\end{array} :
$$

---

[*] This article is based on a lecture given to the graduate student seminar in analysis at Tel Aviv University.

the total cost is 4 in each case. For general measures, it is difficult to demonstrate the existence of any optimal transportation map. In the late seventies, Sudakov [S] outlined an important new approach to the problem which has recently been implemented in a number of different ways: by Caffarelli, Feldman and McCann [CFM], Trudinger and Wang [TW] and Ambrosio [A] and in particular contexts by Ambrosio, Kirchheim and Pratelli [AKP]. I am indebted to Bernd Kirchheim for his guidance on the history of this problem.

One reason for the difficulty is that the condition (1) is highly non-linear in T. This difficulty can be overcome by relaxing the requirement that μ be transported to ν by a *map*. Instead of insisting that all the μ-measure at x should end up at the same place Tx, we may allow this mass to be "smeared out" over many points. The aim is thus to describe for each *pair* of points (x, y), how much mass is moved from x to y. More formally, we look for a measure γ on the product $\mathbf{R}^n \times \mathbf{R}^n$ whose marginal in the x direction is μ and whose marginal in the y direction is ν: so for measurable sets A and B,

$$\gamma(A \times \mathbf{R}^n) = \mu(A)$$

$$\gamma(\mathbf{R}^n \times B) = \nu(B).$$

A transportation measure is then optimal if it minimises

$$\int \|x - y\| \, d\gamma(x, y).$$

In this form, the problem is a linear program (for γ), whose discrete version is familiar from introductory courses on linear programming. The conversion of the transportation problem to a linear program was first described by Kantorovich. In this setting, the existence of optimal transport measures is not so hard to establish: the difficulty is to decide whether there is such a measure which is concentrated on the graph of a function T.

In the context of linear programming it is customary to consider transportation costs that are much more general than simply the distance $\|x - y\|$. The total cost will be

$$\int c(x, y) \, d\gamma(x, y)$$

if c measures the cost per unit mass of transporting from x to y. If c is a strictly convex function of the distance $\|x - y\|$ then examples of non-unique optimal transportation like the one above, do not occur. In [Br], Brenier explained that there is one particular choice of cost function for which not only is the optimal map unique but also it has a particularly special form, which makes it suitable for a wide range of applications. From the geometric point of view, this cost function is undoubtedly the "right" one. The cost in question is the square of the Euclidean distance: $c(x, y) = \|x - y\|^2$.

To see why this choice is so special, consider the problem of transporting a probability measure μ on $\mathbf{R}^2$ (which will be assumed to be absolutely continuous with respect to Lebesgue measure) to the probability measure ν which assigns mass $1/2$ to each of the points $(1,0)$ and $(-1,0)$. The problem is illustrated in Figure 1: the shaded region on the left is supposed to represent the measure μ.

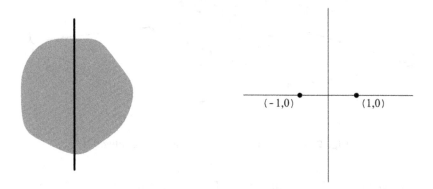

Fig. 1. A simple transport problem

The problem is to find a partition of $\mathbf{R}^2$ into sets A and B with

$$\mu(A) = \mu(B) = 1/2$$

so as to minimise the cost of transporting A to $(-1,0)$ and B to $(1,0)$:

$$\int_A \|x - (-1,0)\|^2 \, d\mu + \int_B \|x - (1,0)\|^2 \, d\mu.$$

I claim that the best thing to do is to divide the measure μ using a line in the direction $(0,1)$, as shown in Figure 1.

To establish the claim, we need to check that given two points (a, u) and (b, v) with $a < b$, it is better to move the leftmost point to $(-1,0)$ and the rightmost point to $(1,0)$, than it is to swap the order. The quadratic cost $\|x - y\|^2$ ensures this because it "keeps separate" the contributions from the first and second coordinates. The costs in the two cases are

$$(a + 1)^2 + u^2 + (b - 1)^2 + v^2$$

and

$$(a - 1)^2 + u^2 + (b + 1)^2 + v^2$$

and the only difference arises because $a < b$.

Because the transport function T maps everything to the right of the line, to the point $(1,0)$ and everything to the left of the line to the point $(-1,0)$, the map T is the gradient of a 'V'-shaped function as shown in Figure 2.

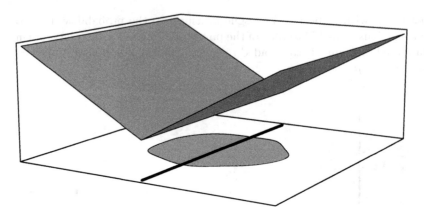

Fig. 2. Representing the transport as a gradient

More generally, the special feature of the Brenier map is that whenever we find a transport map which is optimal with respect to the quadratic cost, it will be the gradient of a convex function. Let's see why such a gradient should be optimal. Suppose that $\phi : \mathbf{R}^n \to \mathbf{R}^n$ is convex, $T = \nabla\phi$ and we have points $x_1, x_2, \ldots, x_k$ whose images under T are $y_1, y_2, \ldots, y_k$ respectively. The aim is to show that if $u_1, u_2, \ldots, u_k$ is a permutation of the x_i then

$$\sum \|x_i - y_i\|^2 \leq \sum \|u_i - y_i\|^2 :$$

that the total cost goes up if we map the points any other way than by T. Plainly the inequality amounts to

$$\sum \langle x_i, y_i \rangle \geq \sum \langle u_i, y_i \rangle. \tag{3}$$

Now, since ϕ is convex, we have that for every x and u

$$\phi(u) \geq \phi(x) + \langle u - x, \nabla\phi|_x \rangle.$$

Hence

$$\sum \langle u_i - x_i, y_i \rangle \leq \sum \left(\phi(u_i) - \phi(x_i) \right)$$

and the expression on the right is zero, since the u_i are a permutation of the x_i. This establishes (3).

Brenier demonstrated the existence of such optimal transport maps under certain conditions on the measures μ and ν: his result was generalised by McCann in [Mc1]. The Brenier map is sometimes called a monotone transportation map by analogy with the 1-dimensional case, in which the derivative of a convex function is monotone increasing. The rest of this article is organised as follows. In the next section we will demonstrate the existence of the

Brenier map under fairly general conditions on the measures, using an argument which is somewhat in the spirit of the example above. In the last section we shall explain how the special structure of the Brenier map fits neatly with a number of geometric inequalities.

There are a number of excellent surveys relating to mass transportation and monotone maps, for example the lecture notes of Villani [V].

1 A Construction of the Brenier Map

The aim of this section is to outline a proof of the following theorem:

Theorem 1. *If μ and ν are probability measures on $\mathbf{R}^n$, ν has compact support and μ assigns no mass to any set of Hausdorff dimension $(n-1)$, then there is a convex function $\phi : \mathbf{R}^n \to \mathbf{R}$, so that $T = \nabla\phi$ transports μ to ν.*

The proof has two main steps. We first establish the existence of ϕ in the case that the second measure ν is discrete and then pass to general measures by approximating them weakly by discrete ones. Suppose then that ν is a convex combination of point masses

$$\nu = \sum_1^m \alpha_i \delta_{u_i}.$$

The convex function we want has the form

$$\phi(x) = \max\{\langle x, u_i \rangle - s_i\}$$

for some suitable choice of real numbers $s_1, s_2, \ldots, s_m$. This function partitions $\mathbf{R}^n$ into m pieces according to whichever linear function is biggest. If A_i is the set where ϕ is given by

$$x \mapsto \langle x, u_i \rangle - s_i$$

then we want to juggle the s_i so as to arrange that $\mu(A_i) = \alpha_i$ for each i. In order to use a fixed point theorem we shall define a map H on the simplex of points $t = (t_1, t_2, \ldots, t_m)$ with non-negative coordinates satisfying $\sum t_i = 1$, by considering the function

$$\phi_t(x) = \max\left\{\langle x, u_i \rangle - \frac{1}{t_i}\right\}.$$

If t has non-zero coordinates, $H(t)$ will be the point $(\mu(A_1), \ldots, \mu(A_m))$ whose coordinates are the measures of the sets on which ϕ is linear. As t_i approaches 0, the ith linear function drops to negative infinity and so $\mu(A_i)$ decreases to zero. In view of the hypothesis on μ, H can be defined continuously on the simplex and maps each face of the simplex into itself. It is a well-known consequence (or reformulation) of Brouwer's fixed point theorem, that such a map is surjective. (If the map omits a point, say in the interior of the simplex, then we can follow it by a projection of the punctured simplex onto its boundary, to obtain a continuous map from the simplex to its boundary which fixes each face. Now by cycling the coordinates of the simplex we obtain a continuous map with no fixed point.) So there is a choice of t for which $H(t)$ is $(\alpha_1, \ldots, \alpha_m)$.

Now suppose that we have a general probability ν. Approximate it weakly by a sequence (ν_k) of discrete measures and choose convex functions ϕ_k whose gradients transport μ to these approximants. Assume that the ϕ_k are pinned down by (for example) $\phi_k(0) = 0$. We may assume that all ν_k are supported on the same compact set from which it is immediate that the ϕ_k are equicontinuous. So we may assume that they have a locally uniform limit, ϕ, say. The function ϕ is convex.

A standard result in convex analysis guarantees that outside a set of Hausdorff dimension $n - 1$, ϕ and all the ϕ_k are differentiable: (for the ϕ_k the differentiability is obvious by their construction). It is quite easy to check that except on the exceptional set,

$$\nabla \phi_k \to \nabla \phi.$$

Let $T_k = \nabla \phi_k$ for each k and $T = \nabla \phi$. By the condition on the support of ϕ, $T_k \to T$, μ-almost everywhere. We want to conclude that T transports μ to ν. This is a standard argument in weak convergence: if f is a bounded continuous function then

$$\int f \, d\nu = \lim \int f \, d\nu_k$$
$$= \lim \int f \circ T_k \, d\mu$$
$$= \int f \circ T \, d\mu \ .$$

Hence T transports μ to ν. □

The preceding theorem establishes the existence of a monotone transportation map T. Since T is the gradient of a convex function ϕ, its derivative is the Hessian of ϕ, so T' is positive semi-definite symmetric. Therefore $\det(T')$ is non-negative. This means that T is essentially 1-1. Therefore, if μ and ν have densities f and g respectively, the condition (1) for the Brenier map, has a local formulation as the familiar change of variables formula

$$f(x) = g(Tx) . \det\left(T'(x)\right). \tag{4}$$

This version of the measure transportation property is particularly useful in a number of geometric applications such as those below.

2 The Brunn–Minkowski Inequality

The aim of this section and the following one is to illustrate why the Brenier map is the perfect tool for a number of geometric applications. The first such applications were found by McCann [Mc2] who gave a proof of the Brunn–Minkowski inequality using displacement convexity, which depends upon mass transportation, and Barthe [B] who used the Brenier map to give a very clear proof of the Brascamp-Lieb inequality [BL], and its generalisations proved in [L], and also of a reverse inequality that had been conjectured by the present author. There have been several other, striking, geometric applications, for example those in [CNV].

In this article I have chosen two somewhat simpler applications: to results which were originally proved in other ways but which illustrate quite well, just why monotone maps are so appropriate. The first is the Brunn–Minkowski inequality.

If A and B are non-empty measurable sets in $\mathbf{R}^n$ and $\lambda \in (0,1)$ then we define the set

$$(1-\lambda)A + \lambda B$$

to be

$$\{(1-\lambda)a + \lambda b : a \in A, b \in B\}.$$

The Brunn–Minkowski inequality states that

$$\mathrm{vol}\big((1-\lambda)A + \lambda B\big)^{1/n} \geq (1-\lambda)\mathrm{vol}(A)^{1/n} + \lambda\mathrm{vol}(B)^{1/n}.$$

By using the arithmetic/geometric mean inequality, we can deduce a multiplicative version of the Brunn–Minkowski inequality, which has a number of advantages (one of them being that it admits a number of simple proofs)

$$\mathrm{vol}\big((1-\lambda)A + \lambda B\big) \geq \mathrm{vol}(A)^{1-\lambda}\mathrm{vol}(B)^{\lambda}.$$

For fixed A, B and λ, this latter inequality is weaker than the Brunn–Minkowski inequality, but it is not hard to check that its truth for *all* A, B and λ implies the formally stronger statement.

Using the Brenier map it is easy to give a short explanation of the multiplicative form of the Brunn–Minkowski inequality, although to make the argument formal one requires regularity results for the Brenier map which are not so easy to derive. (Cafarelli [C1] and [C2] has obtained a powerful regularity theory for these maps which is more than sufficient for the applications described here.)

We may assume that both A and B have finite non-zero measure. Let μ and ν be the restrictions of Lebesgue measure to A and B respectively, rescaled by the volumes of these sets so as to have total measure 1. Let T be the Brenier map transporting μ to ν.

Equation (4) for the transportation shows that for each $x \in A$,

$$\frac{1}{\text{vol}(A)} = \frac{1}{\text{vol}(B)} \cdot \det(T'(x)).$$

Now, let T_λ be the map given by

$$x \mapsto (1-\lambda)x + \lambda T(x).$$

This map transports μ to a measure, of total mass 1, supported on $(1-\lambda)A + \lambda B$. The density f_λ of this measure satisfies

$$\frac{1}{\text{vol}(A)} = f_\lambda(T_\lambda(x)) \cdot \det(T'_\lambda(x)).$$

In order to prove that $(1-\lambda)A + \lambda B$ is large, it suffices to check that the density is small: to be precise, that

$$f_\lambda(T_\lambda(x)) \leq \frac{1}{\text{vol}(A)^{1-\lambda}\text{vol}(B)^\lambda}.$$

This will follow if

$$\det(T'_\lambda(x)) \geq \left(\frac{\text{vol}(B)}{\text{vol}(A)}\right)^\lambda$$
$$= \left(\det(T'(x))\right)^\lambda.$$

This says that for each x,

$$\det\left((1-\lambda)I + \lambda T'(x)\right) \geq \left(\det(T'(x))\right)^\lambda$$

where I is the identity map on $\mathbf{R}^n$.

Now we use the fact that the Brenier map is monotone. At each x, $T'(x)$ is the Hessian of a convex function ϕ and so is a positive semi-definite symmetric matrix. With respect to an appropriate orthonormal basis, it is diagonal: let's say its diagonal entries are $t_1, t_2, \ldots, t_n$. The problem is to show that for these positive t_i

$$\prod(1 - \lambda + \lambda t_i) \geq \left(\prod t_i\right)^\lambda.$$

This inequality is the special case of the Brunn–Minkowski inequality in which the set A is a unit cube and the set B is a cuboid with sides $t_1, \ldots, t_n$, aligned in the same way as the cube. It is immediate because the arithmetic/geometric mean inequality shows that for each i

$$(1 - \lambda) + \lambda t_i \geq t_i^\lambda. \qquad \square$$

As long ago as 1957, Knothe [K] gave a proof of the Brunn–Minkowski inequality which involved a kind of mass transportation. His argument used induction upon dimension and corresponded, in effect, to the construction of a transportation map whose derivative is upper triangular rather than symmetric, as is the derivative of the Brenier map. Since such a choice of map is basis dependent, the proof is geometrically less elegant, but it has the advantage that regularity is not a serious issue.

Recently Alesker, Dar and Milman [ADM] used monotone transportation to give a proof of the Brunn–Minkowski inequality which was "constructive" in the following slightly exotic way. Normally each point of the combination $(1 - \lambda)A + \lambda B$ can be realised as a combination of points (one from A and one from B) in many different ways. In [ADM] the authors construct a map which associates each point of the convex combination with a particular combination that generates it, and which allows them to compare directly the sets themselves, rather than just their volumes.

3 The Marton–Talagrand Inequality

In [M] and subsequent articles, Marton described a new method for demonstrating probabilistic deviation inequalities based upon a comparison between transportation cost and entropy. Her main focus was on Markov chains where the existing methods did not give any estimates and in this setting she considered transportation costs based on an L_1-type cost function rather than the L_2-type discussed in this article. In his article [T], Talagrand pointed out that Marton's method could be applied very elegantly to the isoperimetric inequality for Gaussian measure provided one uses the quadratic cost function. The aim of this section is to explain the Marton–Talagrand inequality using the Brenier map. As in the previous section, the argument will treat regularity of the maps naively. The fact that the Brenier map fits well with the Marton–Talagrand inequality was noticed independently by several people. The first available references seem to be [Blo] and [OV].

Let γ be the standard Gaussian measure on $\mathbf{R}^n$ with density

$$g(x) = \frac{1}{(\sqrt{2\pi})^n} e^{-|x|^2/2}.$$

For a density f on $\mathbf{R}^n$ we define the relative entropy of f (relative to the Gaussian) to be

$$\text{Ent}(f\|\gamma) = \int_{\mathbf{R}^n} f \log(f/g).$$

The Marton–Talagrand inequality compares the relative entropy of f with the cost of transporting γ to the probability with density f. This cost is

$$C(g, f) = \int |x - T(x)|^2 \, d\gamma$$

where T is the Brenier map transporting γ to the measure with density f.

Theorem 2. *With the notation above*

$$\frac{1}{2}C(g,f) \le Ent(f\|\gamma).$$

Proof. Let T be the Brenier map transporting γ to the probability with density f. Then for each x

$$g(x) = f(T(x))T'(x).$$

The relative entropy is

$$\int f(y) \log \left(f(y)/g(y) \right) dy$$

and after the change of variables $y = T(x)$ this becomes

$$\int f(T(x)) \log \left(\frac{f(T(x))}{g(T(x))} \right) T'(x) \, dx = \int g(x) \log \left(\frac{g(x)}{g(T(x))T'(x)} \right).$$

The latter simplifies to

$$\int g(x) \left[-|x|^2/2 + |T(x)|^2/2 - \log T'(x) \right]$$

and we want to show that this expression is at least

$$\frac{1}{2} \int g(x)|x - T(x)|^2.$$

That amounts to showing that

$$\int g(x) \log T'(x) \le \int g(x)\langle x, T(x) \rangle - x \rangle.$$

As in the Brunn–Minkowski argument above, for each x, the derivative $T'(x)$ is a positive semi-definite symmetric matrix with eigenvalues $t_1, \ldots, t_n$ (say). Therefore $\log T'(x) = \sum \log t_i$ and this is at most $\sum (t_i - 1)$ which is the trace of the matrix $T' - I$. This in turn is the divergence (at x) of the map

$$x \mapsto T(x) - x.$$

Using integration by parts and the fact that $\nabla g(x) = -xg(x)$ we can conclude as follows:

$$\int g(x) \log T'(x) \le \int g(x)\mathrm{div}(T(x) - x)$$

$$= -\int \langle \nabla g(x), T(x) - x \rangle$$

$$= \int g(x)\langle x, T(x) \rangle - x \rangle. \qquad \square$$

The original argument of Talagrand, like that of Marton, used induction on dimension, much as in the proof of the Brunn–Minkowski inequality found by Knothe.

To see how the Marton–Talagrand inequality provides a deviation estimate, consider a measurable set B and form a density by restricting the Gaussian density g to B and dividing by $\gamma(B)$: call it f_B. So

$$f_B(x) = \mathbf{1}_B(x)g(x)/\gamma(B).$$

Then the relative entropy of f_B is just

$$\log\left(\frac{1}{\gamma(B)}\right).$$

So the cost of transporting γ to f_B is at most $-2\log\gamma(B)$. Now suppose that A is a set of fairly large Gaussian measure ($1/2$ perhaps) and that all points of B are at least distance ϵ from A. Then in transporting γ to f_B, we must transport all the measure in A, at least a distance ϵ. So the cost is at least $\gamma(A)\epsilon^2$.

Hence

$$\gamma(A)\epsilon^2 \le -2\log\gamma(B)$$

which implies that

$$\gamma(B) \le e^{-\gamma(A)\epsilon^2/2}.$$

Thus the ϵ-neighbourhood of a set of fairly large measure has Gaussian measure, close to 1.

References

[ADM] Alesker, S., Dar, S., Milman, V.: A remarkable measure-preserving transformation between two convex bodies in $\mathbf{R}^n$. Geom. Dedicata, **74**, 201–212 (1999)

[A] Ambrosio, L.: Lecture notes on optimal transport problems. CIME Springer Lecture Notes, to appear

[AKP] Ambrosio, L., Kirchheim, B., Pratelli, A.: Existence of optimal transport maps for crystalline norms. Preprint

[B] Barthe, F.: Inégalités de Brascamp-Lieb et convexité. C.R. Acad. Sci. Paris, **324**, 885–888 (1997)

[Blo] Blower, G.: The Gaussian isoperimetric inequality and transportation. Positivity, **7**, 203–224 (2003)

[BL] Brascamp, H.J., Lieb, E.H.: Best constants in Young's inequality, its converse, and its generalization to more than three functions. Advances in Math., **20**, 151–173 (1976)

[Br] Brenier, J.: Polar factorization and monotone rearrangement of vector-valued functions. Comm. Pure Appl. Math., **44**, 375–417 (1991)

[C1] Cafarelli, L.: Boundary regularity of maps with a convex potential. Comm. Pure Appl. Math., **45**, 1141-1151 (1992)

[C2]   Cafarelli, L., Regularity of mappings with a convex potential. J.A.M.S., **5**, 99–104 (1992)

[CFM]   Cafarelli, L., Feldman, M., McCann, R.J.: Constructing optimal maps for Monge's transport problem as a limit of strictly convex costs. J.A.M.S., **15**, 1–26 (2002)

[CNV]   Cordero-Erausquin, D., Nazaret B., Villani, C.: A mass transportation approach to sharp Sobolev and Gagliardo–Nirenberg inequalities. Advances in Math., to appear

[K]   Knothe, H.: Contributions to the theory of convex bodies. Michigan Math. J., **4**, 39–52 (1957)

[L]   Lieb, E.H.: Gaussian kernels have only Gaussian maximizers. Invent. Math., **102**, 179–208 (1990)

[M]   Marton, K.: Bounding $\bar{d}$-distance by informational divergence: A method to prove measure concentration. The Annals of Probability, **24**, 857–866 (1996)

[Mc1]   McCann, R.J.: Existence and uniqueness of measure preserving maps. Duke Math. J., **80**, 309–323 (1995)

[Mc2]   McCann, R.J.: A convexity principle for interacting gases. Adv. Math., **228**, 153–179 (1997)

[OV]   Otto, F., Villani, C.: Generalisation of an inequality by Talagrand and links with the logarithmic Sobolev inequality. J. Funct. Anal., **173**, 361–400 (2000)

[S]   Sudakov, V.N.: Geometric problems in the theory of infinite-dimensional distributions. Proc. Steklov Inst. Math., **141**, 1–178 (1979)

[T]   Talagrand, M.: Transportation cost for Gaussian and other product measures. Geometric and Functional Analysis, **6**, 587–599 (1996)

[TW]   Trudinger, N.S., Wang, X.J.: On the Monge mass transfer problem. Calc. Var. PDE, **13**, 19–31 (2001)

[V]   Villani, C.: Topics in optimal transportation. Graduate Studies in Mathematics, **58**, Amer. Math. Soc. (2003)

A Continuous Version
of the Brascamp–Lieb Inequalities

F. Barthe

Institut de Mathématiques, Laboratoire de Statistiques et Probabilités-CNRS
UMR C5583, Université Paul Sabatier, 118 route de Narbonne, 31062 Toulouse
Cedex 4, France `barthe@math.ups-tlse.fr`

Summary. We describe continuous extensions of the convexity versions of the
Brascamp–Lieb inequality and its inverse form.

1 Introduction

Throughout this note, $\mathbb{R}^n$ is equipped with its canonical Euclidean structure
$(\mathbb{R}^n, \langle \cdot, \cdot \rangle, |\cdot|)$. The canonical basis is denoted by $(e_1, \ldots, e_n)$ and I_n is the
identity map. Finally $(u \otimes u)(x) = \langle x, u \rangle u$.

Keith Ball's version of the Brascamp–Lieb inequality (see [BrL, B2, B3]
and the survey [B1]) and its inverse form [Ba1, Ba2] involve a decomposition
of the identity

$$c_i > 0, \quad u_i \in \mathbb{R}^n \quad |u_i| = 1, \quad \sum_{i=1}^{m} c_i \, u_i \otimes u_i = I_n, \tag{1}$$

or equivalently an isotropic measure $\mu = \sum_{i=1}^{m} c_i \delta[u_i]$ on the unit sphere
S^{n-1} of $\mathbb{R}^n$ ($\delta[x]$ stands for the Dirac mass at x). They assert that for every
non-negative measurable $f_i : \mathbb{R} \to \mathbb{R}^+$ one has

$$\int_{\mathbb{R}^n} \prod_{i \leq m} f_i^{c_i}(\langle x, u_i \rangle) \, dx \leq \prod_{i \leq m} \left(\int_{\mathbb{R}} f_i \right)^{c_i} \tag{2}$$

and if $h : \mathbb{R}^n \to \mathbb{R}^+$ is measurable and $h(\sum_{i \leq m} c_i \theta_i u_i) \geq \prod_{i \leq m} f_i^{c_i}(\theta_i)$ for all
$(\theta_i)_{i \leq m} \in \mathbb{R}^m$ then

$$\int_{\mathbb{R}^n} h(x) \, dx \geq \prod_{i \leq m} \left(\int_{\mathbb{R}} f_i \right)^{c_i}. \tag{3}$$

In the applications given by Ball, decompositions as (1) appeared when pro-
jecting the unit ball of ℓ_p^n onto a subspace, or when considering a convex
body with the Euclidean unit ball as maximal volume ellipsoid. Many more

isotropic measures were recently investigated, in relation to various optimal positions of convex bodies (see [GP, GM, GMR] and in particular the paper by Giannopoulos and Milman, where an overview is provided). Some of them are not discrete any more. For this reason the Brascamp–Lieb inequality could not be applied directly. It was necessary to first approximate the body by other ones with discrete associated isotropic measure (which required specific tricks in particular situations) and then to apply (2) or (3). Very recently, Lutwak, Yang and Zhang [LYZ2] proved optimal volume inequalities for balls of subspaces of L_p associated to arbitrary isotropic measures, with complete equality conditions. Their argument builds on Ball's approach and our transportation proof for the Brascamp–Lieb inequalities. These authors emphasize that they avoid the use of Brascamp–Lieb inequalities. In the last section of this note we put forward a version of these inequalities for arbitrary isotropic measures, which is implicit in their work. We think that it could be useful for further application and that it provides a nice unified framework for the various transport arguments written in [LYZ2, LYZ1]. In the second section we briefly develop a very simple approximation argument to get their result from the previously known ones, though without equality conditions.

2 The Approximation Argument

A Borel measure μ on the sphere is isotropic if it satisfies

$$\int_{S^{n-1}} u \otimes u \, d\mu(u) = I_n. \tag{4}$$

Note that $\mu(S^{n-1}) = n$ by taking traces. Let $p \geq 1$ and let K be the convex body in $\mathbb{R}^n$ defined by its associated norm

$$\|x\|_K^p = \int_{S^{n-1}} |\langle x, u \rangle|^p d\mu(u). \tag{5}$$

The paper [LYZ2] shows that $\mathrm{Vol}(K) \leq \mathrm{Vol}(B_p^n)$ and $\mathrm{Vol}(K^\circ) \geq \mathrm{Vol}(B_q^n)$, where B_p^n is the unit ball of ℓ_p^n, K° is the polar of K and q is the dual exponent of p. If μ is a sum of finitely many atoms, this recovers one step in Ball's upper bounds of volume ratio [B3], and our corresponding dual results [Ba2]. We recall the proof here: assume that $\mu = \sum_{i=1}^m c_i \delta[u_i]$ (with distinct u_i's), then by standard level sets integration and Brascamp–Lieb

$$\Gamma\left(1 + \frac{n}{p}\right)\mathrm{Vol}(K) = \int_{\mathbb{R}^n} e^{-\|x\|_K^p} dx = \int_{\mathbb{R}^n} \prod_{i \leq m} e^{-c_i |\langle x, u_i \rangle|^p} dx$$

$$\leq \left(\int_{\mathbb{R}} e^{-|t|^p} dt\right)^{\sum_{i \leq m} c_i} = \left(2\Gamma\left(1 + \frac{1}{p}\right)\right)^n.$$

So $\mathrm{Vol}(K) \leq \mathrm{Vol}(B_p^n)$. It follows from [Ba2] that equality occurs only when $(u_i)_{i \leq m}$ is an orthonormal basis, that is, when K is isometric to B_p^n.

For the dual bound, we need the inequality (it is actually an equality)

$$\|x\|_{K^\circ}^q \leq \inf \left\{ \sum_{i \leq m} c_i |\theta_i|^q; \ \sum_{i \leq m} c_i \theta_i u_i = x \right\}.$$

To see this, we write for $x = \sum c_i \theta_i u_i$

$$\langle x, y \rangle = \sum c_i \theta_i \langle u_i, y \rangle \leq \left(\sum c_i |\theta_i|^q \right)^{\frac{1}{q}} \left(\sum c_i |\langle y, u_i \rangle|^p \right)^{\frac{1}{p}}.$$

So the inverse Brascamp–Lieb inequality yields

$$\Gamma\left(1 + \frac{n}{q}\right) \mathrm{Vol}(K^\circ) = \int_{\mathbb{R}^n} e^{-\|x\|_{K^\circ}^q} \, dx \geq \int_{\mathbb{R}^n} \sup_{\sum c_i \theta_i u_i = x} \prod_{i \leq m} e^{-c_i |\theta_i|^q} \, dx$$

$$\geq \left(\int_{\mathbb{R}} e^{-|t|^p} \, dt \right)^{\sum_{i \leq m} c_i} = \left(2\Gamma\left(1 + \frac{1}{q}\right) \right)^n.$$

So $\mathrm{Vol}(K^\circ) \geq \mathrm{Vol}(B_q^n)$. Due to the characterisation of equality cases in (3) [Ba2], equality may occur only when K is isometric to B_p^n.

Next we give an approximation argument to reach bodies K associated to arbitrary isotropic measures. For $\varepsilon > 0$, we choose a (finite) maximal ε-net N in S^{n-1} together with a partition of S^{n-1} into Borel sets $(U_x)_{x \in N}$ with $U_x \subset B(x, \varepsilon)$ (the closed ball of radius ε around x). Given a finite measure μ on S^{n-1} we consider its approximation

$$\mu_\varepsilon = \sum_{x \in N} \mu(U_x) \delta[x].$$

For any continuous $f : S^{n-1} \to \mathbb{R}$,

$$\left| \int f \, d\mu - \int f \, d\mu_\varepsilon \right| = \left| \sum_{x \in N} \int_{U_x} (f - f(x)) \, d\mu \right| \leq \mu(S^{n-1}) \omega_f(\varepsilon),$$

where ω_f is the modulus of continuity of f. So when ε goes to zero, μ_ε tends weakly to μ.

Next, if we start from an isotropic measure μ, the measure μ_ε is close to isotropic. Indeed $Q_\varepsilon = \int u \otimes u \, d\mu_\varepsilon$ considered as a matrix has coefficients $(Q_\varepsilon)_{i,j} = \int \langle u, e_i \rangle \langle u, e_j \rangle \, d\mu_\varepsilon(u)$. Since these numbers involve integrals against $n(n-1)/2$ continuous functions (actually two, up to rotations), there exists a function $\omega_2(\varepsilon)$ with limit zero at zero such that $|(Q_\varepsilon - I_n)_{i,j}| \leq \omega_2(\varepsilon)$. So when ε is small enough, $Q_\varepsilon^{-1/2}$ is defined. Multiplying left and right by $Q_\varepsilon^{-1/2}$, the equation that defines Q_ε yields

$$I_n = \int_{S^{n-1}} Q_\varepsilon^{-\frac{1}{2}} u \otimes Q_\varepsilon^{-\frac{1}{2}} u \, d\mu(u) = \int_{S^{n-1}} v \otimes v \, d\nu_\varepsilon(v),$$

where

$$\nu_\varepsilon = \sum_{x \in N} \mu_\varepsilon(x) |Q_\varepsilon^{-\frac{1}{2}} x|^2 \delta \left[\frac{Q_\varepsilon^{-\frac{1}{2}} x}{|Q_\varepsilon^{-\frac{1}{2}} x|} \right].$$

Finally for any continuous function f on the sphere

$$\left| \int f \, d\nu_\varepsilon - \int f \, d\mu_\varepsilon \right| \le \sum_{x \in N} \mu_\varepsilon(x) \left| |Q_\varepsilon^{-\frac{1}{2}} x|^2 f \left(\frac{Q_\varepsilon^{-\frac{1}{2}} x}{|Q_\varepsilon^{-\frac{1}{2}} x|} \right) - f(x) \right|$$

$$\le \mu(S^{n-1}) \max_{x \in S^{n-1}} \left| |Q_\varepsilon^{-\frac{1}{2}} x|^2 f \left(\frac{Q_\varepsilon^{-\frac{1}{2}} x}{|Q_\varepsilon^{-\frac{1}{2}} x|} \right) - f(x) \right|.$$

Since $\|Q_\varepsilon - I_n\|_{\ell_2^n \to \ell_2^n} \le c\omega_2(\varepsilon)$, the above quantity can be upper-bounded by a function $\omega_3(\varepsilon)$ with limit zero at zero, and depending only on ω_2, on the modulus of continuity ω_f and on $\max |f|$ (we omit the details).

We have built an isotropic finite sum of Dirac masses ν_ε with $|\int f d\mu - \int f d\nu_\varepsilon| \le \omega_{4,f}(\varepsilon)$ and $\lim_{0^+} \omega_{4,f} = 0$ where $\omega_{4,f}$ depends only on the modulus of continuity of f and on its ℓ_∞-norm. This is enough to approximate the body K given in (5). We define K_ε by

$$\|x\|_{K_\varepsilon}^p = \int_{S^{n-1}} |\langle x, u \rangle|^p d\nu_\varepsilon(u), \quad x \in S^{n-1}.$$

By the previous argument, for all x on the sphere, $\lim_{\varepsilon \to 0} \|x\|_{K_\varepsilon} = \|x\|_K$. Moreover, since the functions $u \in S^{n-1} \mapsto |\langle x, u \rangle|^p$ have the same modulus of continuity and the same supremum, the above convergence is uniform on S^{n-1}. It follows that K_ε tends to K in Hausdorff distance. The inequalities $\mathrm{Vol}(K_\varepsilon) \le \mathrm{Vol}(B_p^n)$ and $\mathrm{Vol}(K_\varepsilon^\circ) \ge \mathrm{Vol}(B_q^n)$, recalled in the beginning of this section, easily pass to the limit K.

3 The Continuous Version of the Inequalities

We present extensions of (2) and (3) to arbitrary isotropic distributions of vectors on the sphere. Our statement contains an assumption on the functions. This is not a problem for applications in convexity, where only nicely behaved functions are used. Moreover it is clear that approximation arguments could be combined with our result in order to relax the assumptions. We say that a family of functions $(f_u)_{u \in S^{n-1}}$ verifies Hypothesis (H) if it satisfies the following two conditions:

- There exists a positive continuous function $F : S^{n-1} \times \mathbb{R} \to (0, +\infty)$ and two functions a, b on S^{n-1} with $a < b$ pointwise and a (resp. b) is either real-valued continuous or constant with value in $\{-\infty, +\infty\}$ such that for all $(u, t) \in S^{n-1} \times \mathbb{R}$

$$f_u(t) = \mathbf{1}_{t \in [a(u), b(u)]} F(u, t).$$

- There exists a function $\mathcal{F} \in L_1(\mathbb{R}) \cap L_\infty(\mathbb{R})$ such that for all $u \in S^{n-1}$, one has $0 \leq f_u \leq \mathcal{F}$.

Theorem 1. *Let μ be an isotropic Borel measure on S^{n-1} and for $u \in S^{n-1}$ let $f_u : \mathbb{R} \to \mathbb{R}^+$ such that the family $(f_u)_{u \in S^{n-1}}$ satisfies Hypothesis (H). Then*

$$\int_{\mathbb{R}^n} \exp\left(\int_{S^{n-1}} \log f_u(\langle x, u \rangle) \, d\mu(u)\right) dx \leq \exp\left(\int_{S^{n-1}} \log\left(\int_{\mathbb{R}} f_u\right) d\mu(u)\right).$$
(6)

The reverse inequality states that if h is measurable and verifies for every integrable function θ on the sphere

$$h\left(\int_{S^{n-1}} \theta(u) \, u \, d\mu(u)\right) \geq \exp\left(\int_{S^{n-1}} \log f_u(\theta(u)) \, d\mu(u)\right)$$

then

$$\int_{\mathbb{R}^n} h \geq \exp\left(\int_{S^{n-1}} \log\left(\int_{\mathbb{R}} f_u\right) d\mu(u)\right).$$
(7)

Remark. Note that the logarithmic terms take values in $[-\infty, +\infty)$ and that their integral make sense also in $[-\infty, +\infty)$, since by (H) their positive parts converge.

As in [Ba2] the two inequalities can be derived from a single one, which is proved along the same lines.

Lemma 1. *Let $(f_u)_{u \in S^{n-1}}, (g_u)_{u \in S^{n-1}}$ be two families of functions satisfying (H). Then*

$$\exp\left(\int_{S^{n-1}} \log\left(\int_{\mathbb{R}} g_u\right) d\mu(u)\right) \int_{\mathbb{R}^n} \exp\left(\int_{S^{n-1}} \log f_u(\langle x, u \rangle) \, d\mu(u)\right) dx$$

$$\leq \exp\left(\int_{S^{n-1}} \log\left(\int_{\mathbb{R}} f_u\right) d\mu(u)\right)$$

$$\cdot \int_{\mathbb{R}^n} \sup_{y = \int \theta(u) u \, d\mu(u)} \exp\left(\int_{S^{n-1}} \log g_u(\theta(u)) \, d\mu(u)\right) dy.$$

Proof of Theorem 1. The extension (6) of the Brascamp–Lieb inequality follows from the choice $g_u(t) = \exp(-\pi t^2)$ in Lemma 1. Indeed if $y = \int \theta(u) \, u \, d\mu(u)$ then

$$|y|^2 = \left\langle y, \int_{S^{n-1}} \theta(u) \, u \, d\mu(u) \right\rangle = \int_{S^{n-1}} \theta(u)\langle y, u \rangle \, d\mu(u)$$

$$\leq \left(\int_{S^{n-1}} \theta(u)^2 \, d\mu(u)\right)^{\frac{1}{2}} \left(\int_{S^{n-1}} \langle y, u \rangle^2 \, d\mu(u)\right)^{\frac{1}{2}} = \left(\int_{S^{n-1}} \theta(u)^2 \, d\mu(u)\right)^{\frac{1}{2}} |y|,$$

by the isotropy condition (4). Therefore for $g_u(t) = \exp(-\pi t^2)$,

$$\int_{\mathbb{R}^n}^* \sup_{y = \int \theta(u) \, u \, d\mu(u)} \exp\left(\int_{S^{n-1}} \log g_u(\theta(u)) \, d\mu(u)\right) dy \leq \int_{\mathbb{R}^n} e^{-\pi |y|^2} \, dy = 1.$$

The inverse inequality (7) follows from Lemma 1 by choosing $f_u(t) = \exp(-\pi t^2)$, when

$$\int_{\mathbb{R}^n} \exp\left(\int_{S^{n-1}} \log f_u(\langle x, u \rangle) \, d\mu(u)\right) dx$$

$$= \int_{\mathbb{R}^n} \exp\left(\int_{S^{n-1}} -\pi \langle x, u \rangle^2 \, d\mu(u)\right) dx = \int_{\mathbb{R}^n} e^{-\pi |x|^2} \, dx = 1. \qquad \square$$

Proving Lemma 1 requires the extension to arbitrary isotropic measures of a determinant inequality observed by Ball (see e.g. Proposition 9 in [Ba2]). It is established by Lutwak, Yang and Zhang in [LYZ2] with a neat description of equality cases. We quote their result here

Lemma 2. *If μ is an isotropic Borel measure on S^{n-1} and if $t : S^{n-1} \to (0, \infty)$ is continuous then*

$$\det\left(\int_{S^{n-1}} t(u) \, u \otimes u \, d\mu(u)\right) \geq \exp\left(\int_{S^{n-1}} \log\left(t(u)\right) d\mu(u)\right).$$

There is equality if and only if the quantity $t(v_1) \ldots t(v_n)$ is constant for linearly independent $v_1, \ldots, v_n$ in $\mathrm{Supp}(\mu)$.

Proof of Lemma 1. We may assume that the left hand side in the conclusion of Lemma 1 is positive. Note that the second exponential in this term vanishes outside the set

$$S = \{x \in \mathbb{R}^n; \text{ for } \mu - \text{a.e. } u \in S^{n-1}, \, a(u) \leq \langle x, u \rangle \leq b(u)\}$$
$$= \{x \in \mathbb{R}^n; \text{ for all } u \in \mathrm{supp}(\mu), \, a(u) \leq \langle x, u \rangle \leq b(u)\},$$

where the equality follows from the continuity of a, b. This set is closed and convex. Thus its boundary is Lebesgue-negligible and we may consider the integral on its interior. One easily checks that

$$\mathrm{Int}(S) = \Big\{x \in \mathbb{R}^n; \sup_{u \in \mathrm{supp}(\mu)} \max\left(\langle x, u \rangle - b(u), a(u) - \langle x, u \rangle\right) < 0\Big\}.$$

For every $u \in S^{n-1}$ we define the map $T_u : (a(u), b(u)) \to (c(u), d(u))$ by

$$\frac{\int_{a(u)}^t f_u}{\int_{\mathbb{R}} f_u} = \frac{\int_{c(u)}^{T_u(t)} g_u}{\int_{\mathbb{R}} g_u}.$$

Our hypotheses ensure that $(u, t) \mapsto T_u(t)$ is continuous on the open set $\{(u, t); \ u \in S^{n-1} \text{ and } a(u) < t < b(u)\}$ (note that the second assertion in (H) allows an extensive use of dominated convergence). Moreover for every u, T_u is differentiable, strictly increasing and verifies

$$f_u(t) \int_{\mathbb{R}} g_u = g_u(T_u(t)) \, T'_u(t) \int_{\mathbb{R}} f_u \quad \text{for } t \in (a(u), b(u)).$$

Thus for every $x \in \text{Int}(S)$, the maps $u \mapsto T_u(\langle x, u\rangle)$ and $u \mapsto T'_u(\langle x, u\rangle)$ are continuous and therefore bounded on the sphere. This ensures that the integrals in the following lines are well defined.

We consider the map $T : \text{Int}(S) \subset \mathbb{R}^n \to \mathbb{R}^n$ defined by

$$T(x) = \int_{S^{n-1}} T_u(\langle x, u\rangle) \, u \, d\mu(u).$$

By dominated convergence

$$dT(x) = \int_{S^{n-1}} T'_u(\langle x, u\rangle) \, u \otimes u \, d\mu(u).$$

Note that $dT(x)$ is a symmetric semi-definite map. Its determinant is bounded from below by Lemma 2:

$$\det \big(dT(x)\big) \geq \exp\left(\int_{S^{n-1}} \log T'_u(\langle x, u\rangle) \, d\mu(u) \right). \tag{8}$$

This shows in particular that $dT(x)$ cannot be degenerate, therefore it is positive and T is injective. Finally, if h is measurable and satisfies for every integrable θ

$$h\left(\int_{S^{n-1}} \theta(u) \, u \, d\mu(u) \right) \geq \exp\left(\int_{S^{n-1}} \log g_u(\theta(u)) \, d\mu(u) \right),$$

we combine the previous results as follows:

$$\exp\left(\int_{S^{n-1}} \log\left(\int_{\mathbb{R}} f_u \right) d\mu(u) \right) \int_{\mathbb{R}^n} h(y) \, dy$$

$$\geq \exp\left(\int_{S^{n-1}} \log\left(\int_{\mathbb{R}} f_u \right) d\mu(u) \right) \int_{\text{Int}(S)} h(T(x)) \det\big(dT(x)\big) \, dx$$

$$\geq \exp\left(\int_{S^{n-1}} \log\left(\int_{\mathbb{R}} f_u \right) d\mu(u) \right)$$

$$\cdot \int_{\text{Int}(S)} \exp\left(\int_{S^{n-1}} \log g_u\left(T_u(\langle x, u\rangle)\right) d\mu(u) \right) \det\big(dT(x)\big) \, dx$$

$$\geq \int_{\text{Int}(S)} \exp\left(\int_{S^{n-1}} \log\left(g_u\left(T_u(\langle x, u\rangle)\right) \int_{\mathbb{R}} f_u \right) d\mu(u) \right)$$

$$\cdot \exp\left(\int_{S^{n-1}} \log T'_u(\langle x, u \rangle) \, d\mu(u) \right) dx$$

$$= \int_{\text{Int}(S)} \exp\left(\int_{S^{n-1}} \log\left(g_u\left(T_u(\langle x, u \rangle) \right) T'_u(\langle x, u \rangle) \int_{\mathbb{R}} f_u \right) d\mu(u) \right) dx$$

$$= \int_{\text{Int}(S)} \exp\left(\int_{S^{n-1}} \log\left(f_u(\langle x, u \rangle) \int_{\mathbb{R}} g_u \right) d\mu(u) \right) dx$$

$$= \exp\left(\int_{S^{n-1}} \log\left(\int_{\mathbb{R}} g_u \right) d\mu(u) \right) \int_{\mathbb{R}^n} \exp\left(\int_{S^{n-1}} \log f_u(\langle x, u \rangle) \, d\mu(u) \right) dx.$$

$\square$

In [Ba2] we were able to characterize equality cases in (2) and (3). The description was a bit tedious, but we put forward a convenient result for non-Gaussian functions. Recall that $f : \mathbb{R} \to \mathbb{R}^+$ is Gaussian if there exists $a > 0, b, c \in \mathbb{R}$ such that for every $t \in \mathbb{R}$, $f(t) = \exp(-at^2 + bt + c)$. Next, we extend this result to (6) and (7). The argument follows [Ba2] and [LYZ2].

Theorem 2. *Let μ be an isotropic Borel measure on S^{n-1}. Let $(f_u)_{u \in S^{n-1}}$ satisfy (H) and assume that none of the f_u's is Gaussian. If there is equality in (6) or in (7) then there exists an orthonormal basis $(u_1, \ldots, u_n)$ of $\mathbb{R}^n$ such that $\text{Supp}(\mu) \subset \{u_1, -u_1, \ldots, u_n, -u_n\}$.*

Remarks.

1. This means that for non-Gaussian functions equality occurs only when the directions are orthogonal and all the integrals on $\mathbb{R}^n$ are products of integrals on $\mathbb{R}$, due to Fubini's theorem.
2. Thanks to Theorems 1 and 2, the argument given in the beginning of Section 2 can be processed for arbitrary measures μ on the sphere, thus recovering the results of [LYZ2].

Proof of Theorem 2. Assume $n \geq 2$. Our task is to analyse equality cases in Lemma 1 in two cases:

1. (f_u) are non-Gaussian and $g_u(t) = \exp(-\pi t^2)$ for all $u \in S^{n-1}$, $t \in \mathbb{R}$.
2. $f_u(t) = \exp(-\pi t^2)$ for all $u \in S^{n-1}$, $t \in \mathbb{R}$ and (g_u) are non-Gaussian.

We deal with both cases at a time. First we assume, with the notation of the proof of Lemma 1, that $a = -\infty$ and $b = +\infty$, so that $S = \mathbb{R}^n$.

The isotropy condition prevents μ from being supported in a hyperplane. So let $(u_1, \ldots, u_n)$ be a basis of $\mathbb{R}^n$ included in $\text{Supp}(\mu)$. Let $v \in \text{Supp}(\mu)$. Assume that v is linearly independent of $(u_2, \ldots, u_n)$. Equality in Lemma 1 implies that every step in the proof is an equality. In particular for a.e. x the determinant inequality (8) is. By Lemma 2 and continuity, it follows that for all $x \in \mathbb{R}^n$,

$$T'_v(\langle x, v\rangle)T'_{u_2}(\langle x, u_2\rangle)\ldots T'_{u_n}(\langle x, u_n\rangle)$$
$$= T'_{u_1}(\langle x, u_1\rangle)T'_{u_2}(\langle x, u_2\rangle)\ldots T'_{u_n}(\langle x, u_n\rangle).$$

Since the previous derivatives do not vanish, we conclude that

$$T'_v(\langle x, v\rangle) = T'_{u_1}(\langle x, u_1\rangle), \quad x \in \mathbb{R}^n.$$

If $v \notin \mathbb{R}u_1 \cap \mathrm{Supp}(\mu) = \{-u_1, u_1\}$ then we can find $w \in \mathbb{R}^n$ such that $\langle w, u_1\rangle = 0$ and $\langle w, v\rangle = 1$. For $x = tw$ the latter yields $T'_v(t) = T'_{u_1}(0)$ for all $t \in \mathbb{R}$. Thus T_v is affine, and the relation

$$\frac{f_v(t)}{\int_{\mathbb{R}} f_v} = T'_v(t)\frac{g_v(T_v(t))}{\int_{\mathbb{R}} g_v}$$

contradicts assumption 1 (resp. 2), which tells that in $\{f_v, g_v\}$ there is a Gaussian function and a non-Gaussian function.

As a conclusion, we obtain that for every $v \in \mathrm{Supp}(\mu)$ either $v \in \{-u_1, u_1\}$ or $v \in \mathrm{span}\{u_2, \ldots, u_n\}$. This also holds for any u_i instead of u_1, and this property easily implies that $v = \pm u_j$ for some j. Therefore $\mathrm{Supp}(\mu) \subset \{-u_1, u_1, \ldots, -u_n, u_n\}$. As noted in [LYZ2] this ensures that $(u_i)_{i \leq n}$ is orthonormal. Indeed the isotropy condition reads as

$$I_n = \int_{S^{n-1}} u \otimes u \, d\mu(u) = \sum_{i \leq n} c_i u_i \otimes u_i,$$

where $c_i = \mu(\{-u_i, u_i\})$. Applying this to u_j gives $u_j = \sum_{i \leq n} c_i \langle u_j, u_i\rangle u_i$. By uniqueness of the decomposition in a basis, we get $\langle u_i, u_j\rangle = 0$ if $i \neq j$ and $c_i = 1$. This completes the proof when $a = -\infty, b = \infty$. In particular it settles equality conditions for (7) in general, and for (6) only when the functions f_u are everywhere positive. In principle the latter argument could be refined when this is not the case. For technical reasons we prefer to conclude with an observation of K. Ball, which we establish for general μ in the next lemma. It guarantees that if (f_u) gives equality in (6) and satisfies (H) then $(f_u * g)$ enjoys the same properties, where $g(t) = \exp(-t^2)$, $t \in \mathbb{R}$. If f_u is non-Gaussian then neither is $f_u * g$ (check on Fourier transforms). These new functions are everywhere positive, non-Gaussian and give equality in (6). By the preceding result, μ is supported in the union of an orthonormal basis and the symmetric basis. The proof is therefore complete. □

As discovered by Ball, the convolution of maximisers (i.e. functions for which equality holds) in the Brascamp–Lieb inequality is again a maximiser (see Lemma 2 in [Ba3]). We need the following particular case for general μ. The argument is the same.

Lemma 3. *If $(f_u)_{u \in S^{n-1}}$ satisfies (H) and gives equality in (6), then so does the family $(f_u * g)_{u \in S^{n-1}}$, where $g(t) = \exp(-\pi t^2)$, $t \in \mathbb{R}$.*

Proof. Hypothesis (H) is easy to check. For every $x \in \mathbb{R}^n$ we consider

$$f_u^{(x)}(t) := f_u(\langle x, u \rangle - t) g(t).$$

These functions satisfy (H) as well and the extension of the Brascamp–Lieb inequality gives

$$\int_{\mathbb{R}^n} \exp \left(\int_{S^{n-1}} \log \left(f_u(\langle x, u \rangle - \langle y, u \rangle) g(\langle y, u \rangle) \right) d\mu(u) \right) dy$$

$$\leq \exp \left(\int_{S^{n-1}} \log \left(\int_{\mathbb{R}} f_u(\langle x, u \rangle - t) g(t) \, dt \right) d\mu(u) \right).$$

Set $\Phi(x) = \exp \left(\int \log f_u(\langle x, u \rangle) d\mu(u) \right)$ and $\Gamma(y) = \exp \left(\int \log g(\langle y, u \rangle) d\mu(u) \right) = \exp(-\pi|y|^2)$. The latter inequality may be rewritten as

$$\Phi * \Gamma(x) \leq \exp \left(\int_{S^{n-1}} \log(f_u * g)(\langle x, u \rangle) d\mu(u) \right).$$

Integrating over $x \in \mathbb{R}^n$ yields

$$\int_{\mathbb{R}^n} \exp \left(\int \log(f_u * g)(\langle x, u \rangle) d\mu(u) \right) dx$$

$$\geq \int_{\mathbb{R}^n} \Phi(x) \, dx \int_{\mathbb{R}^n} \Gamma(y) \, dy = \int_{\mathbb{R}^n} \exp \left(\int_{S^{n-1}} \log f_u(\langle x, u \rangle) d\mu(u) \right) dx$$

$$= \exp \left(\int_{S^{n-1}} \log \left(\int_{\mathbb{R}} f_u \right) d\mu(u) \right) = \exp \left(\int_{S^{n-1}} \log \left(\int_{\mathbb{R}} f_u * g \right) d\mu(u) \right).$$

So $(f_u * g)_{u \in S^{n-1}}$ gives equality in the Brascamp–Lieb inequality (6).   □

Acknowledgement. We thank Erwin Lutwak, Deane Yang and Gaoyong Zhang for communicating their preprints to us, Vitali Milman for stimulating questions and Shahar Mendelson for his kind invitation to the Research School of Information Sciences and Engineering of the Australian National University, Canberra, where part of this work was done.

References

[B1]   Ball, K.: Convex geometry and functional analysis. In: Johnson, W.B. et al. (eds) Handbook of the Geometry of Banach Spaces, Vol. 1, 161–194. Elsevier, Amsterdam (2001)

[B2]   Ball, K.M.: Volumes of sections of cubes and related problems. In: Lindenstrauss, J., Milman, V.D. (eds) Israel Seminar on Geometric Aspects of Functional Analysis, Vol. 1376, Lectures Notes in Math. Springer-Verlag (1989)

[B3]   Ball, K.M.: Volume ratio and a reverse isoperimetric inequality. J. London Math. Soc., **44**, no. 2, 351–359 (1991)

[Ba1]   Barthe, F.: Inégalités de Brascamp-Lieb et convexité. C.R. Acad. Sci. Paris Sér. I Math., **324**, 885–888 (1997)

[Ba2]   Barthe, F.: On a reverse form of the Brascamp-Lieb inequality. Invent. Math., **134**, 335–361 (1998)

[Ba3]   Barthe, F.: Optimal Young's inequality and its converse: a simple proof. Geom. Funct. Anal., **8**, 234–242 (1998)

[BrL]   Brascamp, H.J., Lieb, E.H.: Best constants in Young's inequality, its converse and its generalization to more than three functions. Adv. Math., **20**, 151–173 (1976)

[GM]   Giannopoulos, A., Milman, V.: Extremal problems and isotropic positions of convex bodies. Israel J. Math., **117**, 29–60 (2000)

[GMR]   Giannopoulos, A., Milman, V., Rudelson, M.: Convex bodies with minimal mean width. Geometric Aspects of Functional Analysis, Vol. 1745, Lecture Notes in Math., 81–93. Springer, Berlin (2000)

[GP]   Giannopoulos, A., Papadimitrakis, M.: Isotropic surface area measures. Mathematika, **46**, no. 1, 1–13 (1999)

[LYZ1] Lutwak, E., Yang, D., Zhang, G.: Volume inequalities for isotropic measures. Preprint (2003)

[LYZ2] Lutwak, E., Yang, D., Zhang, G.: Volume inequalities for subspaces of L_p. Preprint (2003)

Inverse Brascamp–Lieb Inequalities along the Heat Equation

F. Barthe[1] and D. Cordero-Erausquin[2]

[1] Institut de Mathématiques, Laboratoire de Statistiques et Probabilités, CNRS UMR C5583, Université Paul Sabatier, 118 route de Narbonne, 31062 Toulouse Cedex 4, France `barthe@math.ups-tlse.fr`

[2] Laboratoire d'Analyse et Mathématiques Appliquées, CNRS UMR 8050, Université de Marne la Vallée, Boulevard Descartes, Cité Descartes, Champs sur Marne, 77454 Marne la Vallée Cedex 2, France `cordero@math.univ-mlv.fr`

Summary. Adapting Borell's proof of Ehrhard's inequality for general sets, we provide a semi-group approach to the reverse Brascamp–Lieb inequality, in its "convexity" version.

1 Introduction

We work in the Euclidean space $(\mathbb{R}^n, \langle \cdot, \cdot \rangle, |\cdot|)$. Given m vectors $(u_i)_{i=1}^m$ in $\mathbb{R}^n$ and m numbers $(c_i)_{i=1}^m$, we shall say that they decompose the identity of $\mathbb{R}^n$ if

$$c_i > 0, \quad |u_i| = 1, \quad \sum_{i=1}^m c_i \, u_i \otimes u_i = I_n. \tag{1}$$

This relation ensures that for every $x \in \mathbb{R}^n$,

$$x = \sum_{i=1}^m c_i \langle x, u_i \rangle u_i, \quad \text{and} \quad |x|^2 = \sum_{i=1}^m c_i \langle x, u_i \rangle^2.$$

Such a decomposition appears in John's description of the maximal volume ellipsoid. Keith Ball was able to adapt a family of inequalities by Brascamp and Lieb [BraL] to this setting. He used it in order to derive several optimal estimates of volume ratios and volumes of sections of convex bodies [Ba1, Ba2]. His version of the Brascamp–Lieb inequality is as follows. Consider for $i = 1, \dots, m$, vectors $u_i \in \mathbb{R}^n$, and numbers c_i satisfying the decomposition relation (1); then for all non-negative integrable functions $f_i : \mathbb{R} \to \mathbb{R}^+$ one has

$$\int_{\mathbb{R}^n} \prod_{i \leq m} f_i^{c_i} (\langle x, u_i \rangle) \, dx \leq \prod_{i \leq m} \left(\int_{\mathbb{R}} f_i \right)^{c_i}. \tag{2}$$

He also conjectured a reverse inequality, which was established by the first-named author [Bar1, Bar2], and had also several applications in convex geometry. It asserts that under the same assumptions, if a measurable $h : \mathbb{R}^n \to \mathbb{R}^+$ verifies for all $(\theta_1, \ldots, \theta_m) \in \mathbb{R}^m$

$$h\left(\sum_{i \leq m} c_i \theta_i u_i \right) \geq \prod_{i \leq m} f_i^{c_i}(\theta_i), \tag{3}$$

then

$$\int_{\mathbb{R}^n} h \geq \prod_{i \leq m} \left(\int_{\mathbb{R}} f_i \right)^{c_i}. \tag{4}$$

Up to now the shortest proof of (2) and (4) stands on measure transportation [Bar1]. This tool recently appeared as a powerful challenger to the semi-group interpolation method and provided new approaches and extensions of Sobolev type inequalities and other inequalities related to concentrations. For a presentation of the semi-group and of the mass transport methods, one can consult [L, B] and [V], respectively. The transport method seemed more powerful for Brascamp–Lieb and Brunn–Minkowski type inequalities.

But we shall see that the semi-group approach is still alive! In the next section we adapt a recent argument of Christer Borell [Bor3]. He gave an impressive proof of Ehrhard's inequality for general sets based on the precise study of inequalities along the heat semi-group. We recover by similar arguments the reverse Brascamp–Lieb inequality (4). It should be mentioned that Borell already used (somewhat different) diffusion semi-groups arguments in the setting of Prékopa–Leindler type inequalities [Bor2, Bor1]. It is natural to ask whether there exists a semi-group proof of (2) since the two inequalities (2) and (4) were obtained simultaneously in the transport method. The answer is yes. We learned from Eric Carlen of a major work in preparation by Carlen, Lieb and Loss [CLL] containing a semi-group approach of quite general Brascamp–Lieb inequalities with striking applications to kinetic theory. The (surprisingly) short argument we present in the third section is nothing else but a particular case of their work where "geometry" provides welcome simplifications.

2 Proof of the Inverse Brascamp–Lieb Inequality

We work with the heat semi-group defined for $f : \mathbb{R}^n \to \mathbb{R}^+$ and $t \geq 0$ by

$$P_t f(x) = \int_{\mathbb{R}^n} f(x + \sqrt{t}\, z) \, d\gamma_n(z),$$

where γ_n is the standard Gaussian measure on $\mathbb{R}^n$, with density $(2\pi)^{-n/2}$ $\exp(-|z|^2/2)$, $z \in \mathbb{R}^n$ with respect to Lebesgue's measure. Under appropriate assumptions, $g(t, x) = P_t f(x)$ solves the equation

$$\begin{cases} \dfrac{\partial g}{\partial t} = \dfrac{1}{2}\Delta g, \\ g(0, \cdot) = f. \end{cases}$$

In this equation and in the rest of the paper, the Laplacian and the gradient act on space variables only. Note that the function $(t, x) \mapsto F^{(t)}(x) = \log P_t f(x)$ satisfies

$$\begin{cases} 2\dfrac{\partial F^{(t)}}{\partial t}(x) = \Delta F^{(t)}(x) + |\nabla F^{(t)}(x)|^2, \\ F^{(0)} = \log f. \end{cases} \tag{5}$$

Let $(u_i)_{i=1}^m$ be vectors in $\mathbb{R}^n$ and $(c_i)_{i=1}^m$ be non-negative numbers, inducing a decomposition of the identity (1). Let $(f_i)_{i=1}^m$ be non-negative integrable functions on $\mathbb{R}$ and let h on $\mathbb{R}^n$ satisfy (3). Introduce $H^{(t)} = \log P_t h : \mathbb{R}^n \to \mathbb{R}$ and $F_i^{(t)} = \log P_t f_i : \mathbb{R} \to \mathbb{R}$, where we use the same notation for the n-dimensional and the one-dimensional semi-groups. We define

$$C(t, \theta_1, \dots, \theta_m) := H^{(t)}\left(\sum_{i=1}^m c_i \theta_i u_i\right) - \sum_{i=1}^m c_i F_i^{(t)}(\theta_i) \tag{6}$$

$$= H^{(t)}(\Theta) - \sum_{i=1}^m c_i F_i^{(t)}(\theta_i)$$

with the notation $\Theta = \Theta(\theta_1, \dots, \theta_m) = \sum_{i=1}^m c_i \theta_i\, u_i \in \mathbb{R}^n$. The evolution equations (5) satisfied by $F_i^{(t)}$ and $H^{(t)}$ give

$$2\dfrac{\partial C}{\partial t}(t, \theta_1, \dots, \theta_m) = \Delta H^{(t)}(\Theta) - \sum_{i=1}^m c_i (F_i^{(t)})''(\theta_i)$$

$$+ |\nabla H^{(t)}|^2(\Theta) - \sum_{i=1}^m c_i (F_i^{(t)})'^2(\theta_i).$$

For simplicity, we shall write $H = H^{(t)}$ and $F_i = F_i^{(t)}$, keeping in mind that we are working at a fixed t. Following Borell we try to express the right hand side in terms of C. We start with the second order terms. From the definition (6), we get

$$\dfrac{\partial^2 C}{\partial \theta_i \partial \theta_j}(t, \theta_1, \dots, \theta_m) = c_i c_j \langle D^2 H(\Theta) u_i, u_j \rangle - \delta_{i,j}\, c_i F''(\theta_i).$$

By the decomposition of the identity (1), we have $D^2 H = \sum c_i (D^2 H u_i) \otimes u_i$ and

$$\Delta H = \operatorname{tr} D^2 H = \sum_{i \le m} c_i \langle D^2 H u_i, u_i \rangle = \sum_{i,j \le m} c_i c_j \langle u_i, u_j \rangle \langle D^2 H u_i, u_j \rangle.$$

Therefore if we denote by $\mathcal{E}$ the degenerate elliptic operator

$$\mathcal{E} = \sum_{i,j \leq m} \langle u_i, u_j \rangle \frac{\partial^2}{\partial \theta_i \partial \theta_j},$$

we have, since $|u_i| = 1$,

$$\mathcal{E}C(t, \theta_1, \ldots, \theta_m) = \Delta H(\Theta) - \sum_{i=1}^{m} c_i F_i''(\theta_i). \tag{7}$$

Next we turn to first order terms. The gradient of C has coordinates

$$\frac{\partial C}{\partial \theta_i}(t, \theta_1, \ldots, \theta_m) = c_i \big(\langle \nabla H(\Theta), u_i \rangle - F_i'(\theta_i) \big).$$

We introduce the vector valued function $b : \mathbb{R} \times \mathbb{R}^m \to \mathbb{R}^m$ with coordinates

$$b_i(t, \theta_1, \ldots, \theta_m) = \langle \nabla H(\Theta), u_i \rangle + F_i'(\theta_i), \tag{8}$$

so that

$$\langle b, \nabla C \rangle(t, \theta_1, \ldots, \theta_m) = \sum_{i \leq m} c_i \left(\langle \nabla H(\Theta), u_i \rangle^2 - F_i'^2(\theta_i) \right)$$

$$= |\nabla H|^2(\Theta) - \sum_{i \leq m} c_i F_i'^2(\theta_i),$$

where we have used again the decomposition of identity (1). Finally, the evolution equation for C can be rewritten as

$$2 \frac{\partial C}{\partial t} = \mathcal{E}C + \langle b, \nabla C \rangle. \tag{9}$$

Note that we are now working in the m-dimensional Euclidean space. It follows from the theory of parabolic-elliptic evolution equations that when $C(0, \cdot)$ is a non-negative function, then for all $t > 0$, $C(t, \cdot)$ remains non-negative. Alternately, an explicit expression of C can be given in terms of stochastic processes, on which this property can be read. See the remark after the proof.

By the hypothesis (3) we know that $C(0, \cdot) \geq 0$. At time $\sigma > 0$, $C(\sigma, 0, \ldots, 0) \geq 0$ reads as

$$P_\sigma h(0) \geq \prod_{i \leq m} (P_\sigma f_i)^{c_i}(0),$$

or equivalently,

$$\int_{\mathbb{R}^n} h(z) \frac{e^{-|z|^2/(2\sigma)} \, dz}{(\sqrt{2\pi\sigma})^n} \geq \prod_{i \leq m} \left(\int_{\mathbb{R}} f_i(s) \frac{e^{-|s|^2/(2\sigma)} \, ds}{\sqrt{2\pi\sigma}} \right)^{c_i}.$$

Since (1) implies $\sum c_i = n$, we find, letting $\sigma \to +\infty$, $\int_{\mathbb{R}^n} h \geq \prod_{i \leq m} (\int_{\mathbb{R}} f_i)^{c_i}$. $\qquad \square$

Remark. In order to justify that a solution of (9) preserves the non-negativity (maximum principle), or in order to express the solution in terms of a stochastic process, one has to be a little bit careful. It is convenient to assume first that the functions h and $(f_i)_{i \leq m}$ are "nice" enough (this will guarantee that b defined in (8) is also nice), and to conclude by approximation. This procedure is described in detail in Borell's papers. Assuming for instance that $b(t, \cdot)$ is Lipschitz with Lipschitz constant uniformly bounded in $t \in [0, T]$, the solution of (9) at time T can be expressed as

$$C(T, \theta_1, \ldots, \theta_m) = \mathbb{E}\left[C(0, W(T))\right]. \tag{10}$$

Here $W(t)$ is the m-dimensional (degenerate) process with initial value $W(0) = (\theta_1, \ldots, \theta_m)$ satisfying

$$dW(t) = \sigma dB(t) + \frac{1}{2}b(T - t, W(t))\, dt, \qquad t \in [0, T]$$

where B is a standard n-dimensional Brownian motion and σ is the $m \times n$ matrix whose i-th row is given by the vector u_i. It follows from (10) that $C(T, \cdot) \geq 0$ when $C(0, \cdot) \geq 0$.

3 The Brascamp–Lieb Inequality

As explained in the introduction, this section is a particular case of the recent work of Carlen, Lieb and Loss.

Instead of working with the heat semi-group, it will be more convenient to introduce the Ornstein–Uhlenbeck semi-group with generator $L := \Delta - \langle x, \nabla \rangle$:

$$P_t f(x) = \int_{\mathbb{R}^n} f\left(e^{-t}x + \sqrt{1 - e^{-2t}}\, z\right) d\gamma_n(z).$$

The function $g(t, x) = P_t f(x)$ solves the equation

$$\begin{cases} \dfrac{\partial g}{\partial t}(t, x) = \Delta g(t, x) - \langle x, \nabla g(t, x) \rangle = Lg(t, x), \\ g(0, \cdot) = f. \end{cases}$$

Thus, $F^{(t)}(x) := \log P_t f(x)$ satisfies

$$\begin{cases} \dfrac{\partial F^{(t)}}{\partial t}(x) = LF^{(t)}(x) + |\nabla F^{(t)}(x)|^2, \\ F(0, \cdot) = \log f. \end{cases}$$

We recall that under appropriate assumptions, $P_t f \longrightarrow \int f \, d\gamma_n$ almost surely and that

$$\int (Lu)v \, d\gamma_n = -\int \langle \nabla u, \nabla v \rangle \, d\gamma_n \tag{11}$$

whenever it makes sense.

We introduce the *one-dimensional* semi-groups $P_t(f_i)$ and $F_i^{(t)} = \log P_t f_i$, together with

$$\alpha(t) := \int \prod_{i \leq m} (P_t f_i)^{c_i}\left(\langle x, u_i \rangle\right) d\gamma_n(x) = \int e^{\sum_{i \leq m} c_i F_i^{(t)}(\langle x, u_i \rangle)} d\gamma_n(x).$$

We have

$$\alpha'(t) = \int \sum_{i \leq m} c_i \left(LF_i(\langle x, u_i \rangle) + (F_i')^2(\langle x, u_i \rangle) \right) e^{\sum c_i F_i(\langle x, u_i \rangle)} d\gamma_n(x),$$

where we have written F_i instead of $F_i^{(t)}$ (after differentiating in t, we are working at a fixed t). For a function F on $\mathbb{R}$, the function $\tilde{F}(x) := F(\langle x, u_i \rangle)$ on $\mathbb{R}^n$ verifies $L\tilde{F}(x) = LF(\langle x, u_i \rangle)$ since $|u_i| = 1$ (here we used the same notation for the 1 and n dimensional generators). Thus we can use the integration by parts formula (11) in $\mathbb{R}^n$ and get

$$\alpha'(t) = \int \left(-\sum_{i,j \leq m} c_i c_j \langle u_i, u_j \rangle F_i'(\langle x, u_i \rangle) F_j'(\langle x, u_j \rangle) + \sum_{i \leq m} c_i (F_i')^2(\langle x, u_i \rangle) \right)$$
$$\cdot e^{\sum c_i F_i(\langle x, u_i \rangle)} d\gamma_n(x).$$

This can be rewritten as

$$\alpha'(t) = \int \left(-\left| \sum_{i \leq m} c_i F_i'(\langle x, u_i \rangle) u_i \right|^2 + \sum_{i \leq m} c_i (F_i')^2(\langle x, u_i \rangle) \right)$$
$$\cdot e^{\sum c_i F_i(\langle x, u_i \rangle)} d\gamma_n(x).$$

We can deduce that $\alpha'(t) \geq 0$ from the inequality

$$\left| \sum_{i \leq m} c_i \theta_i u_i \right|^2 \leq \sum_{i \leq m} c_i \theta_i^2, \qquad \forall (\theta_i)_{i \leq m} \in \mathbb{R}^m,$$

which is an easy consequence of the decomposition of the identity (1). To see this, set $v = \sum_{i \leq m} c_i \theta_i u_i$ and write

$$|v|^2 = \left\langle v, \sum_{i \leq m} c_i \theta_i u_i \right\rangle = \sum_{i \leq m} c_i \theta_i \langle v, u_i \rangle$$
$$\leq \left(\sum_{i \leq m} c_i \theta_i^2 \right)^{\frac{1}{2}} \left(\sum_{i \leq m} c_i \langle v, u_i \rangle^2 \right)^{\frac{1}{2}} = \left(\sum_{i \leq m} c_i \theta_i^2 \right)^{\frac{1}{2}} |v|.$$

The inequality $\alpha(0) \leq \alpha(+\infty)$ reads as

$$\int_{\mathbb{R}^n} \prod_{i \leq m} f_i^{c_i}(\langle x, u_i \rangle) d\gamma_n(x) \leq \prod_{i \leq m} \left(\int_{\mathbb{R}} f_i \, d\gamma_1 \right)^{c_i}.$$

Setting $g_i(s) := f_i(s)e^{-|s|^2/2}$ and using (1) (and $\sum c_i = n$) we end up with the classical form

$$\int_{\mathbb{R}^n} \prod_{i \leq m} g_i^{c_i}\left(\langle x, u_i \rangle\right) dx \leq \prod_{i \leq m} \left(\int_{\mathbb{R}} g_i\right)^{c_i}.$$ $\square$

References

[B]   Bakry, D.: L'hypercontractivité et son utilisation en théorie des semigroupes (French) [Hypercontractivity and its use in semigroup theory]. In: Lectures on Probability Theory (Saint-Flour, 1992), 1–114, Lecture Notes in Math., **1581**. Springer, Berlin (1994)

[Ba1]   Ball, K.M.: Volumes of sections of cubes and related problems. In: Lindenstrauss, J., Milman, V.D. (eds) Israel Seminar on Geometric Aspects of Functional Analysis, Lectures Notes in Mathematics, **1376**. Springer-Verlag (1989)

[Ba2]   Ball, K.M.: Volume ratio and a reverse isoperimetric inequality. J. London Math. Soc., **44**, no. 2, 351–359 (1991)

[Bar1]   Barthe, F.: Inégalités de Brascamp-Lieb et convexité. C.R. Acad. Sci. Paris Sér. I Math., **324**, no. 8, 885–888 (1997)

[Bar2]   Barthe, F.: On a reverse form of the Brascamp–Lieb inequality. Invent. Math., **134**, no. 2, 335–361 (1998)

[Bor1]   Borell, C.: Geometric properties of some familiar diffusions in R^n. Ann. Probab., **21**, no. 1, 482–489 (1993)

[Bor2]   Borell, C.: Diffusion equations and geometric inequalities. Potential Anal., **12**, no. 1, 49–71 (2000)

[Bor3]   Borell, C.: The Ehrhard inequality. Preprint (2003)

[BraL]   Brascamp, H.J., Lieb, E.H.: Best constants in Young's inequality, its converse and its generalization to more than three functions. Adv. Math., **20**, 151–173 (1976)

[CLL]   Carlen, E., Lieb, E., Loss, M.: In preparation (2003)

[L]   Ledoux, M.: The concentration of measure phenomenon. Mathematical Surveys and Monographs, **89**. American Mathematical Society, Providence, RI (2001)

[V]   Villani, C.: Topics in optimal transportation. Graduate Studies in Mathematics, **58**. American Mathematical Society, Providence, RI (2003)

Pinched Exponential Volume Growth Implies an Infinite Dimensional Isoperimetric Inequality

I. Benjamini[1] and O. Schramm[2]

[1] Department of Mathematics, Weizmann Institute of Science, Rehovot, Israel
`itai@wisdom.weizmann.ac.il`
[2] Microsoft Research, One Microsoft Way, Redmond, WA 98052, USA
`schramm@microsoft.com`

Summary. Let G be a graph which satisfies $c^{-1} a^r \leq |B(v,r)| \leq c a^r$, for some constants $c, a > 1$, every vertex v and every radius r. We prove that this implies the isoperimetric inequality $|\partial A| \geq C|A|/\log|A|$ for some constant $C = C(a,c)$ and every nonempty finite set of vertices A.

A graph $G = (V(G), E(G))$ has pinched growth $f(r)$ if there are two constants $0 < c < C < \infty$ so that every ball $B(v,r)$ of radius r centered around a vertex $v \in V(G)$ satisfies

$$c\, f(r) < |B(v,r)| < C\, f(r).$$

For example, Cayley graphs and vertex-transitive graphs have pinched growth.

It is easy to come up with an example of a tree for which every ball satisfies $|B(v,r)| \geq 2^{r/2}$, yet there are arbitrarily large finite subsets of G with one boundary vertex. For example, start with $\mathbb{N}$ (an infinite one-sided path) and connect every vertex n to the root of a binary tree of depth n. This tree does not have pinched growth.

We will see that, perhaps surprisingly, the additional assumption of pinched exponential growth (that is, pinched growth a^r, for some $a > 1$) implies an infinite dimensional isoperimetric inequality. For a set $A \subset V(G)$ of vertices denote by ∂A the (vertex) boundary of A, consisting of vertices outside of A which have a neighbor in A.

Theorem 1. *Let G be an infinite graph with pinched growth a^r, where $a > 1$. Then there is a constant $c > 0$ such that for every nonempty finite set of vertices $A \subset V(G)$,*

$$|\partial A| \geq c|A|/\log|A|. \tag{1}$$

We say that G satisfies an s-dimensional isoperimetric inequality if there is a $c > 0$ such that $|\partial A| \geq c|A|^{(s-1)/s}$ holds for every finite $A \subset V(G)$. Thus, (1) may be considered an infinite dimensional isoperimetric inequality.

Coulhon and Saloff-Coste [CS-C] proved that when G is a Cayley graph of an infinite, finitely-generated, group, the isoperimetric inequality

$$|\partial A| \geq \frac{|A|}{4\, m\, \phi(2|A|)} \tag{2}$$

holds for every finite $A \subset V(G)$, where m is the number of neighbors every vertex has and $\phi(n) = \inf\{r \geq 1 : |B(v,r)| \geq n\}$ (here, $v \in V(G)$ is arbitrary). This result implies Theorem 1 for the case where G is a Cayley graph, even when the upper bound in the pinched growth condition is dropped. The tree example discussed above shows that Theorem 1 is not valid without the upper bound. Thus, the (short and elegant) proof of (2) from [CS-C] does not generalize to give Theorem 1, and, in fact, the proof below does not seem related to the arguments from [CS-C].

It is worthwhile to note that (2) is also interesting for Cayley graphs with sub-exponential growth. For example, it shows that $\mathbb{Z}^d$ satisfies a d-dimensional isoperimetric inequality.

Another related result, with some remote similarity in the proof, is due to Babai and Szegedy [BaS]. They prove that for a finite vertex transitive graph G, and $A \subset V(G)$, $0 < |A| < |G|/2$,

$$|\partial A| \geq |A|/(1 + \operatorname{diam} G).$$

The isoperimetric inequality (1) is sharp up to the constant, since there are groups with pinched growth a^r where (1) cannot be improved. Examples include the lamplighter on $\mathbb{Z}$ [LPP]. See [dlHa] for a discussion of growth rates of groups and many related open problems.

Regarding pinched polynomial growth, it is known that for every $d > 1$ there is a tree with pinched growth r^d containing arbitrarily large sets A with $|\partial A| = 1$, see, e.g., [BS2].

Problem 1. Does every graph of a pinched exponential growth contain a tree with pinched exponential growth?

In [BS1] it was shown that every graph satisfying the linear isoperimetric inequality $|\partial A| \geq c|A|$ ($c > 0$) contains a tree satisfying such an inequality, possibly with a different constant. The question whether one can find a *spanning* tree with a linear isoperimetric inequality was asked earlier [DSS]. It follows from Theorem 1 that a tree with pinched exponential growth satisfies the linear isoperimetric inequality. (If a tree satisfies $|\partial A| \geq 3$ for every vertex set A of size at least k, then every path of k vertices in the tree must contain a branch point, a point whose removal will give at least 3 infinite components. Consequently the tree contains a modified infinite binary tree, where every edge is subdivided into at most k edges.) Consequently, Problem 1 is equivalent to the question whether every graph with pinched exponential growth contains a tree satisfying a linear isoperimetric inequality.

As a warm up for the proof of Theorem 1, here is an easy argument showing that when G has pinched growth a^r it satisfies a two-dimensional isoperimetric inequality. Let $A \subset V(G)$ be finite. Let v be a vertex of A that is farthest from ∂A, and let r be the distance from v to ∂A . Note that $B(v, 2r) \subset \bigcup_{u \in \partial A} B(u, r)$. This gives, $a^{2r} \leq O(1) |\partial A| a^r$, and therefore $O(1) |\partial A| \geq a^r$. On the other hand, $\bigcup_{u \in \partial A} B(u, r) \supset A$, which gives $O(1) |\partial A| a^r \geq |A|$. Hence, $O(1) |\partial A|^2 \geq |A|$.

Proof of Theorem 1. For vertices v, u set $z(v, u) := a^{-d(v,u)}$, where $d(v, u)$ is the graph distance between v and u in G.

We estimate in two ways the quantity

$$Z = Z_A := \sum_{v \in A} \sum_{u \in \partial A} z(v, u) .$$

Fix $v \in A$. For every $w \notin A$, fix some geodesic path from v to w, and let w' be the first vertex in ∂A on this path. Let R be sufficiently large so that $|B(v, R)| \geq 2 |A|$, and set $W := B(v, R) \setminus A$. Then

$$\left| \{(w, w') : w \in W\} \right| = |W| \geq a^R / O(1) .$$

On the other hand, we may estimate the left hand side by considering all possible $u \in \partial A$ as candidates for w'. If $w \in W$, then $d(v, w') + d(w', w) \leq R$. Thus, each u is equal to w' for at most $O(1) a^{R - d(v,u)}$ vertices $w \in W$. This gives

$$\left| \{(w, w') : w \in W\} \right| \leq O(1) \sum_{u \in \partial A} a^{R - d(u,v)} = O(1) a^R \sum_{u \in \partial A} z(v, u) .$$

Combining these two estimates yields $O(1) \sum_{u \in \partial A} z(v, u) \geq 1$. By summing over v, this implies

$$O(1) Z \geq |A| . \tag{3}$$

Now fix $u \in \partial A$, set $m_r := \left| \{v \in A : d(v, u) = r\} \right|$, and consider

$$Z(u) := \sum_{v \in A} z(v, u) = \sum_r m_r a^{-r} . \tag{4}$$

For $r \leq \log |A| / \log a$, we use the inequality $m_r \leq |B(u, r)| = O(1) a^r$, while for $r > \log |A| / \log a$, we use $m_r \leq |A|$. We apply these estimates to (4), and get $Z(u) \leq O(1) \log |A|$, which gives $Z = \sum_{u \in \partial A} Z(u) \leq O(1) |\partial A| \log |A|$. Together with (3), this gives (1). $\qquad \square$

Next, we present a slightly different version of Theorem 1, which also applies to finite graphs.

Theorem 2. *Let G be a finite or infinite graph, $c > 0$, $a > 1$, $R \in \mathbb{N}$, and suppose that $c^{-1} a^r \leq |B(v, r)| \leq c a^r$ holds for all $r = 1, 2, \ldots, R$ and for all $v \in V(G)$. Then there is a constant $C = C(a, c)$, depending only on a and c, such that*

$$C \, |\partial A| \geq |A| / \log |A|$$

holds for every finite $A \subset V(G)$ with $|A| \leq C^{-1} a^R$.

The proof is the same. A careful inspection of the proof shows that one only needs the inequality $c^{-1} a^r \leq |B(v, r)|$ to be valid for $v \in A$ and the inequality $|B(v, r)| \leq c a^r$ only for $v \in \partial A$.

Acknowledgement. We thank Thierry Coulhon and Iftach Haitner for useful discussions.

References

[BaS]   Babai, L., Szegedy, M.: Local expansion of symmetrical graphs. Combin. Probab. Comput., **1**, no. 1, 1–11 (1992)

[BS1]   Benjamini, I., Schramm, O.: Every graph with a positive Cheeger constant contains a tree with a positive Cheeger constant. Geom. Funct. Anal., **7**, no. 3, 403–419 (1997)

[BS2]   Benjamini, I., Schramm, O.: Recurrence of distributional limits of finite planar graphs. Electron. J. Probab., **6**, no. 23 (2001) 13 pp. (electronic)

[CS-C] Coulhon, T., Saloff-Coste, L.: Isope'rime'trie pour les groupes et les varie'te's (French), [Isoperimetry for groups and manifolds]. Rev. Mat. Iberoamericana, **9**, no. 2, 293–314 (1993)

[DSS]   Deuber, W., Simonovits, M., Sós, V.: A note on paradoxical metric spaces. Studia Sci. Math. Hungar., **30**, no. 1-2, 17–23 (1995)

[dlHa] de la Harpe, P.: Topics in geometric group theory. In: Chicago Lectures in Mathematics. University of Chicago Press, Chicago, IL (2000) vi+310 pp

[LPP]   Lyons, R., Pemantle, R., Peres, Y.: Random walks on the lamplighter group. Ann. Probab., **24**, no. 4, 1993–2006 (1996)

On Localization for Lattice Schrödinger Operators Involving Bernoulli Variables

J. Bourgain

Institute for Advanced Study, Princeton, NJ 08540, USA
bourgain@math.ias.edu

1 Introduction

In this paper we attempt to progress on various problems related to the Anderson–Bernoulli lattice Schrödinger operators.

The first (and main) part of the paper aims to develop a method of establishing localization results in dimension $D > 1$. For $D = 1$, the model

$$H_\lambda = \lambda \varepsilon_n \delta_{n,n'} + \Delta \qquad (1.1)$$

is well known to exhibit Anderson-localization (hence pure point spectrum) for all values of $\lambda \neq 0$, almost surely in $\varepsilon = (\varepsilon_n)_{n \in \mathbb{Z}} \in \{1, -1\}^{\mathbb{Z}}$. There are two proofs known for this fact. The first is the Furstenberg–Lepage approach through the transfer matrix (see [C-K-M]). The second is based on the supersymmetric formalism that was exploited in [C-V-V] in the Bernoulli context. Neither of them could so far be extended to treat the Bernoulli model in $D > 1$ (the transfer-matrix approach is 1D, naturally; the supersymmetric approach does apply but leads to substantial difficulties). Among the spectral properties, the easiest one to establish here is the presence of a point spectrum near the edge of the spectrum. This fact should be true in arbitrary dimension.

The main difficulty of establishing a Wegner-type estimate following the usual multiscale strategy in the case of Bernoulli variables, is that proving the required probabilistic statements on the Green's function $G(\varepsilon; E)$ may not be achieved by varying a single variable (as in the case of a continuous distribution). In the present setting, 'rare-event' bounds are obtained considering the dependence of eigenvalues on a large collection of variables $(\varepsilon_n)_{n \in \mathcal{S}}$. The key result used here is a new probabilistic estimate (Lemma 2.1) that we believe is of independent interest. To carry out this type of analysis, one needs further information on sites $n \in \mathbb{Z}^D$ where the eigenfunction $\xi = (\xi_n)_{n \in \mathbb{Z}^D}$ is not 'too small' (as is clear from the first order eigenvalue variation). While this issue is easily taken care of in $D = 1$ (again using the transfer matrix), for $D > 1$ our understanding of the set

$$\{n \in \mathbb{Z}^D \mid |n| \sim N \text{ and } |\xi_n| > e^{-CN}\} \tag{1.2}$$

for large N, is rather poor. In order to avoid this issue, we have been forced to adopt in Theorem 3.1 and section 4 a modification of the original Bernoulli model. For instance, the continuum version considered in section 4 is given by

$$-\Delta + \lambda \sum_{n \in \mathbb{Z}^D} \varepsilon_n \phi(x - n) \tag{1.3}$$

with ϕ *exponentially decaying* for $|x| \to \infty$ (but not compactly supported as in the usual Bernoulli-setting).

The second part of the paper deals with another unsolved issue on the IDS (Integrated Density of States) of (1.1). Given $\lambda \neq 0, \mathcal{N} = \mathcal{N}(E)$ is known to have only a limited smoothness in the energy E and this smoothness is expected to improve for $\lambda \to 0$ (which is also heuristically reasonable). However, at this point, one does not even seem to be able to establish a Lipschitz property of $\mathcal{N}$, taking λ small enough. In fact, the known arguments implying some Hölder-continuity produce an exponent $\alpha(\lambda)$ which deteriorates for $\lambda \to 0$. In this paper, we use the Figotin–Pastur formula and elementary martingale theory to obtain a uniform (and explicit) Hölder estimate in λ (keeping the energy E away from the edges, and zero).

The author is grateful to Tom Spencer for several discussions on the Anderson–Bernoulli problems and in particular on the strategy to prove Wegner estimates.

2 A Distributional Inequality

We prove the following

Lemma 2.1. *Let* $f = f(\varepsilon_1, \ldots, \varepsilon_n)$ *be a bounded function on* $\{1, -1\}^n$ *and denote*

$$I_j = f|_{\varepsilon_j = 1} - f|_{\varepsilon_j = -1}$$

as the j-influence, which is a function of $\varepsilon_{j'}, j' \neq j$.

Let

$$2 < b < a \qquad \frac{\log a}{\log b} < constant \tag{2.2}$$

and assume

$$a^{-j} < I_j < b^{-j} \qquad (1 \leq j \leq n). \tag{2.3}$$

Let $\delta > a^{-n}$. *Then for all* E

$$\text{mes}\,[|f - E| < \delta] < e^{-c(\log \frac{\log \delta^{-1}}{\log a})^2}. \tag{2.4}$$

Proof. Take

$$j_1 \sim \frac{\log \frac{1}{\delta}}{\log a}; \quad a^{-j_1} > 4\delta$$

and define the decreasing sequence

$$j_2 = \gamma j_1$$

$$\vdots$$

$$j_s = \gamma j_{s-1}$$

with

$$\gamma = \frac{1}{10} \frac{\log b}{\log a}$$

and $j_s \sim j_1^{1/2}$. Thus

$$s \sim \log \left(\frac{\log \frac{1}{\delta}}{\log a} \right).$$

Consider the consecutive intervals

$$J_s < J_{s-1} < \cdots < J_1 \text{ where } J_{s'} = \left[\frac{1}{2} j_{s'}, j_{s'} \right]$$

and denote $\bar{\varepsilon}_{s'} = (\varepsilon_{s'})_{j \in J_{s'}}$.

Consider $f = F(\bar{\varepsilon}_1, \ldots, \bar{\varepsilon}_s)$ as a function of $\bar{\varepsilon}_1, \ldots, \bar{\varepsilon}_s$ (other variables fixed).

By (2.3)

$$\left| F(\bar{\varepsilon}_1, \ldots, \bar{\varepsilon}_{s'}, \bar{\varepsilon}_{s'+1}, \ldots, \bar{\varepsilon}_s) - F(\bar{\varepsilon}_1', \ldots, \bar{\varepsilon}_{s'}', \bar{\varepsilon}_{s'+1}, \ldots, \bar{\varepsilon}_s) \right|$$
$$< \sum_{t \leq s'} \sum_{j \in J_t} \|I_j\|_\infty < \sum_{t \leq s'} \sum_{j \in J_t} b^{-j} < 2.b^{-\frac{1}{2} j_{s'}}$$

and also, if $\bar{\varepsilon}_{s'} \overset{>}{\neq} \bar{\varepsilon}_{s'}'$ ($>$ refers to the natural ordering)

$$F(\bar{\varepsilon}_1, \ldots, \bar{\varepsilon}_{s'-1}, \bar{\varepsilon}_{s'}, \bar{\varepsilon}_{s'+1}, \ldots, \bar{\varepsilon}_s) - F(\bar{\varepsilon}_1, \ldots, \bar{\varepsilon}_{s'-1}, \bar{\varepsilon}_{s'}', \bar{\varepsilon}_{s'+1}, \ldots, \bar{\varepsilon}_s)$$
$$> a^{-j_{s'}}$$

$$F(\bar{\varepsilon}_1, \ldots, \bar{\varepsilon}_{s'-1}, \bar{\varepsilon}_{s'}, \bar{\varepsilon}_{s'+1}, \ldots, \bar{\varepsilon}_s) - F(\bar{\varepsilon}_1', \ldots, \bar{\varepsilon}_{s'-1}', \bar{\varepsilon}_{s'}', \bar{\varepsilon}_{s'+1}, \ldots, \bar{\varepsilon}_s)$$
$$> F(\bar{\varepsilon}_1, \ldots, \bar{\varepsilon}_{s'-1}, \bar{\varepsilon}_{s'}, \bar{\varepsilon}_{s'+1}, \ldots, \bar{\varepsilon}_s) - F(\bar{\varepsilon}_1, \cdots, \bar{\varepsilon}_{s'-1}, \bar{\varepsilon}_{s'}', \bar{\varepsilon}_{s'+1}, \ldots, \bar{\varepsilon}_s)$$
$$- \left| F(\bar{\varepsilon}_1, \ldots, \bar{\varepsilon}_{s'-1}, \bar{\varepsilon}_{s'}', \bar{\varepsilon}_{s'+1}, \ldots, \bar{\varepsilon}_s) - F(\bar{\varepsilon}_1', \ldots, \bar{\varepsilon}_{s'-1}', \bar{\varepsilon}_{s'}', \bar{\varepsilon}_{s'+1}, \ldots, \bar{\varepsilon}_s) \right|$$
$$> a^{-j_{s'}} - 2b^{\frac{1}{2} j_{s'-1}}$$
$$> a^{-j_{s'}} - 2b^{-\frac{1}{2\gamma} j_{s'}}$$
$$> \frac{1}{2} a^{-j_{s'}} \tag{2.5}$$

by the choice of γ. Hence, for all $\bar{\varepsilon}_{s'+1}, \ldots, \bar{\varepsilon}_s$

$$\min_{\bar{\varepsilon}_1,\ldots,\bar{\varepsilon}_{s'-1}} |F(\bar{\varepsilon}_1,\ldots,\bar{\varepsilon}_{s'-1},\bar{\varepsilon}_{s'},\bar{\varepsilon}_{s'+1},\ldots,\bar{\varepsilon}_s) - E| < \frac{1}{4}a^{-j_{s'}} \qquad (2.6)$$

may only happen for non-comparable elements $\bar{\varepsilon}_{s'} = (\varepsilon_j)_{j \in J_{s'}}$.

Thus, by Sperner's lemma

$$\operatorname{mes}_{\bar{\varepsilon}_{s'}} \left[(2.6) \right] \leq |J_{s'}|^{-1/2} \sim j_{s'}^{-1/2}. \qquad (2.7)$$

Estimate

$$\operatorname{mes} \left[|f - E| < \delta \right]$$
$$\leq \operatorname{mes} \left[|F - E| < \delta \right]$$
$$\leq \operatorname{mes}_{\bar{\varepsilon}_s} \left[\min_{\bar{\varepsilon}_1,\ldots,\bar{\varepsilon}_{s-1}} |F(\bar{\varepsilon}_1,\ldots,\bar{\varepsilon}_{s-1},\bar{\varepsilon}_s) - E| < \frac{1}{4}a^{-j_s} \right].$$
$$\max_{\bar{\varepsilon}_s} \operatorname{mes}_{\bar{\varepsilon}_{s-1}} \left[\min_{\bar{\varepsilon}_1,\ldots,\bar{\varepsilon}_{s-2}} |F - E| < \frac{1}{4}a^{-j_{s-1}} \right].$$
$$\vdots$$
$$\max_{\bar{\varepsilon}_s,\ldots,\bar{\varepsilon}_2} \operatorname{mes}_{\bar{\varepsilon}_1} \left[|F - E| < \frac{1}{4}a^{-j_1} \right]$$
$$\leq \prod_{s'=1}^{s} j_{s'}^{-1/2} \leq j_1^{-s/4} < \exp\left[-c\left(\log \frac{\log \frac{1}{\delta}}{\log a} \right)^2 \right] \quad \text{(by (2.7))}.$$

This proves the lemma. $\qquad \square$

Remark. If we bound

$$\frac{\log a}{\log b} < K \qquad (2.8)$$

instead of (2.2), inequality (2.4) clearly becomes

$$\operatorname{mes} \left[|f - E| < \delta \right] < \exp -c\left\{ \frac{(\log \frac{\log \delta^{-1}}{\log a})^2}{\log K} \right\}. \qquad (2.9)$$

3 Application to Certain Lattice Schrödinger Operators

We will prove the following localization result

Theorem 3.1. *Let* $d \geq 1$ *be arbitrary and* $H(\varepsilon)$ *the following lattice Schrödinger operator on* $\mathbb{Z}^d$

$$H(\varepsilon) = D(\varepsilon) + \Delta \qquad (3.2)$$

with $D(\varepsilon)$ *a diagonal given by*

$$D(\varepsilon)_n = \lambda \sum_{m \in \mathbb{Z}^d} 2^{-|n-m|} \varepsilon_m \ \ (\lambda > 0) \ , \tag{3.3}$$

where

$$\varepsilon = (\varepsilon_n)_{n \in \mathbb{Z}^d} \in \{1, -1\}^{\mathbb{Z}^d}$$

and Δ the lattice Laplacian on $\mathbb{Z}^d$, i.e.

$$\Delta(n, n') = 1 \ \ if \ \max_{1 \le j \le d} |n_j - n'_j| = 1$$
$$= 0 \ otherwise.$$

Then ε a.s., $H(\varepsilon)$ exhibits A.L. at the edge of the spectrum; thus for

$$\big| |E| - E_* \big| < \delta \tag{3.4}$$

$$E_* = 2d + \lambda \sum_{m \in \mathbb{Z}^d} 2^{-|m|} \tag{3.5}$$

and $\delta > 0$ small enough.

Remark. The usual Bernoulli model is given by

$$H_0(\varepsilon) = \lambda \varepsilon_n \delta_{n,n'} + \Delta, \ \varepsilon \in \{1, -1\}^{\mathbb{Z}^d}. \tag{3.6}$$

It is known that for $d = 1$, $H(\varepsilon)$ has a.s. exponentially localized states on the entire spectrum. A proof based on the Furstenberg–Lepage method appears in [C-K-M] and uses the super-symmetric formalism in [S-V-V]. Theorem 3.1 results from an attempt to produce point spectrum for (3.6) when $d > 1$ (which is expected to appear at least at the edge). In fact, minor adjustment of the proof of Theorem 3.1 permits us to show A.L. at the spectrum edge for $d = 1$ and quasi one-dimensional models. These results are of course of little interest on their own.

Notice that

$$H(\varepsilon) = \sum_{n \in \mathbb{Z}^d} 2^{-|n|} S_{-n} H_0(\varepsilon) S_n \ \ \ (S_n = \text{n-shift}).$$

The choice of this particular modification of (3.6) will become clear later on.

To obtain Theorem 3.1, we establish the following Wegner-type estimate.

Proposition 3.7. *Denote*

$$G_N(E; \varepsilon) = \Big[R_{-N,N]^d} \big(H(\varepsilon) - E + io \big) R_{[-N,N]^d} \Big]^{-1}$$

(R_Λ denoting the restriction to $\Lambda \subset \mathbb{Z}^d$).
Then, for fixed E satisfying (3.4)

$$\|G_N(E; \varepsilon)\| < e^{N^{9/10}} \tag{3.8}$$

and

$$|G_N(E;\varepsilon)(n,n')| < e^{-\frac{N}{(\log N)^{10}}} \quad for \; |n-n'| > \frac{N}{10} \qquad (3.8')$$

for $\varepsilon \notin \Omega_N(E)$, *where*

$$\text{mes } \Omega_N(E) < e^{-c\frac{(\log N)^2}{\log\log N}}. \qquad (3.9)$$

Deducing Theorem 3.1 from Proposition 3.7 is standard.

Proposition 3.7 is proven inductively on the scale N. In this process, we involve an additional property (3.10). Denote $H(x)$ as the obvious extension of $H(\varepsilon)$ to $[1,-1]^{\mathbb{Z}^d}$.

(3.10). *There is a set* $\Omega'_N(E) \subset \{1,-1\}^{\mathbb{Z}^d \setminus \{0\}}$ *satisfying*

$$\text{mes } \Omega'_N(E) < \frac{1}{100}$$

and if $\varepsilon \notin \Omega'_N(E), t \in [-1,1]$ *arbitrary,*

$$G_N\big(E; x_0 = t, x_j = \varepsilon_j (j \neq 0)\big)$$

satisfies (3.8), (3.8').

At an initial (large) scale N_0, properties (3.7), (3.10) are easily obtained from assumption (3.4).

Indeed, write

$$G_{N_0}(E) = \big(D_{N_0}(\varepsilon) - E + \Delta_{N_0}\big)^{-1}$$
$$= \big(D_{N_0}(\varepsilon) - E - 1\big)^{-1} \sum_{s \geq 0} (-1)^s \Big[(\Delta_{N_0} + 1)(D_{N_0}(\varepsilon) - E - 1)^{-1}\Big]^s. \qquad (3.11)$$

To control the Neumann series, we claim that if

$$E > E_* - \frac{1}{(\log N_0)^8} \qquad (3.12)$$

then

$$\big\|(\Delta_{N_0}+1)(E+1-D(\varepsilon))^{-1}(\Delta_{N_0}+1)(E+1-D(\varepsilon))^{-1}\big\| < 1 - \frac{1}{(\log N_0)^8} \qquad (3.13)$$

except for ε in a set of measure $< e^{-(\log N_0)^2}$.

We sketch the argument. Notice first that

$$\|1 + \Delta\| = 2d + 1$$
$$E + 1 - D(\varepsilon)_n > E_* + 1 - D(\varepsilon)_n - \delta > 2d + 1 - \delta \; ;$$

hence

$$\left\|\left[E+1-D(\varepsilon)\right]^{-1}\right\| < \frac{1}{2d+1-\delta}.$$

Assume

$$\left\|(\Delta_{N_0}+1)(E+1-D(\varepsilon))^{-1}(\Delta_{N_0}+1)(E+1-D(\varepsilon))^{-1}\right\| > 1-\delta$$

implying

$$\left\|(\Delta_{N_0}+1)(E+1-D(\varepsilon))^{-1}(\Delta_{N_0}+1)\right\| > (1-\delta)(2d+1-\delta).$$

Take $\xi \in [e_n \mid |n_j| \le N_0], \|\xi\| = 1$ s.t.

$$\left\langle (E+1-D(\varepsilon))^{-1}(\Delta_{N_0}+1)\xi, (\Delta_{N_0}+1)\xi \right\rangle > (1-\delta)(2d+1-\delta). \quad (3.14)$$

Then, since

$$\|(\Delta+1)\xi\| \geqq \|(\Delta_{N_0}+1)\xi\| > (1-\delta)^{1/2}(2d+1-\delta) > 2d+1-C\delta$$

we have

$$\sum_n \left(\xi_n + \sum_{\substack{i=1,\ldots,d \\ \nu_i=\pm 1}} \xi_{n+\nu_i e_i} \right)^2 > (2d+1)^2 - C\delta$$

$$\langle \xi, S_{e_i}\xi \rangle = \sum \xi_n \xi_{n+e_i} > 1 - C\delta \qquad \text{(for all } i=1,\ldots,d).$$

Therefore

$$\|\xi - S_{e_i}\xi\| < C\sqrt{\delta}, \|(2d+1)\xi - (\Delta_{N_0}+1)\xi\| < C\sqrt{\delta}. \quad (3.15)$$

Returning to (3.14), it follows that

$$\left\langle (E+1-D(\varepsilon))^{-1}\xi, \xi \right\rangle > \frac{1}{2d+1} - C\sqrt{\delta}$$

$$\left\|(2d+1)(E+1-D(\varepsilon))^{-1}\xi - \xi\right\| < C\sqrt{\delta}$$

$$C\sqrt{\delta} > \|2d\xi - (E-D(\varepsilon))\xi\| = \|(E-2d)\xi - D(\varepsilon)\xi\|. \quad (3.16)$$

For $r = 0, 1, \ldots, R = 0\left(\frac{1}{\sqrt{\delta}}\right)$, (3.15), (3.16) imply

$$\|\xi - S_{re_i}\xi\| < CR\sqrt{\delta}$$

$$\|(E-2d)\xi - S_{-re_i}D(\varepsilon)S_{re_i}\xi\| < CR\sqrt{\delta}$$

$$\left\| \sum_{r=1}^R S_{-re_i}D(\varepsilon)S_{re_i}\xi \right\| > \frac{R}{2}. \quad (3.17)$$

Clearly the left side of (3.17) is at most

$$\max_{|n|\lesssim N_0} \left| \sum_{r=1}^{R} D(\varepsilon)_{n+re_i} \right| \leq \max_{|n|\lesssim N_0} \sum_{m} 2^{-|n-m|} \left| \sum_{r=1}^{R} \varepsilon_{m+re_i} \right|$$

and therefore (3.17) holds with probability at most

$$e^{-c\frac{R}{\log N_0}}. \tag{3.18}$$

With $\delta = (\log N_0)^{-8}$, we may take $R \sim (\log N_0)^4$, so that $(3.18) < e^{-c(\log N_0)^3}$.
    This proves the claim.
    If (3.13) holds, (3.11) implies the bounds

$$\|G_{N_0}(E)\| < (\log N_0)^{10}$$

and

$$|G_{N_0}(E)(n,n')| < e^{-\frac{N_0}{(\log N_0)^{10}}} \quad \text{for } |n-n'| > \frac{N_0}{10}.$$

Next, we discuss the inductive step from scale N to scale

$$N_1 = N^{4/3}.$$

One thing to point out first is that the restriction $H_N(\varepsilon) = R_{[-N,N]^d} H(\varepsilon)$
$R_{[-N,N]^d}$ does not only depend on the variables $(\varepsilon_j)_{j\in[-N,N]^d}$ as in the usual
Bernoulli case.
    However, taking (3.3) into account

$$H_N(\varepsilon) = H_N\left(\varepsilon_j | j \in \left[-\frac{11}{10}N, \frac{11}{10}N\right]^d\right) + 0\left(2^{-\frac{N}{10}}\right)$$

and, consequently, properties (3.7), (3.10) may be considered as only depen-
dent on the variable $(\varepsilon_j)_{j\in[-\frac{11}{10}N,\frac{11}{10}N]^d}$ say. This fact will repeatedly be exploited in the sequel.
    Next, it should also be observed that (3.10) at scale N_1 follows from (3.10)
at scale N, property (3.7) at scale N and the resolvent identity (together with
the previous observation).
    Assuming (3.8), (3.8') valid for $\varepsilon \notin \Omega_N(E)$, with

$$\text{mes } \Omega_N(E) < \delta_N \tag{3.19}$$

we study the Green's function at scale $N_1 = N^{4/3}$. For notational simplicity,
let $d = 1$. But the argument readily applies to arbitrary dimension.
    Define

$$S = \left\{ \frac{5}{4}rN \,\middle|\, r \in \mathbb{Z}, |r| < \frac{4}{5}\frac{N_1}{N} \right\}$$

and cover $[-N_1, N_1]$ by the intervals $m + [-N, N], m \in S$

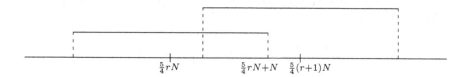

$$\frac{5}{4}rN \qquad \frac{5}{4}rN+N \quad \frac{5}{4}(r+1)N$$

From the comment on essential independence and (3.19) and for ε outside a set of measure $< N_1^2\delta_N^2$, we ensure that there is at most one $r_0 \in \mathbb{Z}$ s.t. if $|r - r_0| \geq 2$, then $G_{J_r}\left(E; x_{\frac{5}{4}rN} = \pm 1, x_j = \varepsilon_j (j \neq \frac{5}{4}rN)\right), J_r$ = the interval $\frac{5}{4}rN + [-N, N]$, will satisfy (3.8), (3.8′).

Moreover, by (3.10) and excluding another ε-set of measure $< 2^{-(\log N)^2}$, there is a subset

$$\mathcal{R} \subset \mathbb{Z} \cap \left[-\frac{4}{5}\frac{N_1}{N}, \frac{4}{5}\frac{N_1}{N}\right]$$

with the following properties

$$|r - r_0| > 10 \quad \text{for} \quad r \in \mathcal{R} \tag{3.20}$$

$$\max_{|r'| < \frac{4}{5}\frac{N_1}{N}} \operatorname{dist}(r', \mathcal{R}) < (\log N)^2 \tag{3.21}$$

$$G_{J_r}\left(E; x_{\frac{5}{4}rN} = t, x_j = \varepsilon_j\left(j \neq \frac{5}{4}rN\right)\right) \tag{3.22}$$

satisfying (3.8), (3.8′) for all $t \in [-1, 1]$.

Denote

$$\mathcal{S}' = \left\{\frac{5}{4}rN \,\middle|\, r \in \mathcal{R}\right\}. \tag{3.23}$$

It is important to notice that the set $\mathcal{S}' \subset \mathcal{S}$ is determined by the variables $(\varepsilon_n)_{n \notin \mathcal{S}}$ only.

For $n \in \mathcal{S} \setminus \mathcal{S}'$, choose ε_n arbitrary.

Defining

$$\Lambda = \bigcup_{|r-r_0|>5} J_r$$

it follows from the preceding and the resolvent identity that

$$G_\Lambda\left(E; x_j = \varepsilon_j (j \notin \mathcal{S}'), x_j = t_j (j \in \mathcal{S}')\right)$$

and $t_j \in [-1, 1]$ arbitrary, will satisfy

$$\|G_\Lambda\| < e^{N^{9/10}} \tag{3.24}$$

$$|G_\Lambda(n, n')| < e^{-\frac{1}{10}\frac{|n-n'|}{(\log N)^{10}}} \quad \text{for} \quad |n - n'| > N. \tag{3.24′}$$

Let $E_\tau(t) = E_\tau(t_j | j \in \mathcal{S}')$ be a continuous eigenvalue parametrization of $H_{N_1}\left(\varepsilon_j (j \notin \mathcal{S}'), t_j (j \in \mathcal{S}')\right)$, where $(\varepsilon_j)_{j \notin \mathcal{S}'}$ are fixed. Thus $E_\tau(t)$ is piecewise analytic in each variable $t_j \in [-1, 1]$ separately. From first order eigenvalue variation and (3.3)

$$|E(t) - E(t')| \le |t - t'| \left(\sum_{j \in \mathcal{S}'} \max_{t_j'' \in [t_j, t_j']} \left| \left\langle \frac{\partial H}{\partial t_j} \xi, \xi \right\rangle \right| \right)$$

$$\le |t - t'| \left\{ \lambda \sum_{j \in \mathcal{S}', n \in \mathbb{Z}} 2^{-|n-j|} \max_{t''} |\langle \xi(t''), e_n \rangle|^2 \right\} \quad (3.25)$$

where $\xi(t)$ denotes the corresponding normalized eigenfunction.

Assume

$$|E(t) - E| < e^{-N^{10/11}} \quad (3.26)$$

and

$$|t' - t| < e^{-N^{10/11}}. \quad (3.27)$$

Since (3.24), (3.24') remain clearly preserved under $e^{-N^{\frac{10}{11}}}$ perturbation of the energy, it follows that

$$G_\Lambda\big(E(t), \varepsilon_j(j \notin \mathcal{S}'), t_j(j \in \mathcal{S}')\big)$$

satisfies (3.24), (3.24') and hence

$$|\xi_n(t)| < e^{-\frac{1}{(\log N)^{10}} [\text{dist}\,(n, \frac{5}{4} r_0 N) - 6N]}. \quad (3.28)$$

Thus in particular, by (3.20)

$$|\xi_n(t)| < e^{-\frac{N}{(\log N)^{10}}} \quad \text{for dist}\,(n, \mathcal{S}') < N. \quad (3.29)$$

If (3.27) and $t_j'' \in [t_j, t_j']$, $|t - t''| < e^{-N^{10/11}}$, and hence $|E(t) - E(t'')| < Ne^{-N^{10/11}}$. Thus $|E - E(t'')| < Ne^{-N^{10/11}}$ and (3.29) implies

$$|\xi_n(t'')| < e^{-N^{1-}} \quad \text{if dist}\,(n, \mathcal{S}') < N. \quad (3.30)$$

Substitution of (3.30) in (3.25) therefore gives

$$|E(t) - E(t')| \le |t - t'| \big(N_1 2^{-N} + N_1 e^{-N^{1-}} \big)$$
$$\le e^{-N^{1-}} |t - t'|. \quad (3.31)$$

The preceding inequality thus shows that if

$$\min_t |E - E_\tau(t)| < e^{-N^{10/11}} \quad (3.32)$$

then

$$\max_t |E - E_\tau(t)| < 2e^{-N^{10/11}}. \quad (3.33)$$

Here t refers to $(t_j)_{j \in \mathcal{S}'} \in [-1, 1]^{\mathcal{S}'}$.

Consider the eigenvalues $E_\tau(t)$ satisfying (3.32), hence (3.33) and also (3.28) for the corresponding eigenfunction. Choose $r_\ell \in \mathcal{R}$ s.t.

$$|r_\ell - r_0| \sim \ell(\log N)^2 \tag{3.34}$$

which is possible by (3.21).

Consider the function

$$f\left(\varepsilon'_\ell \,\middle|\, 1 \le \ell \le \frac{N_1}{N(\log N)^3}\right)$$
$$= E_\tau\left(x_{\frac{5}{4}r_\ell N} = \varepsilon'_\ell, x_j = \varepsilon_j \text{ for } j \in \mathcal{S}'\backslash\left\{\frac{5}{4}r_\ell N\right\}\right)$$

with the $(\varepsilon_j)_{j \in \mathcal{S}'\backslash\{\frac{5}{4}r_\ell N\}}$ fixed.

In order to apply the distributional result from section 2, we need to estimate the influence I_ℓ from above and below.

Denote ∂_ℓ as the derivative w.r.t. $t_j, j = \frac{5}{4}r_\ell N$.

Thus

$$I_\ell = \int_{-1}^{1} \partial_\ell E_\tau$$
$$= \lambda \int_{-1}^{1} \left\{\sum_{n \in \mathbb{Z}} e^{-|n - \frac{5}{4}r_\ell N|} \langle \xi(t_{\frac{5}{4}r_\ell N} = s, t_j = \varepsilon_j \text{ otherwise}), e_n\rangle^2\right\} ds. \tag{3.35}$$

From (3.28) and distinguishing the cases $|n - \frac{5}{4}r_\ell N| \le \frac{1}{2}|r_\ell - r_0|N, |n - \frac{5}{4}r_\ell N| > \frac{1}{2}|r_\ell - r_0|N$, it follows that

$$I_\ell < e^{-\frac{1}{(\log N)^{10}}(\frac{1}{2}|r_\ell - r_0|N - 6N)} < e^{-\frac{N}{(\log N)^9}\ell}. \tag{3.36}$$

On the other hand, (3.28) also implies that for all t

$$\sum_{|n - \frac{5}{4}r_0 N| > 10N} \xi_n(t)^2 < e^{-N^{1-}}$$

and hence

$$\sum_{|n - \frac{5}{4}r_0 N| \le 10N} \xi_n(t)^2 > \frac{1}{2}. \tag{3.37}$$

It follows that the integrand in (3.35) is at least

$$\frac{1}{2} \min_{|n - \frac{5}{4}r_0 N| \le 10N} e^{-|n - \frac{5}{4}r_\ell N|} > e^{-(\frac{5}{4}|r_\ell - r_0|N + 10N)}$$

$$> e^{-C\ell(\log N)^2 N}. \tag{3.38}$$

(This is a crucial point where the modification (3.3) of the Bernoulli model is exploited.)

Hence, also

$$I_\ell > e^{-CN(\log N)^2 \ell}. \tag{3.39}$$

Summarizing (3.36), (3.39), we get

$$a^{-\ell} < I_\ell < b^{-\ell} \quad \left(1 \leq \ell \leq n \equiv \frac{N_1}{N(\log N)^3}\right)$$

with

$$a = e^{CN(\log N)^2} \tag{3.40}$$

$$b = e^{\frac{N}{(\log N)^9}} \; ; \tag{3.41}$$

hence

$$\frac{\log a}{\log b} < (\log N)^{12} = K. \tag{3.42}$$

Applying inequality (2.9) with

$$\delta = e^{-N.N_1^{\frac{1}{10}}} > a^{-n} \tag{3.43}$$

implies that

$$\operatorname{mes}_{\varepsilon'} \left[|f - E| < e^{-N.N_1^{\frac{1}{10}}}\right] < e^{-c\frac{(\log N_1)^2}{\log\log N}}. \tag{3.44}$$

Thus if E_τ satisfies (3.32), then

$$\operatorname{mes}_{(\varepsilon_j)_{j \in \mathcal{S}'}} \left[|E_\tau(\varepsilon) - E| < e^{-NN_1^{\frac{1}{10}}}\right] < e^{-c\frac{(\log N_1)^2}{\log\log N}}$$

and, therefore, clearly

$$\operatorname{mes}_{(\varepsilon_j)_{j \in \mathcal{S}'}} \left[\operatorname{dist}\left(E, \operatorname{Spec} H_{N_1}(\varepsilon)\right) < e^{-N.N_1^{\frac{1}{10}}}\right] < e^{-c\frac{(\log N_1)^2}{\log\log N}}. \tag{3.45}$$

Recall that $(\varepsilon_j)_{j \in \mathcal{S} \backslash \mathcal{S}'}$ was arbitrary and $(\varepsilon_j)_{j \notin \mathcal{S}}$ restricted to the complement of a set of measure $< N_1^2 \delta_N^2$.

Hence

$$\operatorname{mes}_\varepsilon \left[\operatorname{dist}\left(E, \operatorname{Spec} H_{N_1}(\varepsilon)\right) < e^{-N.N_1^{\frac{1}{10}}}\right] < N_1^2 \delta_N^2 + e^{-c\frac{(\log N_1)^2}{\log\log N}} \equiv \delta'_{N_1}. \tag{3.46}$$

In particular, $\|G_{N_1}(E, \varepsilon)\| < e^{N.N_1^{\frac{1}{10}}} < e^{N_1^{9/10}}$ for ε outside a set of measure $< \delta'_{N_1}$.

Applying (3.46) with $H_{N_1}(\varepsilon)$ replaced by $H_\Lambda(\varepsilon)$ with $\Lambda \subset [-N_1, N_1]$, an arbitrary interval of size $\frac{N_1}{100}$, it follows from (3.24), (3.24′), the bounds

$$\|G_\Lambda(E; \varepsilon)\| < e^{N.N_1^{\frac{1}{10}}} \tag{3.47}$$

and the resolvent identity, that

$$|G_{N_1}(E;\varepsilon)(n,n')| < e^{-\frac{1}{20}\frac{|n-n'|}{(\log N)^{10}}} < e^{-\frac{|n-n'|}{(\log N_1)^{10}}} \text{ for } |n-n'| > \frac{N_1}{10} \quad (3.48)$$

if $\varepsilon \notin \Omega_{N_1}(E)$, where

$$\text{mes } \Omega_{N_1}(E) < N_1^2 \delta'_{\frac{N_1}{100}} < \delta_{N_1} \equiv N_1^4 \left(\delta_N^2 + e^{-c\frac{(\log N_1)^2}{\log \log N_1}}\right). \quad (3.49)$$

Since $N_1 = N^{4/3}$, (3.49) is consistent with (3.9).

This completes the inductive step and proves Proposition 3.7. $\qquad \square$

Remark. Consider a 1-D Schrödinger operator $H = v_n \delta_{nn'} + \Delta$ with bounded potential and a solution

$$(H - E)\xi = 0, \quad \xi_0 = 1. \quad (3.50)$$

It follows from the transfer matrix formulation that if $\Omega = \{n, n+1\}$, then

$$|\xi_n| + |\xi_{n+1}| > C^{-|n|}. \quad (3.51)$$

Considering the 1D Bernoulli model, this fact permits us by previous reasoning to get lower bounds on the influences by varying a pair of consecutive sites. Thus

$$E_\tau(t_{\frac{5}{4}r_\ell N} = 1, t_{\frac{5}{4}r_\ell N+1} = 1) - E_\tau(t_{\frac{5}{4}r_\ell N} = -1, t_{\frac{5}{4}r_\ell N+1} = -1)$$

$$= \lambda \int_{-1}^{1} (|\xi_{\frac{5}{4}r_\ell N}|^2 + |\xi_{\frac{5}{4}r_\ell N+1}|^2)(t_{\frac{5}{4}r_\ell N} = s, t_{\frac{5}{4}r_\ell N+1} = s)ds$$

$$> C^{-|r_\ell - r_0|N} \quad (3.52)$$

and hence inequality (3.39).

The problem one encounters with the Bernoulli model, $D > 1$, following this approach, is to vary sites on sets $\Omega \subset \mathbb{Z}^D$ that are large enough to control $\max_{n \in \Omega} |\xi_n|$ from below, if ξ satisfies (3.50).

4 A Continuum Model

There are continuous analogues of the discrete model described in §3. Let ϕ be a function on $\mathbb{R}^d$ satisfying

$$\phi(x) \sim e^{-|x|} \quad (4.1)$$

$$\hat{\phi}(0) = 1 \quad (4.2)$$

$$\hat{\phi}(n) = 0 \text{ for } n \in \mathbb{Z}^d \backslash \{0\}. \quad (4.3)$$

(Simply let $\phi = e^{-|x|} * \chi_{[-\frac{1}{2},\frac{1}{2}]^d}$ for instance.)

Thus by the Poisson formula and (4.2), (4.3)

$$\sum_{n \in \mathbb{Z}^d} \phi(x-n) = 1. \tag{4.4}$$

Consider then the following random Schrödinger operator on $\mathbb{R}^d$

$$H(\varepsilon) = -\Delta + V_\varepsilon(x) \tag{4.5}$$

with Δ the Laplacian and

$$V_\varepsilon(x) = \sum_{n \in \mathbb{Z}^d} \varepsilon_n \phi(x-n). \tag{4.6}$$

Thus by (4.4)

$$\|V_\varepsilon\|_\infty \le 1 \tag{4.7}$$

and a.s.

$$\inf \operatorname{Spec} H(\varepsilon) = \inf V_\varepsilon = -1. \tag{4.8}$$

Proposition 4.9. $H(\varepsilon)$ *satisfies a.s. the Anderson localization near the edge* $E \approx -1$.

Write

$$
\begin{aligned}
&G(E + io; \varepsilon) \\
&= \left(H(\varepsilon) - E - io\right)^{-1} \\
&= (-\Delta - E)^{-1/2}(I + (-\Delta - E)^{-1/2}V(\varepsilon)(-\Delta - E)^{-1/2} + io)^{-1}(-\Delta - E)^{-1/2}
\end{aligned}
\tag{4.9}
$$

which reduces the problem to the control of the inverse of

$$I + io + A_\varepsilon \tag{4.10}$$

$$A_\varepsilon = (-\Delta - E)^{-1/2} V(\varepsilon)(-\Delta - E)^{-1/2}. \tag{4.11}$$

Notice that since

$$(-\Delta - E)^{-1/2}\xi(x) = \int_{\mathbb{R}^d} \frac{\hat{\xi}(\lambda)}{(-E + |\lambda|^2)^{1/2}} e^{2\pi i x.\lambda} d\lambda, \tag{4.12}$$

$$\|(-\Delta - E)^{-1/2}\| = |E|^{-1/2}. \tag{4.13}$$

Denote R_N as the restriction to $[-N, N]^d$.

One has

$$(1 + R_N A_\varepsilon R_N)^{-1} = \sum_{s \ge 0} (-1)^s (R_N A_\varepsilon R_N)^s. \tag{4.14}$$

Fix $N = N_0$ a large integer. We claim that if the energy E satisfies

$$|1 + E| < \delta \equiv (\log N_0)^{-10^3} \tag{4.15}$$

then

$$\|R_{N_0} A_\varepsilon R_{N_0}\| < 1 - \delta \tag{4.16}$$

for $\varepsilon \notin \Omega_{N_0}(E)$,

$$\text{mes } \Omega_{N_0}(E) < e^{-(\log N_0)^{10}}. \tag{4.17}$$

Assume indeed $\xi = R_{N_0}\xi$, $\|\xi\|_2 = 1$ and

$$|\langle A_\varepsilon \xi, \xi \rangle| = |\langle V_\varepsilon (-\Delta - E)^{-1/2}\xi, (-\Delta - E)^{-1/2}\xi \rangle| > 1 - \delta. \tag{4.18}$$

From (4.7), (4.13), (4.15), (4.18)

$$\left(\int \frac{|\hat{\xi}(\lambda)|^2}{|\lambda|^2 - E} d\lambda \right)^{1/2} = \|(-\Delta - E)^{-1/2}\xi\| > |E|^{1/2}(1 - \delta) > 1 - 2\delta$$

$$\int_{|\lambda| > \delta^{1/4}} |\hat{\xi}(\lambda)|^2 < C\delta^{1/2} \tag{4.19}$$

$$\|(-\Delta - E)^{-1/2}\xi - \xi\|_2 < C\delta^{1/2}. \tag{4.20}$$

Returning to (4.18), (4.20) implies

$$|\langle V_\varepsilon \xi, \xi \rangle| > 1 - C\delta^{1/2}$$

and by (4.19)

$$\|V_\varepsilon * \widehat{\psi}_{\delta^{1/5}}\|_{L^\infty(|x| < N_0)} > \frac{1}{2}. \tag{4.21}$$

We denote here $\psi_\gamma(\lambda) = \psi(\gamma^{-1}\lambda), 0 \leq \psi \leq 1$ as a smooth bump function, where $\psi = 1$ on $|\lambda| \leq 1$, and $\psi = 0$ on $|\lambda| > 2$.

Denoting τ_y as the translation operator for $y \in \mathbb{R}^d$, we have

$$\left\| \widehat{\psi}_{\delta^{1/5}} - \tau_y \widehat{\psi}_{\delta^{1/5}} \right\|_1 < \delta^{1/10} \quad \text{if } |y| < \delta^{-1/10}. \tag{4.22}$$

Taking $K \sim \delta^{-\frac{1}{10}}$, (4.21), (4.22) imply therefore

$$\frac{1}{4} < \left\| V_\varepsilon * \left(K^{-d} \sum_{k_i \in \mathbb{Z}, 0 \leq k_i < K} \tau_k \widehat{\psi}_{\delta^{1/5}} \right) \right\|_{L^\infty(|x| < N_0)}$$

$$\leq \left\| K^{-d} \left(\sum_{0 \leq k_i \leq K} \tau_k V_\varepsilon \right) \right\|_\infty$$

$$\leq K^{-d} \left\| \sum_{n \in \mathbb{Z}^d, 0 \leq k_i \leq K} \varepsilon_n \phi(x - k - n) \right\|_{L^\infty(|x| < N_0)}$$

$$< \max_{|n| < 2N_0} K^{-d} \left| \sum_{0 < k_i < K} \varepsilon_{n-k} \right| + e^{-\frac{1}{2}N_0}. \tag{4.23}$$

For ε outside a set Ω_{N_0} satisfying (4.17), (4.23) is clearly bounded by

$$CK^{-d}K^{d/2}(\log N_0) < C\delta^{d/20}(\log N_0) < \frac{1}{\log N_0}$$

(a contradiction).

Hence, if $\varepsilon \notin \Omega_{N_0}$

$$\|R_{N_0} A_\varepsilon R_{N_0}\| < 1 - (\log N_0)^{-10^3} \qquad (4.24)$$

implying convergence of the Neumann series (4.14) and

$$\|(1 + R_{N_0} A_\varepsilon R_{N_0})^{-1}\| < (\log N_0)^{10^3}. \qquad (4.25)$$

Since $|A_\varepsilon(x,y)| < e^{-c|x-y|}$, we get moreover for $\varepsilon \notin \Omega_N$

$$|(1 + R_{N_0} A_\varepsilon R_{N_0})^{-1}(x,y)| < e^{-\frac{N_0}{(\log N_0)^{10^4}}} \text{ for } |x-y| > \frac{N_0}{10}. \qquad (4.26)$$

Next, we prove inductively on the scale N that

$$\|(1 + R_N A_\varepsilon R_N)^{-1}\| < e^{N^{9/10}} \qquad (4.27)$$

and

$$|(1 + R_N A_\varepsilon R_N)^{-1}(x,y)| < e^{-\frac{N}{(\log N)^{10^4}}} \text{ for } |x-y| > \frac{N}{10} \qquad (4.27')$$

for $\varepsilon \notin \Omega_N$, with

$$\text{mes } \Omega_N < e^{-c\frac{(\log N)^2}{\log\log N}} \qquad (4.28)$$

(and also the analogue of (3.10)).

The preceding inequality yields this at scale N_0. The inductive step from scale N to scale N_1 is a modification of the proof of Proposition 3.7. We indicate the main points.

Using the notation from Proposition 3.7, assume the singular N-interval $\subset [-N_1, N_1]^d$ centered at 0, i.e. $r_0 = 0$. We consider now continuous eigenvalue functions $E_\tau(t)$ of

$$A_{N_1}(x_j = \varepsilon_j(j \notin S'), x_j = t_j(j \in S')); \quad A_{N_1} = R_{N_1} A R_{N_1} \qquad (4.29)$$

such that

$$\min_t |1 + E_\tau(t)| < e^{-N^{\frac{10}{11}}}. \qquad (4.30)$$

Thus again

$$\frac{\partial E_\tau}{\partial t_j} = \left\langle \frac{\partial A}{\partial t_j} \xi, \xi \right\rangle \qquad (4.31)$$

$\left(\xi = \xi_t \text{ is the corresponding normalized eigenfunction of } A_{N_1}(t)\right)$.

Since

$$|\xi_t(x)| < e^{-c\frac{|x|}{(\log N)^{10^4}}} \quad \text{for} \quad |x| > CN \tag{4.32}$$

we get

$$(4.31) = \int |(-\Delta - E)^{-1/2}\xi|^2 \, e^{-|x-j|}dx$$

$$< \int_{|x-j|<\frac{|j|}{2}} + \int_{|x-j|>\frac{|j|}{2}}$$

$$\leq \|(-\Delta - E)^{-1/2}\xi\|_{L^2[|x|>\frac{|j|}{2}]} + e^{-\frac{|j|}{2}}$$

$$\leq \|\xi\|_{L^2[|x|>\frac{|j|}{4}]} + e^{-c|j|}$$

$$< e^{-c\frac{|j|}{(\log N)^{10^4}}+CN} \tag{4.33}$$

by (4.32).

In order to apply Lemma 2.1, we also need a lower bound on (4.31). Write

$$(4.31) > e^{-\frac{|j|}{2}} \int_{|x|<\frac{|j|}{2}} |(-\Delta - E)^{-1/2}\xi|^2$$

$$= e^{-\frac{|j|}{2}}\left(\|(-\Delta - E)^{-1/2}\xi\|_2^2 - e^{-|j|(\log N)^{-10^4}}\right)$$

$$> \frac{1}{2}e^{-\frac{|j|}{2}}. \tag{4.34}$$

Indeed, since

$$(-\Delta - E)^{-1/2}V(-\Delta - E)^{-1/2}\xi \approx -\xi \,,$$

$$\|(-\Delta - E)^{-1/2}\xi\|_2^2 \geq \int |V| \, |(-\Delta - E)^{-1/2}\xi|^2 > \frac{1}{2} \,.$$

Following the argument in Proposition 3.7, one then again obtains the estimate

$$\text{mes}_\varepsilon\left[\text{dist}\left(-1, \text{Spec } A_{N_1}(\varepsilon)\right) < e^{-N.N_1^{\frac{1}{10}}}\right] < e^{-c\frac{(\log N_1)^2}{\log\log N}} + N_1^d\delta_N^2. \tag{4.35}$$

5 Remark on the IDS of the 1-D Bernoulli Model with Weak Disorder

Consider the 1D Bernoulli model

$$H_\lambda(\varepsilon) = \lambda\varepsilon_n\delta_{nn'} + \Delta, \, \lambda > 0. \tag{5.1}$$

It is known that the integrated density of states (IDS) is Hölder continuous

$$|N(E) - N(E')| < C|E - E'|^\alpha \tag{5.2}$$

for some exponent $\alpha = \alpha(\lambda) > 0$ (see [S-V-W], for instance).

For $\lambda \to \infty, \alpha(\lambda) \to 0$ (see [S-T]). On the other hand, if $\lambda \to 0$ one should reasonably expect the smoothness properties of $N(E)$ to improve. The exponent $\alpha(\lambda)$ in [S-V-W] deteriorates, however, for $\lambda \to 0$. We show here the following

Proposition 5.3. *There is an absolute constant $c_1 > 0$ s.t.*

$$|N(E) - N(E')| < C(\lambda)|E - E'|^{c_1} \tag{5.3}$$

if E, E' satisfy $\delta < |E|, |E'| < 2 - \delta (\delta > 0)$ and $0 < \lambda < \lambda_0(\delta)$.

To prove this, we use the same approach as in [BS] based on the Figotin–Pastur formula and the avalanche principle from [G-S]. But the question of smoothness of $N_\lambda(E)$ for $\lambda \to 0$ remains.

By Thouless' formula

$$L(E) = \int \log |E - E'| dN(E') , \tag{5.4}$$

the problem is equivalent to Hölder's regularity of the Lyapounov exponent

$$L(E) = \lim_{N \to \infty} L_N(E)$$

where in our case

$$L_N(E) = \frac{1}{N} \int \log \|M_N(E; \varepsilon)\| d\varepsilon \tag{5.5}$$

$$M_N(E; \varepsilon) = \prod_{n=N}^{1} \begin{pmatrix} \lambda\varepsilon_n - E & -1 \\ 1 & 0 \end{pmatrix} . \tag{5.6}$$

Denoting

$$E = 2 \cos \kappa \tag{5.7}$$

$$V_n = -\frac{\varepsilon_n}{\sin \kappa} \tag{5.8}$$

the Figotin–Pastur formula gives

$$\frac{1}{N} \log \|M_N(E, \varepsilon)\| = \frac{1}{2N} \sum_{1}^{N} \log \left(1 + \lambda V_n \sin 2(\varphi_n + \kappa) + \lambda^2 V_n^2 \sin^2(\varphi_n + \kappa)\right) \tag{5.9}$$

with

$$\zeta_n = e^{2i\varphi_n} \tag{5.10}$$

recursively given by

$$\zeta_{n+1} = \mu\zeta_n + i\frac{\lambda}{2}V_n \frac{(\mu\zeta_n - 1)^2}{1 - \frac{i\lambda}{2}V_n(\mu\zeta_n - 1)} \tag{5.11}$$

and
$$\mu = e^{2i\kappa}. \tag{5.12}$$

Notice that by (5.8), (5.11), ζ_n only depends on $\varepsilon_{n'}$ for $n' \leq n - 1$.

Expanding (5.9), we obtain

$$(5.9) = \frac{\lambda^2}{8N} \sum_1^N V_n^2 \tag{5.13}$$

$$+ \frac{\lambda}{2N} \sum_1^N V_n \sin 2(\varphi_n + \kappa) \tag{5.14}$$

$$- \frac{\lambda^2}{4N} \sum_1^N V_n^2 \cos 2(\varphi_n + \kappa) \tag{5.15}$$

$$+ \frac{\lambda^2}{8N} \sum_1^N V_n^2 \cos 4(\varphi_n + \kappa) \tag{5.16}$$

$$+ 0(\lambda^3)$$

and

$$(5.13) = \frac{\lambda^2}{8 \sin^2 \kappa} = \frac{\lambda^2}{2(4 - E^2)}. \tag{5.17}$$

By (5.11)

$$|1 - \mu| \left| \sum_1^N \zeta_n \right| < 1 + 0(\lambda N)$$

$$\left| \sum_1^N \xi_n \right| < \frac{0(\lambda N)}{\sin^2 \kappa} \tag{5.18}$$

and similarly

$$\left| \sum_1^N \zeta_n^2 \right| < \frac{0(\lambda N)}{\sin^2 2\kappa}. \tag{5.19}$$

(Since we assume $\delta < |E| < 2 - \delta$, κ stays away from $0, \frac{\pi}{2}, \pi$.)

Writing $\cos 2(\varphi_n + \kappa) = \frac{1}{2}(\mu \zeta_n + \bar{\mu}\bar{\zeta}_n)$, $\cos 4(\varphi_n + \kappa) = \frac{1}{2}(\mu^2 \zeta_n^2 + \bar{\mu}^2 \bar{\zeta}_n^2)$,
(5.18), (5.19) imply

$$(5.15), (5.16) = 0(\lambda^3). \tag{5.20}$$

Write

$$(5.14) = \frac{-\lambda}{2N \sin \kappa} \sum_1^N \tilde{\varepsilon}_n \sin 2(\varphi_n + \kappa)$$

$$= -\frac{\lambda}{2N \sin \kappa} \sum_1^N \varepsilon_n d_n(\varepsilon_{n'}; n' < n) \tag{5.21}$$

which is a martingale difference sequence, with

$$\sum_1^N |d_n|^2 = \sum_1^N \sin^2 2(\varphi_n + \kappa) < \frac{N}{2} + \frac{1}{2}\left|\sum_1^N \mu^2 \zeta_n^2 + \bar{\mu}^2 \bar{\zeta}_n^2\right|$$
$$< \left(\frac{1}{2} + 0(\lambda)\right)N. \tag{5.22}$$

In conclusion

$$\frac{1}{N}\log\|M_N(E;\varepsilon)\| = \frac{\lambda^2}{8\sin^2\kappa} - \frac{\lambda}{2N\sin\kappa}\sum_1^N \varepsilon_n d_n + 0(\lambda^3) \tag{5.23}$$

$$L(E) = \frac{\lambda^2}{8\sin^2\kappa} + 0(\lambda^3). \tag{5.24}$$

From Rubin's martingale inequality (see [C-W-W]) and (5.22), we get for $a > 0$

$$\text{mes}\left[\varepsilon\middle|\left|\frac{1}{N}\log\|M_N(E;\varepsilon)\| - L(E)\right| > aL(E)\right] < e^{-(\frac{a^2\lambda^2}{16\sin^2\kappa} + 0(\lambda^3))N}$$
$$< e^{-(\frac{a^2}{2}L(E) + 0(\lambda^3))N}. \tag{5.25}$$

The next ingredient is the 'avalanche principle' for $SL_2(\mathbb{R})$-matrices (see [G-S] or [B-S]).

Lemma 5.26. *Let $A_1, \ldots, A_n$ be a sequence in $SL_2(\mathbb{R})$ and $\mu \gg n$ satisfy the conditions*

$$\min_{1\le j\le n}\|A_j\| > \mu \tag{5.27}$$

$$\max_{1\le j\le n}\left|\log\|A_j\| + \log\|A_{j+1}\| - \log\|A_{j+1}A_j\|\right| < \frac{1}{2}\log\mu. \tag{5.28}$$

Then

$$\left|\log\|A_n\cdots A_1\| + \sum_2^{n-1}\log\|A_j\| - \sum_{j=1}^{n-1}\log\|A_{j+1}A_j\|\right| < C\frac{n}{\mu}. \tag{5.29}$$

One then combines the LDT (5.25) with (5.26), taking

$$A_j = \prod_{(j-1)\ell}^{j\ell}\begin{pmatrix} \lambda\varepsilon_n - E & -1 \\ 1 & 0 \end{pmatrix}$$

and $\ell \in \mathbb{Z}_+$ large enough. Thus, if

$$\mu = e^{\frac{1}{2}\ell L(E)}, n = e^{\frac{1}{10^3}\ell L(E)}, \tag{5.30}$$

it follows from (5.25) with $N = \ell, 2\ell$ and $a = \frac{1}{10}$ that

$$\text{mes}\left[\varepsilon|\ (5.27)\ \text{or}\ (5.18)\ \text{fail}\right] < Cne^{-\left(\frac{1}{200}L(E)+0(\lambda^3)\right)\ell} < e^{-\frac{1}{300}L(E)\ell}. \tag{5.31}$$

Denoting Ω as this exceptional set and integrating (5.29) over its complement Ω^c, it follows that

$$\left|n\ell L_{n\ell}(E) + (n-2)\ell L_\ell(E) - 2(n-1)\ell L_{2\ell}(E)\right| < C\frac{n}{\mu} + Cn\ell(\text{mes }\Omega)$$

$$\left|L_{n\ell}(E) + \frac{n-2}{n}L_\ell(E) - 2\frac{n-1}{n}L_{2\ell}(E)\right| < e^{-\frac{1}{300}L(E)\ell}. \tag{5.32}$$

Following [G-S], one then derives further that

$$\left|L(E) + \frac{n-2}{n}L_\ell(E) - 2\frac{n-1}{n}L_{2\ell}(E)\right| < e^{-\frac{1}{400}L(E)\ell}. \tag{5.33}$$

Finally, we need an estimate on

$$|\partial_E L_\ell(E)| \leq \frac{1}{\ell}\int \frac{\|\partial_E M_\ell(E;\varepsilon)\|}{\|M_\ell(E,\varepsilon)\|}d\varepsilon. \tag{5.34}$$

Estimating

$$\|\partial_E M_\ell(E;\varepsilon)\| \leq \sum_{m=1}^{\ell}\|M_{m-1}(E;\varepsilon_1,\ldots,\varepsilon_{m-1})\|\ \|M_{\ell-m}(E,\varepsilon_{m+1},\ldots,\varepsilon_\ell)\| \tag{5.35}$$

and recalling (5.23), it follows that

$$\frac{\|\partial_E M_\ell(E;\varepsilon)\|}{\|M_\ell(E;\varepsilon)\|} \leq \sum_{m=1}^{\ell}\exp\left\{\frac{\lambda}{2\sin\kappa}\left[\sum_1^{\ell}\varepsilon_n d_n - \sum_1^{m-1}\varepsilon_n d'_n - \sum_{m+1}^{\ell}\varepsilon_n d'_n\right]+0(\lambda^3\ell)\right\}. \tag{5.36}$$

Hence, again by Rubin's inequality

$$|\partial_E L_\ell(E)| < e^{\ell\left(\frac{\lambda^2}{2\sin^2\kappa}+0(\lambda^3)\right)} < e^{\ell(4L(E)+0(\lambda^3))}. \tag{5.37}$$

From (5.33), (5.37) and $\delta < |E|, |E'| < 2 - \delta$, we therefore obtain

$$|L(E) - L(E')| < e^{-\frac{1}{400}L(E)\ell} + e^{5L(E)\ell}|E - E'|. \tag{5.38}$$

Appropriate choice of ℓ clearly implies (4.3') for some numerical exponent c_1.

Remark. Instead of applying the avalanche principle, we may alternatively proceed as follows. One has that

$$\bar{\partial}N = \mathbb{E}\left[G(0,0;z)\right] \qquad (z = E + i\eta) \tag{5.39}$$

where $\bar{\partial}N$ denotes $\frac{\partial}{\partial \bar{z}}$ of the harmonic extension of $N(E)$ to the upper half-plane. Estimate by the resolvent-identity, for $\Lambda = [-a, b] (a, b \in \mathbb{Z}_+)$

$$|G(0,0;z)| \le |G_\Lambda(0,0;z)| + \frac{1}{|\eta|}\big(|G_\Lambda(-a,0;z)| + |G_\Lambda(0,b,z)|\big)$$

$$\le \|G_\Lambda(E)\| + \frac{1}{|\eta|}(|G_\Lambda(-a,0;z)| + |G_\Lambda(0,b;z)|). \qquad (5.40)$$

Fix N and take $a, b \in \{N-1, N\}$ s.t.

$$|\det[H_\Lambda - E]| \sim \|M_{[-N,N]}(E)\|.$$

Hence, by Cramer's rule

$$\|(5.39)\| \lesssim \sum_{0<m<2N} \mathbb{E}\left[\frac{\|M_{[0,m]}(E;\varepsilon)\| \, \|M_{[m,2N]}(E;\varepsilon)\|}{\|M_{[0,2N]}(E;\varepsilon)\|}\right]$$

$$+ \eta^{-1}\mathbb{E}\left[\frac{\|M_{[0,N]}(E,\varepsilon)\|}{\|M_{[0,2N]}(E,\varepsilon)\|}\right]. \qquad (5.41)$$

Using (5.23), more careful analysis shows that

$$(5.41) \lesssim e^{0(\lambda^3 N)} \sum_{0<m<2N} \mathbb{E}\big[e^{\frac{\lambda}{\sin\kappa}\sum_1^m \varepsilon_n d'_n} \wedge e^{\frac{\lambda}{\sin\kappa}\sum_{m+1}^{2N}\varepsilon_n d_n}\big] \qquad (5.42)$$

$$+ \eta^{-1}e^{0(\lambda^3 N)}e^{-NL(E)}\mathbb{E}\big[e^{\frac{\lambda}{2\sin\kappa}\sum_1^N \varepsilon_n d_n}\big] \qquad (5.43)$$

where $d'_n = d'_n(\varepsilon_{n'}; n' > n)$, $\sum_1^m |d'_n|^2 = \frac{m}{2}(1+0(\lambda^3))$ and $d_n = d_n(\varepsilon_{n'}; n' < n)$,
$\sum_{m+1}^{2N} |d_n|^2 = \frac{2N-m}{2}(1 + 0(\lambda^3))$.
    Hence

$$(5.41) < e^{0(\lambda^3 N)}\big[e^{2L(E)N} + \eta^{-1}e^{-\frac{1}{2}NL(E)}\big]. \qquad (5.44)$$

Appropriate choice of N then implies that

$$|(5.39)| \ll \eta^{-\frac{4}{5}-0(\lambda)}. \qquad (5.45)$$

By (5.39), this implies Hölder-regularity of $N(E)$ with exponent $\alpha > \frac{1}{5} - 0(\lambda)$.

References

[B-S]   Bourgain, J., Schlag, W.: Anderson localization for Schrödinger operators on $\mathbb{Z}$ with strongly mixing potentials. Comm. Math. Phys., **215**, 143–175 (2000)

[C-K-M] Carmona, R., Klein, A., Martinelli, F.: Anderson localization for Bernoulli and other singular potentials. Comm. Math. Phys., **108**, no 1, 41–66 (1987)

[G-S]   Goldstein, G., Schlag, W.: Hölder continuity of the integrated density of states for quasi-periodic Schrödinger equations and averages of shifts of subharmonic functions. Annals of Math. (2), **154**, no. 1, 155–203 (2001)

[S-V-W] Shubin, C., Vakilian, R., Wolff, T.: Some harmonic analysis questions suggested by Anderson–Bernoulli models. GAFA, **8**, no. 5, 932–964 (1998)

[S-T]   Simon, B., Taylor, M.: Harmonic analysis on $SL(2, \mathbb{R})$ and smoothness of the density of states in the one-dimensional Anderson model. Comm. Math. Phys., **101**, no. 1, 1–19 (1985)

Symmetrization and Isotropic Constants of Convex Bodies

J. Bourgain[1], B. Klartag[2]⋆ and V. Milman[2]⋆

[1] Institute for Advanced Study, Princeton, NJ 08540, USA
   bourgain@math.ias.edu
[2] School of Mathematical Sciences, Tel Aviv University, Tel Aviv 69978, Israel
   klartagb@post.tau.ac.il, milman@post.tau.ac.il

Summary. We investigate the effect of a Steiner type symmetrization on the isotropic constant of a convex body. We reduce the problem of bounding the isotropic constant of an arbitrary convex body, to the problem of bounding the isotropic constant of a finite volume ratio body. We also add two observations concerning the slicing problem. The first is the equivalence of the problem to a reverse Brunn–Minkowski inequality in isotropic position. The second is the essential monotonicity in n of $L_n = \sup_{K \subset \mathbb{R}^n} L_K$ where the supremum is taken over all convex bodies in $\mathbb{R}^n$, and L_K is the isotropic constant of K.

1 Introduction

Let $K \subset \mathbb{R}^n$ be a convex body whose barycenter is at the origin (i.e. $b(K) = \int_K \vec{x} \, dx = 0$). The inertia matrix of K is the matrix M_K whose entries are $M_{i,j} = \int_K x_i x_j dx$. The isotropic constant of K, denoted by L_K, is defined as

$$L_K^2 = \frac{\det(M_K)^{\frac{1}{n}}}{\mathrm{Vol}(K)^{1+\frac{2}{n}}}.$$

The isotropic constant is invariant under linear transformations of the body. If M_K is a scalar matrix and $\mathrm{Vol}(K) = 1$, we say that K is isotropic, or that K is in isotropic position. In this case, for any $\theta \in \mathbb{R}^n$,

$$\int_K \langle x, \theta \rangle^2 dx = L_K^2 |\theta|^2$$

where $| \cdot |$ is the standard Euclidean norm in $\mathbb{R}^n$. Any convex body K has a unique affine image of volume one which is in isotropic position. We refer the

⋆ Supported in part by the Israel Science Foundation and by the Minkowski Center for Geometry at Tel Aviv University.

reader to [MP] for more information concerning the isotropic position and the isotropic constant.

A major unsolved problem asks whether there exists a numerical constant C such that $L_K < C$ for every convex body in any finite dimension. This problem is called the slicing problem or the hyperplane conjecture. A positive answer to this question has many interesting consequences, see [MP]. One of these is that every convex body of volume one, has an $n-1$ dimensional section whose $n-1$ dimensional volume is greater than some constant $c > 0$. The current best estimate is $L_K < cn^{1/4} \log n$, for an arbitrary convex body $K \subset \mathbb{R}^n$ (see [Bou], or the presentation in [D]. See [Pa] for the non-symmetric case). For certain classes of convex bodies the question is affirmatively answered, such as for unconditional bodies (as observed by Bourgain, see [MP]), zonoids, duals of zonoids (see [Ba2], also for the connection with the Gordon-Lewis constant), duals to bodies with finite volume ratio (see [MP]), and more (e.g. [J]). Here, we present a reduction of the general problem to the boundness of the isotropic constant of a certain class of convex bodies: those which have a finite volume ratio. For $K \subset \mathbb{R}^n$, the volume ratio of K is defined as,

$$v.r.(K) = \sup_{\mathcal{E} \subset K} \left(\frac{\text{Vol}(K)}{\text{Vol}(\mathcal{E})} \right)^{\frac{1}{n}}$$

where the supremum is over all ellipsoids contained in K. Here we prove the following conditional proposition:

Proposition 1.1. *There exists $v > 1$ such that the following holds:*

If there exists $c_1 > 0$ such that for any n and for any $K \subset \mathbb{R}^n$, the inequality $v.r.(K) < v$ implies that $L_K < c_1$,

then there exists $c_2 > 0$ such that for any n and for any $K \subset \mathbb{R}^n$ we have $L_K < c_2$.

Next, we shall state a qualitative version of Proposition 1.1. Denote $L_n = \sup_{K \subset \mathbb{R}^n} L_K$ where the supremum is over all convex sets in $\mathbb{R}^n$, and define

$$L_n(a) = \sup \left\{ L_K \ ; \ K \subset \mathbb{R}^n \ , \ v.r.(K) \le a \right\}.$$

Then we can bound L_n by a function of $L_n(a)$ for a suitable $a > 1$. As a matter of fact, this function is almost linear:

Proposition 1.2. *For any $\delta > 0$, there exist numbers $v(\delta) > 1$, $c(\delta) > 0$ such that for any n,*

$$L_n < c(\delta) \, L_n\big(v(\delta)\big)^{1+\delta}.$$

A proof of these propositions, using a symmetrization technique, is presented in Section 4. The technique itself is presented in Section 2. We prove the following proposition in Section 3.

Proposition 1.3. *If $m < n$, then $L_m < cL_n$ where c is a numerical constant.*

As observed by K. Ball (see [MP]), the hyperplane conjecture implies that a reverse Brunn–Minkowski inequality holds in the isotropic position. Answering a question posed by K. Ball to one of the authors, we show that the slicing problem is actually equivalent to a reverse Brunn–Minkowski inequality in the isotropic position. The following conditional statement is proved in Section 5:

Proposition 1.4. *Assume that there exists a constant $C > 0$, such that for any n, and for any two isotropic convex bodies $K, T \subset \mathbb{R}^n$,*

$$\mathrm{Vol}(K + T)^{1/n} \leq C \left(\mathrm{Vol}(K)^{1/n} + \mathrm{Vol}(T)^{1/n} \right). \tag{1}$$

Then it follows that for any convex body $K \subset \mathbb{R}^n$,

$$L_K < C'(C)$$

where $C'(C)$ is a number that depends solely on C.

Actually, Proposition 1.4 is correct even if we restrict T to be a Euclidean ball, as is evident from the proof. Note that as proved in [M2], inequality (1) which is a reverse Brunn–Minkowski inequality, holds when K and T are in a special position called M-position (see definition in Section 3). However, the connection of an M-position with the isotropic position is not yet clear.

Throughout the paper we denote by $c, c', \tilde{c}, C$, etc. some positive universal constants whose value is not necessarily the same on different appearances. Whenever we write $A \approx B$, we mean that there exist universal constants $c, c' > 0$ such that $cA < B < c'A$. Also, $\mathrm{Vol}(T)$ denotes the volume of a set $T \subset \mathbb{R}^n$, relative to its affine hull.

The paper [BKM] serves as an extended introduction to this paper.

2 Symmetrization

2.1 Definition

Let $K \subset \mathbb{R}^n$ be a convex body, let $E \subset \mathbb{R}^n$ be a subspace of dimension k, and let $T \subset E$ be a k-dimensional convex body, whose barycenter is at the origin. We define the "(T, E)-symmetrization" of K as the unique body K' such that:

(i) for any $x \in E^{\perp}$, $\mathrm{Vol}(K \cap (x + E)) = \mathrm{Vol}(K' \cap (x + E))$;
(ii) for any $x \in E^{\perp}$ the body $K' \cap (x+E)$ is homothetic to T, and its barycenter lies in $E^{\perp}$.

In other words, we replace any section of K which is parallel to E, with a homothetic copy of T of the appropriate volume. This procedure of symmetrization is known in convexity, see [BF], page 79. For completeness, we shall next prove that this symmetrization preserves convexity, as follows from the Brunn–Minkowski inequality.

Lemma 2.1. K' *is a convex body.*

Proof. For any $z \in E^\perp$, the section $(z+E) \cap K'$ is convex, as a homothetic copy of T. Let $x, y \in \mathrm{Proj}_{E^\perp}(K')$ be any points, where $\mathrm{Proj}_{E^\perp}$ is the orthogonal projection onto $E^\perp$ in $\mathbb{R}^n$. We will show that

$$\mathrm{conv}\big((x+E) \cap K', (y+E) \cap K'\big)$$
$$= \bigcup_{0 \leq \lambda \leq 1} \lambda\big[(x+E) \cap K'\big] + (1-\lambda)\big[(y+E) \cap K'\big] \subset K'.$$

For $z \in E^\perp$, denote $v(z) = \mathrm{Vol}((z+E) \cap K') = \mathrm{Vol}((z+E) \cap K)$. Since K is convex, by Brunn–Minkowski (see e.g. [BF]),

$$v\big(\lambda x + (1-\lambda)y\big)^{1/k} \geq \lambda v(x)^{1/k} + (1-\lambda)v(y)^{1/k} \tag{2}$$

where $k = \dim(E)$. Since $(z+E) \cap K' = z + \big(\frac{v(z)}{\mathrm{Vol}(T)}\big)^{1/k} T$ for any point $z \in E^\perp$, inequality (2) entails that

$$\big(\lambda x + (1-\lambda)y + E\big) \cap K' \supset \lambda\big[(x+E) \cap K'\big] + (1-\lambda)\big[(y+E) \cap K'\big]$$

and the lemma is proved. $\qquad\square$

2.2 The Effect of a Symmetrization on the Isotropic Constant

Let us determine the eigenvectors of the inertia matrix $M_{K'}$. These eigenvectors are also called axes of inertia of the body K'. If K is an arbitrary body of volume one with its barycenter at zero, and $\{e_1, .., e_n\}$ are its axes of inertia, then since $L_K^2 = \det(M_K)^{1/n}$,

$$L_K^2 = \left(\prod_{i=1}^n \int_K \langle x, e_i \rangle^2 dx \right)^{\frac{1}{n}}.$$

Lemma 2.2. *Assume that K is isotropic. Let $e_1, .., e_k$ be axes of inertia of the body $T \subset E$, and let $e_{k+1}, .., e_n$ be any orthonormal basis of $E^\perp$. Then the orthonormal basis $\{e_1, .., e_n\}$ is a basis of inertia axes of K'.*

Proof. By property (i) from the symmetrization definition, for any $v \in E^\perp$

$$\int_{K'} \langle x, v \rangle^2 dx = \int_K \langle x, v \rangle^2 dx = L_K^2 |v|^2 \tag{3}$$

since K is isotropic. By property (ii), for any $v \in E^\perp, u \in E$,

$$\int_{K'} \langle x, v \rangle \langle x, u \rangle dx = \int_{\mathrm{Proj}_{E^\perp}(K')} \langle y, v \rangle \int_{K' \cap [y+E]} \langle z, u \rangle dz dy = 0$$

since the barycenter of T is at zero. Hence, E and $E^\perp$ are invariant subspaces of $M_{K'}$. According to (3), the operator $M_{K'}$ restricted to $E^\perp$ is simply a multiple of the identity. Therefore any orthogonal basis $e_{k+1}, .., e_n$ of $E^\perp$ is a basis of eigenvectors of $M_{K'}$. All that remains is to select k axes of inertia in E. Let $e_1, .., e_k$ be axes of inertia of the k-dimensional body T. It is straightforward to verify that for any $u_1, u_2 \in E$,

$$\int_{K'} \langle x, u_1 \rangle \langle x, u_2 \rangle dx = c(K, E, T) \int_T \langle x, u_1 \rangle \langle x, u_2 \rangle dx$$

where $c(K, E, T) = \dfrac{\int_{\mathrm{Proj}_{E^\perp}(K)} \mathrm{Vol}(K \cap (x+E))^{1+2/k} dx}{\mathrm{Vol}(T)^{1+2/k}}$ depends only on K, E, T. Therefore $e_1, .., e_k$ are also axes of inertia of K'. $\qquad\square$

We postpone the proof of the following lemma to Section 6.

Lemma 2.3. *Let f be a non-negative function on $\mathbb{R}^n$, such that $f^{1/k}$ is concave on its support, and $\int_{\mathbb{R}^n} f(x) dx = 1$. Denote $M = \max_{x \in \mathbb{R}^n} f(x)$. Then,*

$$\frac{(k+1)(k+2)}{(n+k+1)(n+k+2)} M^{2/k} \leq \int_{\mathbb{R}^n} f(x)^{1+\frac{2}{k}} dx \leq M^{2/k}.$$

Now we can estimate L_2 norms of some linear functionals over K'.

Lemma 2.4. *Let $K \subset \mathbb{R}^n$ be a convex body of volume one whose barycenter is at the origin. Let $E \subset \mathbb{R}^n$ be a subspace with $\dim(E) = k$, and let $T \subset E$ be a k-dimensional convex body of volume one with zero as a barycenter. Denote by K' the "(T, E)-symmetrization" of K. Then for any $v \in E$,*

$$\int_{K'} \langle x, v \rangle^2 dx \geq \left(\frac{k+1}{n+1} \right)^2 \mathrm{Vol}(K \cap E)^{2/k} \int_T \langle x, v \rangle^2 dx$$

and

$$\int_{K'} \langle x, v \rangle^2 dx \leq \left(\frac{n+1}{k+1} \right)^2 \mathrm{Vol}(K \cap E)^{2/k} \int_T \langle x, v \rangle^2 dx.$$

Proof.

$$\int_{K'} \langle x, v \rangle^2 dx = \int_{\mathrm{Proj}_{E^\perp}(K')} \int_{K' \cap (E+x)} \langle y, v \rangle^2 dy dx$$

$$= \int_{\mathrm{Proj}_{E^\perp}(K')} \mathrm{Vol}(K' \cap (E+x))^{1+\frac{2}{k}} dx \int_T \langle y, v \rangle^2 dy.$$

Denote $g(x) = \mathrm{Vol}(K' \cap (x + E)) = \mathrm{Vol}(K \cap (x + E))$. Then by the Brunn–Minkowski inequality, $g^{1/k}$ is concave on its support in $E^{\perp}$ and $\int g = \mathrm{Vol}(K) = 1$. By Lemma 2.3,

$$\frac{(k+1)(k+2)}{(n+1)(n+2)} M^{2/k} \int_T \langle y, v \rangle^2 dy \leq \int_{K'} \langle x, v \rangle^2 dx \leq M^{2/k} \int_T \langle y, v \rangle^2 dy$$

where $M = \max_{x \in E^{\perp}} g(x)$. Since the barycenter of K is at the origin, by Theorem 1 in [F],

$$g(0) \leq M \leq \left(\frac{n+1}{k+1} \right)^k g(0)$$

and since $g(0) = \mathrm{Vol}(K \cap E)$, we get

$$\left(\frac{k+1}{n+1} \right)^2 \leq \frac{(k+1)(k+2)}{(n+1)(n+2)} \leq \frac{\int_{K'} \langle x, v \rangle^2 dx}{\mathrm{Vol}(K \cap E)^{\frac{2}{k}} \int_T \langle x, v \rangle^2 dx} \leq \left(\frac{n+1}{k+1} \right)^2.$$

$\square$

The following theorem connects the isotropic constant of the symmetrized body with the isotropic constants of K, T.

Theorem 2.5. *Let K be an isotropic body of volume one, E a subspace of dimension k, T a k-dimensional convex body with its barycenter at the origin, and K' the "(T, E)-symmetrization" of K. Then*

$$L_{K'} \approx L_K^{1 - \frac{k}{n}} L_T^{k/n} \mathrm{Vol}(K \cap E)^{1/n}.$$

In fact, the ratio of these two quantities is always between $\left(\frac{k+1}{n+1} \right)^{k/n}$ and $\left(\frac{n+1}{k+1} \right)^{k/n}$.

Proof. We may assume that $\mathrm{Vol}(T) = 1$. Let $\{e_1, .., e_n\}$ be selected according to Lemma 2.2. Then,

$$L_{K'} = \left(\prod_{i=1}^n \sqrt{\int_{K'} \langle x, e_i \rangle^2 dx} \right)^{1/n} = L_K^{1 - \frac{k}{n}} \left(\prod_{i=1}^k \sqrt{\int_{K'} \langle x, e_i \rangle^2 dx} \right)^{1/n}$$

where the right-most equality follows from (3). By Lemma 2.4,

$$L_{K'} \geq L_K^{1 - \frac{k}{n}} \left(\prod_{i=1}^k \sqrt{\left(\frac{k+1}{n+1} \right)^2 \mathrm{Vol}(K \cap E)^{\frac{2}{k}} \int_T \langle x, e_i \rangle^2 dx} \right)^{1/n}$$

$$= L_K^{1 - \frac{k}{n}} \left(\frac{k+1}{n+1} \right)^{\frac{k}{n}} \mathrm{Vol}(K \cap E)^{\frac{1}{n}} L_T^{\frac{k}{n}}$$

since the vectors $e_1, .., e_k$ are inertia axes of T. Therefore,

$$L_{K'} > cL_K^{1-\frac{k}{n}} L_T^{k/n} \mathrm{Vol}(K \cap E)^{1/n}.$$

Regarding the inverse inequality, according to the opposite inequality in Lemma 2.4 we get,

$$L'_K \leq \left(\frac{n+1}{k+1}\right)^{\frac{k}{n}} L_K^{1-\frac{k}{n}} \mathrm{Vol}(K \cap E)^{\frac{1}{n}} L_T^{\frac{k}{n}}$$
$$< cL_K^{1-\frac{k}{n}} L_T^{k/n} \mathrm{Vol}(K \cap E)^{1/n}$$

for a different constant c. $\qquad\qquad\square$

3 Use of an M-Ellipsoid

We will need to use a special ellipsoid associated with an arbitrary convex body, called an M-ellipsoid. An M-ellipsoid is defined by the following theorem (see [M2], or chapter 7 in the book [P]):

Theorem 3.1. *Let $K \subset \mathbb{R}^n$ be a convex body. Then there exists an ellipsoid $\mathcal{E}$ with $\mathrm{Vol}(\mathcal{E}) = \mathrm{Vol}(K)$ such that*

$$N(K, \mathcal{E}) = \min\{\sharp A; K \subset A + \mathcal{E}\} < e^{cn}$$

where $\sharp A$ is the number of elements in the set A, and c is a numerical constant. We say that $\mathcal{E}$ is an M-ellipsoid of K (with constant c).

An M-ellipsoid may replace K in various volume computations. For example, assume that $\mathcal{E}$ is an M-ellipsoid of K. If $E \subset \mathbb{R}^n$ is a subspace, and Proj_E is the orthogonal projection onto E in $\mathbb{R}^n$, then by Theorem 3.1,

$$\mathrm{Vol}\big(\mathrm{Proj}_E(K)\big)^{1/n} \leq \big(e^{cn} \mathrm{Vol}\big(\mathrm{Proj}_E(\mathcal{E})\big)\big)^{1/n} = c' \mathrm{Vol}\big(\mathrm{Proj}_E(\mathcal{E})\big)^{1/n}.$$

Lemma 3.2. *Let $K \subset \mathbb{R}^n$ be a convex body of volume one whose barycenter is at the origin. Let $E \subset \mathbb{R}^n$ be a subspace of any dimension. Then,*

$$\mathrm{Vol}(K \cap E)^{1/n} > c\frac{1}{\mathrm{Vol}(\mathrm{Proj}_{E^\perp}(K))^{1/n}}.$$

Proof. Denote $m = \dim(E)$ and $M = \max_{x \in E^\perp} \mathrm{Vol}(K \cap (E + x))$. By Fubini and by Theorem 1 in [F],

$$\mathrm{Vol}(K) \leq M\mathrm{Vol}(\mathrm{Proj}_{E^\perp} K) \leq \left(\frac{n+1}{m+1}\right)^m \mathrm{Vol}(\mathrm{Proj}_{E^\perp} K)\mathrm{Vol}(K \cap E).$$

Since $\mathrm{Vol}(K) = 1$, we obtain

$$\mathrm{Vol}(K \cap E)^{1/n} > \left(\frac{m+1}{n+1}\right)^{\frac{m}{n}} \frac{1}{\mathrm{Vol}(\mathrm{Proj}_{E^\perp}(K))^{1/n}}$$

and since $\left(\frac{m+1}{n+1}\right)^{\frac{m}{n}} > c$, the lemma follows.   □

Proof of Proposition 1.3. First assume that $m \geq \frac{n}{2}$. Recall that $L_n = \sup_{C \subset \mathbb{R}^n} L_C$ where the supremum is taken over all isotropic convex bodies in $\mathbb{R}^n$. This supremum is attained by a compactness argument (the collection of all convex sets modulu affine transformations is compact). Define K to be one of the bodies where the supremum is attained; i.e.

$$L_K = L_n$$

and K is isotropic and of volume one. Let $\mathcal{E}$ be an M-ellipsoid of K. Since $\mathcal{E}$ is an ellipsoid of volume one, it has at least one projection onto a subspace $E^\perp$ of dimension $n - m$, such that

$$\mathrm{Vol}(\mathrm{Proj}_{E^\perp} K)^{1/n} < c\,\mathrm{Vol}(\mathrm{Proj}_{E^\perp} \mathcal{E})^{1/n} < C.$$

By Lemma 3.2,
$$\mathrm{Vol}(K \cap E)^{1/n} > c'.$$

Let T be an m-dimensional body such that $L_T = L_m$ and T is of volume one and isotropic. Denote by K' the "(T, E)-symmetrization" of K. Then $L_K = L_n \geq L_{K'}$, and by Theorem 2.5,

$$L_K \geq L_{K'} > cL_K^{1-\frac{m}{n}} L_T^{\frac{m}{n}} \mathrm{Vol}(K \cap E)^{1/n} > \tilde{c} L_K^{1-\frac{m}{n}} L_T^{\frac{m}{n}}$$

or equivalently,
$$L_n = L_K > \tilde{c}^{\frac{n}{m}} L_T = \tilde{c}^{\frac{n}{m}} L_m.$$

Since we assumed that $\frac{n}{m} \leq 2$, we get $L_m < c' L_n$. Regarding the case in which $m < \frac{n}{2}$, note that $L_m \leq L_{2m}$, since the $2m$-dimensional body which is the cartesian product of T with itself, has the same isotropic constant as T. If s is the maximal integer such that $2^s m \leq n$, then clearly $2^s m > \frac{n}{2}$, and therefore

$$L_m \leq L_{2^s m} < c' L_n.$$ □

Remark 3.3. In the proof of Proposition 1.3 we showed that for every convex body $K \subset \mathbb{R}^n$ of volume one, and for any $1 \leq k \leq n$, there exists a k-dimensional subspace E such that $\mathrm{Vol}(K \cap E)^{1/n} > c$. This fact is a direct consequence of the existence of an M-ellipsoid, but may not be very trivial to obtain directly.

We would like to mention an additional property attributed to a body $K \subset \mathbb{R}^n$, which has the largest possible isotropic constant. For this purpose, we will quote a useful result which appears in [Ba1] and in [MP]. Our formulation is closer to the one in [MP] (Lemma 3.10, and Proposition 3.11 there). Although results in that paper are stated only for centrally-symmetric bodies, the symmetry assumption is rarely used. The generalization to non-symmetric bodies is straightforward, and reads as follows:

Lemma 3.4. *Let $K \subset \mathbb{R}^n$ be an isotropic convex body of volume one. Let $1 \le k \le n$ and let E be a k-codimensional subspace. Define C as the unit ball of the (non-symmetric) norm defined on $E^\perp$ as*

$$\|\theta\| = |\theta|^{1+\frac{p}{p+1}} \Big/ \Big(\int_{K \cap E(\theta)} |\langle x, \theta \rangle|^p dx \Big)^{\frac{1}{p+1}}$$

for $p = k+1$, where $E(\theta) = \{x + t\theta; x \in E, t > 0\}$ is a half of a $k-1$-codimensional subspace. Then indeed C is convex, and

$$\frac{L_C}{L_K} \approx \mathrm{Vol}(K \cap E)^{1/k}.$$

Corollary 3.5. *Let $K \subset \mathbb{R}^n$ be a convex isotropic body of volume one, such that $L_K = L_n$. Then for any subspace $E \subset \mathbb{R}^n$ of codimension k,*

$$\mathrm{Vol}(K \cap E)^{1/k} < c$$

where c is a numerical constant.

Proof. By Lemma 3.4,

$$\mathrm{Vol}(K \cap E)^{\frac{1}{k}} \approx \frac{L_C}{L_K} = \frac{L_C}{L_n} \le \frac{L_{n-k}}{L_n} < c$$

where the last inequality follows from Proposition 1.3. □

4 Proof of the Reduction to Bodies with Finite Volume Ratio

In this section, assume that $K \subset \mathbb{R}^n$ is a convex isotropic body of volume one, such that $L_K = L_n$. Apriori, an M-ellipsoid of K may be very different from a Euclidean ball. We shall see that Corollary 3.5 imposes stringent conditions on the axes of an M-ellipsoid.

4.1 Controlling the Axes of an M-Ellipsoid

Denote by κ_m the volume of a unit Euclidean ball in $\mathbb{R}^m$. It is well known that $\kappa_m^{1/m} \approx \frac{1}{\sqrt{m}}$. Let $\mathcal{E} = \{x \in \mathbb{R}^n; \sum_i \frac{x_i^2}{n\lambda_i^2} \le 1\}$ be an M-ellipsoid of K, whose existence is guaranteed in Theorem 3.1. The axes of this ellipsoid are of lengths $\sqrt{n}\lambda_1, .., \sqrt{n}\lambda_n$, and $(\prod_{i=1}^n \lambda_i)^{1/n} \approx 1$, since $1 = \text{Vol}(\mathcal{E}) = \kappa_n \prod \sqrt{n}\lambda_i$. Assume that the λ_i's are ordered, i.e. $\lambda_1 \leq ... \leq \lambda_n$. For convenience, and without loss of generality, we assume that n is divisible by four.

Claim 4.1. $\lambda_{n/2} < c$, for some numerical constant c.

Proof. Let $E \subset \mathbb{R}^n$ be any subspace of any dimension. By Lemma 3.2 and Corollary 3.5,

$$\text{Vol}\big(\text{Proj}_E(K)\big)^{1/n} > \frac{c}{\text{Vol}(K \cap E^\perp)^{1/n}} > c'.$$

Let $E = sp\{e_1, .., e_{n/2}\}$, the linear space spanned by $e_1, .., e_{n/2}$. Then,

$$c < \text{Vol}\big(\text{Proj}_E(K)\big)^{1/n} \leq N(K, \mathcal{E})^{1/n} \left(\kappa_{n/2} \prod_{i=1}^{n/2} \sqrt{n}\lambda_i\right)^{1/n}$$

because $\text{Vol}(\text{Proj}_E(\mathcal{E})) = \kappa_{n/2} \prod_1^{n/2} \sqrt{n}\lambda_i$. Since $(\kappa_{n/2}\sqrt{n}^{n/2})^{1/n} \approx 1$, we get that

$$\left(\prod_{i=1}^{n/2} \lambda_i\right)^{2/n} > c.$$

Hence we obtain,

$$\lambda_{n/2} \leq \left(\prod_{i=\frac{n}{2}+1}^n \lambda_i\right)^{2/n} = \left(\prod_{i=1}^n \lambda_i\right)^{2/n} \left(\prod_{i=1}^{n/2} \lambda_i\right)^{-2/n} < \tilde{c}. \qquad (4)$$

$\square$

4.2 Finite Volume Ratio

The following lemma, whose proof involves the notion of an M-ellipsoid, originally appears in [M1]. It can also be deduced from the proof of Corollary 7.9 in [P].

Lemma 4.2. *Let $K \subset \mathbb{R}^n$ be a convex body. Let $0 < \lambda < 1$. Then there exists a subspace G of dimension $\lfloor \lambda n \rfloor$ such that if $P : \mathbb{R}^n \to \mathbb{R}^n$ is a projection (i.e. P is linear and $P^2 = P$) such that $\ker(P) = G$, then $P(K)$ has a volume ratio smaller than $c(\lambda)$, where $c(\lambda)$ is some function which depends solely on λ.*

The central theme underlying the proof which follows, is the connection between an M-ellipsoid and the isotropy ellipsoid of a body with the largest possible isotropic constant. This connection arises when we project K onto the subspace $E = sp\{e_1, .., e_{n/2}\}$, together with its covering ellipsoid. According to (4) we get that $\text{Proj}_E(\mathcal{E}) \subset c\sqrt{n}D$, so in fact the normalized Euclidean ball is an M-ellipsoid for $\text{Proj}_E(K)$. In other words, the isotropy ellipsoid and the selected M-ellipsoid of K are equivalent in a large projection. Therefore, we may combine the properties of an M-ellipsoid with the properties of the isotropy ellipsoid, to create a finite volume ratio body.

Apply Lemma 4.2 to the body $\mathrm{Proj}_E(K)$. There exists a subspace $F \subset E$ such that $\dim(F) = n/4$ and

$$v.r.\big(\mathrm{Proj}_F(K)\big) = v.r.\big(\mathrm{Proj}_F(\mathrm{Proj}_E(K))\big) < C.$$

Indeed, F is the orthogonal complement in E, to the subspace G from Lemma 4.2. Denote K' as the $(D_{F^\perp}, F^\perp)$-symmetrization of K, where $D_{F^\perp}$ is the standard Euclidean ball in $F^\perp$. Then,

$$K' \cap F = \mathrm{Proj}_F(K') = \mathrm{Proj}_F(K)$$

is a finite volume ratio body, i.e. there exists an ellipsoid $\mathcal{F} \subset K' \cap F$ such that $\left(\frac{\mathrm{Vol}(K' \cap F)}{\mathrm{Vol}(\mathcal{F})}\right)^{4/n} < C$. We claim that K' has a bounded volume ratio. Indeed, the ellipsoid

$$\mathcal{E}' = \big\{\lambda x + \mu y; \lambda^2 + \mu^2 \leq 1, x \in \mathcal{F}, y \in K' \cap F^\perp\big\}$$

satisfies

$$\frac{1}{\sqrt{2}}\mathcal{E}' \subset \mathrm{conv}\{\mathcal{F}, K' \cap F^\perp\} \subset K',$$

$$\mathrm{Vol}(\mathcal{E}')^{1/n} \geq \frac{1}{\sqrt{2}C}\mathrm{Vol}\big(\mathrm{Proj}_F(K')\big)^{1/n}\mathrm{Vol}(K' \cap F^\perp)^{1/n} \geq \frac{1}{\sqrt{2}C},$$

by Lemma 3.2. Hence $\mathcal{E}'$ is evidence of the finite volume ratio property of K'. Note also that according to Claim 4.1,

$$\mathrm{Vol}\big(\mathrm{Proj}_F(K)\big)^{1/n} \leq N(K, \mathcal{E})^{1/n}\mathrm{Vol}\big(\mathrm{Proj}_F(\sqrt{n}\lambda_{n/2}D)\big)^{1/n} < c.$$

Hence by Lemma 3.2 and Theorem 2.5

$$L_{K'} \approx L_n^{1/4}\mathrm{Vol}(K \cap F^\perp)^{1/n} > c\frac{L_n^{1/4}}{\mathrm{Vol}(\mathrm{Proj}_F(K))^{1/n}} > c'L_n^{1/4}$$

and therefore,

$$L_n < c(L_0)^4$$

where $L_0 = L_n(\tilde{c})$ is the largest possible L_K among all convex bodies in $\mathbb{R}^n$, having volume ratio not larger than $\tilde{c}$, and Proposition 1.1 is proved. □

Remark 4.3. Regarding the connection between v, L_n and $L_n(v)$, we have formally proved for some $v > 1$ that $L_n \lesssim (L_n(v))^4$ for all n. However, by adjusting the dimensions of the subspaces E and F, we can reduce the power of $L_n(v)$, at the expense of increasing the volume ratio constant, v. The dependence obtained using this method is quite poor: for any $0 < \theta < 1$,

$$L_n \leq e^{\frac{c}{1-\theta}}L(e^{\frac{c}{1-\theta}})^{\frac{1}{\theta}}.$$

5 The Isotropic Position and an M-Ellipsoid

Proof of Proposition 1.4. Denote $\mathcal{D}_m = \{x \in \mathbb{R}^m; |x| \le \kappa_m^{-1/m}\}$, a Euclidean ball of volume one. Let $K \subset \mathbb{R}^n$ be a convex isotropic body of volume one. Denote

$$K' = \left\{ (x_1, x_2); x_1 \in \sqrt{\frac{L_{\mathcal{D}_n}}{L_K}} K, x_2 \in \sqrt{\frac{L_K}{L_{\mathcal{D}_n}}} \mathcal{D}_n \right\} \subset \mathbb{R}^{2n}.$$

Let $E \subset \mathbb{R}^{2n}$ be the subspace spanned by the first n standard unit vectors, and let $F = E^{\perp}$. We claim that K' is an isotropic body. By a reasoning similar to that in Lemma 2.2, the subspaces E and F are invariant under the action of the matrix $M_{K'}$. In addition, $M_{K'}$ acts as a multiple of the identity in both subspaces. Let us show that it is the same multiple of the identity in both subspaces, and hence $M_{K'}$ is a scalar matrix. For any $v \in E$,

$$\int_{K'} \langle x, v \rangle^2 dx = \frac{L_{\mathcal{D}_n}}{L_K} \int_K \langle x, v \rangle^2 dx = L_{\mathcal{D}_n} L_K.$$

Also, for any $v \in F$,

$$\int_{K'} \langle x, v \rangle^2 dx = \frac{L_K}{L_{\mathcal{D}_n}} \int_{\mathcal{D}_n} \langle x, v \rangle^2 dx = L_K L_{\mathcal{D}_n}.$$

Therefore K' is isotropic. According to our assumption, a reverse Brunn–Minkowski inequality holds. Hence by (1),

$$\mathrm{Vol}(K' + \mathcal{D}_{2n})^{1/2n} < C \left(\mathrm{Vol}(K')^{1/2n} + \mathrm{Vol}(\mathcal{D}_{2n})^{1/2n} \right) = 2C. \qquad (5)$$

But $\sqrt{\frac{L_K}{L_{\mathcal{D}_n}}} \mathcal{D}_n + \mathcal{D}_{2n} \subset K' + \mathcal{D}_{2n}$. Hence,

$$\mathrm{Vol}(K' + \mathcal{D}_{2n})^{1/2n} > \mathrm{Vol}\left(\sqrt{\frac{L_K}{L_{\mathcal{D}_n}}} \mathcal{D}_n + \mathcal{D}_{2n} \right)^{1/2n} > c \left(\frac{L_K}{L_{\mathcal{D}_n}} \right)^{1/4}. \qquad (6)$$

Combining (5) and (6), and using the fact that $L_{\mathcal{D}_n} < c'$ we get

$$L_K < (\tilde{c}C)^4$$

and since K is arbitrary, the isotropic constant of an arbitrary convex body K in $\mathbb{R}^n$ is universally bounded. $\qquad \square$

Remark. The proof of Proposition 1.1 uses the close relation between an M-ellipsoid and the isotropy ellipsoid of the body whose isotropic constant is as large as possible. As follows from Proposition 1.4, if we could deduce such a relation between an M-ellipsoid and the isotropy ellipsoid of an arbitrary convex body $K \subset \mathbb{R}^n$, then a universal bound for the isotropic constant will follow.

6 Appendix: Concave Functions

This section proves Lemma 2.3 in a way similar to the proofs presented in [Bal], [F]. The following lemma reflects the fact that among all concave functions on the line, the linear function is extremal.

Lemma 6.1. *Let* $f : [0, \infty) \to [0, \infty)$ *be a compactly supported function such that* $f^{1/k}$ *is concave on its support and* $a = f(0) > 0$. *Let* $n > 0$ *and choose* b *such that*

$$\int_0^\infty f(x)x^n dx = \int_0^\infty \left(a^{1/k} - bx\right)_+^k x^n dx$$

where $x_+ = \max\{x, 0\}$. *Then for any* $p > 1$

$$\int_0^\infty f(x)^p x^n dx \geq \int_0^\infty \left(a^{1/k} - bx\right)_+^{pk} x^n dx. \tag{7}$$

Proof. Since f has a compact support, $\int_0^\infty f(x)x^n dx < \infty$, so $b > 0$. Denote $h(x) = a^{1/k} - f(x)^{1/k}$. Then h is a convex function and $h(0) = 0$. Therefore $\tilde{h}(x) = \frac{h(x)}{x}$ is increasing. Since

$$\int_0^\infty \left(a^{1/k} - x\tilde{h}\right)_+^k x^n dx = \int_0^\infty \left(a^{1/k} - bx\right)_+^k x^n dx$$

it is impossible that $\tilde{h}$ is always smaller or always larger than b. The function $\tilde{h}$ is increasing, so there exists $x_0 \in [0, \infty)$ such that $\tilde{h} \leq b$ on $[0, x_0]$ and $\tilde{h} \geq b$ on $[x_0, \infty)$. Denote $g(x) = \left(a^{1/k} - bx\right)_+^k$. In order to obtain (7) we need to prove that

$$p \int_0^\infty \int_0^{f(x)} y^{p-1} dy x^n dx \geq p \int_0^\infty \int_0^{g(x)} y^{p-1} dy x^n dx.$$

Since $(g(x) - f(x))(x - x_0) \geq 0$, and g^{p-1} is a decreasing function,

$$\int_0^{x_0} \int_{g(x)}^{f(x)} y^{p-1} dy x^n dx \geq \int_0^{x_0} \int_{g(x)}^{f(x)} g(x_0)^{p-1} dy x^n dx, \tag{8}$$

$$\int_{x_0}^\infty \int_{f(x)}^{g(x)} y^{p-1} dy x^n dx \leq \int_{x_0}^\infty \int_{f(x)}^{g(x)} g(x_0)^{p-1} dy x^n dx. \tag{9}$$

Subtracting (9) from (8), we obtain

$$\int_0^\infty \int_{g(x)}^{f(x)} y^{p-1} dy x^n dx \geq g(x_0)^{p-1} \int_0^\infty \left(f(x) - g(x)\right) x^n dx = 0$$

and the lemma is proven. □

Proof of Lemma 2.3. The inequality on the right has nothing to do with log-concavity: Since $\int_{\mathbb{R}^n} f = 1$,

$$\int_{\mathbb{R}^n} f^{1+\frac{2}{k}} = \int_{\mathbb{R}^n} f \cdot f^{\frac{2}{k}} \le \int_{\mathbb{R}^n} f \cdot M^{\frac{2}{k}} = M^{\frac{2}{k}}.$$

Let us prove the left-most inequality. By translating f if necessary, we may assume that $f(0) = M$. We shall begin by integrating in polar coordinates:

$$\int_{\mathbb{R}^n} f(x)^{1+\frac{2}{k}} dx = \int_{S^{n-1}} \int_0^\infty f(r\theta)^{1+\frac{2}{k}} r^{n-1} dr d\theta.$$

Fix $\theta \in S^{n-1}$, and denote $g(r) = f(r\theta)$. Then $g^{1/k}$ is concave, as a restriction of a concave function to a straight line. Also, f must have a compact support, as it has a finite mass and $f^{1/k}$ is concave. Therefore g is compactly supported and by Lemma 6.1 for $p = 1 + \frac{2}{k}$,

$$\int_0^\infty g(x)^{1+\frac{2}{k}} x^{n-1} dx \ge \int_0^\infty \left(a^{1/k} - bx\right)_+^{k+2} x^{n-1} dx \tag{10}$$

where $a = g(0)$ and b is chosen as in Lemma 6.1, i.e. $\int_0^\infty g(x) x^{n-1} dx = \int_0^\infty \left(a^{1/k} - bx\right)_+^k x^{n-1} dx$. An elementary calculation yields that

$$\frac{\int_0^\infty \left(a^{1/k} - bx\right)_+^{k+2} x^{n-1} dx}{\int_0^\infty \left(a^{1/k} - bx\right)_+^k x^{n-1} dx} = a^{2/k} \frac{(k+1)(k+2)}{(n+k+1)(n+k+2)} \tag{11}$$

where we used the fact that $\int_0^1 x^a (1-x)^b dx = \frac{a!b!}{(a+b+1)!}$. Denote $c_{n,k} = \frac{(k+1)(k+2)}{(n+k+1)(n+k+2)}$. Combining (10) and (11) we obtain

$$\int_0^\infty g(x)^{1+\frac{2}{k}} x^{n-1} dx \ge a^{2/k} c_{n,k} \int_0^\infty \left(a^{1/k} - bx\right)_+^k x^{n-1} dx$$

$$= c_{n,k} g(0)^{2/k} \int_0^\infty g(x) x^{n-1} dx$$

or in other words, for every $\theta \in S^{n-1}$,

$$\int_0^\infty f(r\theta)^{1+\frac{2}{k}} r^{n-1} dr \ge c_{n,k} f(0)^{2/k} \int_0^\infty f(r\theta) r^{n-1} dr.$$

By integrating this inequality over the sphere S^{n-1},

$$\int_{\mathbb{R}^n} f(x)^{1+\frac{2}{k}} dx \ge c_{n,k} f(0)^{2/k} \int_{\mathbb{R}^n} f(x) dx = c_{n,k} f(0)^{2/k}. \qquad \square$$

References

[Ba1]   Ball, K.M.: Logarithmically concave functions and sections of convex sets in $\mathbb{R}^n$. Studia Math., **88**, 69–84 (1988)

[Ba2]   Ball, K.M.: Normed spaces with a weak-Gordon-Lewis property. Proc. of Funct. Anal., University of Texas and Austin (1987–1989), Lecture Notes in Math., **1470**, Springer, 36–47 (1991)

[BF]   Bonnesen, T., Fenchel, W.: Theory of Convex Bodies. Translated from German and edited by L. Boron, C. Christenson and B. Smith. BCS Associates, Moscow, Idaho, USA (1987)

[Bou]   Bourgain, J.: On the distribution of polynomials on high dimensional convex sets. Geometric Aspects of Functional Analysis (1989–90), Lecture Notes in Math., **1469**, Springer, Berlin, 127–137 (1991)

[BKM]   Bourgain, J., Klartag, B., Milman,V.D.: A reduction of the slicing problem to finite volume ratio bodies, C.R. Acad. Sci. Paris, Ser. I, **336**, 331–334 (2003)

[D]   Dar, S.: Remarks on Bourgain's problem on slicing of convex bodies. Geometric Aspects of Functional Analysis, Operator Theory: Advances and Applications, **77**, 61–66 (1995)

[F]   Fradelizi, M.: Sections of convex bodies through their centroid. Arch. Math **69**, 515–522 (1997)

[J]   Junge, M.: Hyperplane conjecture for quotient spaces of L_p. Forum Math. **6**, no. 5, 617–635 (1994)

[M1]   Milman, V.D.: Geometrical inequalities and mixed volumes in the local theory of Banach spaces. Colloquium in honor of Laurent Schwartz, vol. 1 (Palaiseau, 1983). Astérisque, no. 131, 373–400 (1985)

[M2]   Milman, V.D.: Inégalité de Brunn–Minkowski inverse et applications à le théorie locale des espaces normés. C.R. Acad. Sci. Paris, Ser. I, **302**, 25–28 (1986)

[MP]   Milman, V.D., Pajor, A.: Isotropic position and inertia ellipsoids and zonoids of the unit ball of a normed n-dimensional space. Geometric Aspects of Functional Analysis (1987–88), Lecture Notes in Math., **1376**, Springer, Berlin, 64–104 (1989)

[Pa]   Paouris, G.: On the isotropic constant of non-symmetric convex bodies. Geometric Aspects of Functional Analysis, Lecture Notes in Math., **1745**, Springer, Berlin, 239–243 (2000)

[P]   Pisier, G.: The Volume of Convex Bodies and Banach Space Geometry. Cambridge Tracts in Mathematics, Cambridge Univ. Press, **94** (1997)

On the Multivariable Version of Ball's Slicing Cube Theorem

E. Gluskin

School of Mathematical Sciences, Tel Aviv University, Tel Aviv 69978, Israel
gluskin@post.tau.ac.il

In [B] K. Ball proved that for $k \leq n/2$ the maximal volume of a k-codimensional section of the n-dimensional unit cube is equal to $(\sqrt{2})^k$. He presented also some simpler arguments based on the Brascamp–Lieb [B-L] inequality, which gave a slightly worse bound $(\sqrt{e})^k$. Really, for this application of the Brascamp–Lieb inequality, Ball proved that the constant in this inequality is equal to one provided the parameters of the inequality satisfy some special condition (which appeared very often in convex geometry). In this investigation of the Brascamp–Lieb inequality, Barthe [Ba] particularly formulated and proved the multidimensional version of Ball's case of the Brascamp–Lieb inequality (see Lemma below). In this note we combine Barthe's result with Ball's initial arguments to obtain the upper bound for the Lebesgue measure of the sections of the product bodies.

We use the following (mostly standard) notation.

For a Euclidean space H, mes_H or du for $u \in H$ is the Lebesgue measure on H; $\|u\|$ or $\|u\|_H$, $u \in H$, is the Euclidean norm. For a function $f : \mathbb{R}^n \to \mathbb{R}$ and $1 \leq p < \infty$ the standard L_p-norm of f is denoted:

$$\|f\|_p = \left(\int_{\mathbb{R}^n} |f(u)|^p du \right)^{1/p}.$$

For $V \subset \mathbb{R}^n$ we denote by χ_V the indicator function of the set V:

$$\chi_V(u) = \begin{cases} 1, & \text{for } u \in V \\ 0, & \text{otherwise}. \end{cases}$$

Let H_1, H_2 be two Euclidean spaces and $B : H_1 \to H_2$ be a linear operator. As usual the adjoint operator of B is denoted by B^* while $\|B\|$ stands for the operator norm of B.

The identity operator of the space H we denote by I_H ($I_H u = u$, for any $u \in H$).

For given positive integers n, m and k and for a given real $k \times m$ matrix $C = (c_{i,j})\,_{i=1}^{k},\ _{j=1}^{m}$, we denote by H_C or just by H the following linear subspace of the space $\underbrace{\mathbb{R}^{nm} = \mathbb{R}^n \oplus \mathbb{R}^n \oplus \cdots \oplus \mathbb{R}^n}_{m-\text{times}}$

$$H = H_C = \left\{ (x_1, \ldots, x_m) \in \mathbb{R}^{nm} : \sum_{j=1}^{m} c_{ij} x_j = 0 \text{ for } i = 1, \ldots, k \right\}.$$

(Certainly, H being a subspace of a Euclidean space is a Euclidean space itself.) Under this notation the following result holds.

Proposition. *For any positive integer m, n, k, for any real $k \times m$ matrix $C = (c_{ij})\,_{i=1}^{k},\ _{j=1}^{m}$ and for any compact subsets $V_1, \ldots, V_m \in \mathbb{R}^n$ of Lebesgue measure one, the following inequality holds:*

$$\text{mes}_H \{ H_C \cap (V_1 \times V_2 \times \cdots \times V_m) \} \le e^{kn/2}.$$

As was mentioned in the Introduction, the proof followed Ball's arguments using the following result of Barthe.

Lemma [Ba]. *Let a family of linear operators $B_i : H \to \mathbb{R}^{n_i}$, $i = 1, \ldots, m$, satisfy the following properties:*

a) *For any i, $i = 1, \ldots, m$, the adjoint operator $B_i^* : (\mathbb{R}^{n_i})^* \to H^*$ is an isometric embedding*

$$\|B_i^* u\| = \|u\|, \quad \forall u \in (\mathbb{R}^{n_i})^*.$$

b) *For some positive $p_1, \ldots, p_k$, one has*

$$\sum_{i=1}^{k} \frac{1}{p_i} B_i^* B_i = I_H.$$

Then for any positive measurable functions $f_i : \mathbb{R}^{n_i} \to \mathbb{R}_+$ the following inequality holds:

$$\int_H f_1(B_1 u) f_2(B_2 u) \cdots f_k(B_k u) du \le \|f_1\|_{p_1} \cdots \|f_k\|_{p_k}.$$

Proof of the Proposition. Let us denote by J the natural embedding of H_C into $\mathbb{R}^{nm}$ and let $T_j : \mathbb{R}^{nm} \to \mathbb{R}^n$ be the linear operator, which assigns to any vector $(x_1, \ldots, x_m) \in \mathbb{R}^{nm}$ its j-th coordinate $x_j \in \mathbb{R}^n$. Notice that for any one-dimensional subspace $E \subset \mathbb{R}^n$, the restriction of the operator $T_j J$ on the subspace

$$L = \left\{ (x_1, \ldots, x_m) \in H : x_i \in E, \ i = 1, 2, \ldots, m \right\}$$

is an operator of rank one or zero and its norm does not depend on the special choice of the subspace E. It follows that $(T_j J)^*$ is a multiple of an isometrical embedding. Let us denote it by $\alpha_j = \|T_j J\|$. For such j that $\alpha_j \neq 0$ the operator $\frac{1}{\alpha_j}(T_j J)^*$ is an isometric embedding from $\mathbb{R}^n$ into H_C. Next, we have

$$\sum_{j=1}^{m}(T_j J)^* T_j J = J^*\left(\sum_{j=1}^{m} T_j^* T_j\right) J = J^* J .$$

Since $J^* J = I_H$, one sees that the family of operators

$$B_j = \frac{1}{\alpha_j} T_j J , \quad j = 1,\ldots,m, \quad \alpha_j \neq 0$$

and the numbers $p_j = 1/\alpha_j^2$, satisfy all the conditions of the lemma. Following [B], we now apply the lemma to the following family of functions

$$f_j(x) = \chi_{V_j}\left(\frac{1}{\sqrt{p_j}}x\right) = \chi_{\sqrt{p_j}V_j}(x) .$$

We get

$$\int_{H_C} \prod_{j:\alpha_j \neq 0} f_j(B_i u)\,du \leq \prod_{j:\alpha_j \neq 0} \|f_j\|_{p_j} = \prod_{j:\alpha_j \neq 0} \left(\mathrm{mes}_n\left(\sqrt{p_j}V_j\right)\right)^{1/p_j}$$

$$= \exp\left(\frac{n}{2}\sum_{j:\alpha_j \neq 0}\frac{1}{p_j}\ln p_j\right) .$$

By the definition of B_j and f_j, the function $f_j(B_j z)$ is a characteristic function of the set

$$H_C \cap (\mathbb{R}^n \times \mathbb{R}^n \times \cdots \times V_j \times \cdots \times \mathbb{R}^n) .$$

Then for any index j, such that $\alpha_j = 0$ one has $x_j = 0$ for any $(x_1,\ldots,x_m) \in H_C$. Thus the product $\prod_{j:\alpha_j \neq 0} f_j(B_j u)$ is the characteristic function of the set $H \cap (V_1 \times V_2 \times \cdots \times V_m)$ and the last inequality gives

$$\mathrm{mes}_H\left(H \cap V_1 \times \cdots \times V_m\right) \leq \exp\left(\frac{n}{2}\sum_{j:\alpha_j \neq 0}\frac{1}{p_j}\ln p_j\right) . \tag{1}$$

By taking the trace of both parts of the equality

$$\sum_{j:\alpha_j \neq 0}\frac{1}{p_j}B_j^* B_j = I_H ,$$

we get

$$\sum_{j:\alpha_j \neq 0}\frac{1}{p_j}\mathrm{Tr}(B_j^* B_j) = \dim H_C \geq n(m - k) .$$

Since $B_j^* : \mathbb{R}^n \to H$ is an isometry for $\alpha_j \neq 0$, we have $B_j B_j^* = I_{\mathbb{R}^n}$ and

$$\text{Tr} B_j^* B_j = \text{Tr} B_j B_j^* = n$$

for such j, that $\alpha_j \neq 0$.

Finally we get

$$m \geq \sum_{j:\alpha_j \neq 0} \frac{1}{p_j} \geq m - k .$$

Denote $\sum_{j:\alpha_j \neq 0} \frac{1}{p_j} = \lambda$. In this notation, the last inequality stands for $0 \leq m - \lambda \leq k$. It is well known (see e.g. [Ya]) that for any positive λ and integer $\ell \leq m$,

$$\max \left\{ \sum_{i=1}^{\ell} \frac{1}{p_i} \ln p_i ; \sum_{i=1}^{\ell} \frac{1}{p_i} = \lambda, \ p_i > 0, \ i = 1, \dots, \ell \right\} = \lambda \ln \frac{\ell}{\lambda} \leq \lambda \ln \frac{m}{\lambda} .$$

Furthermore, $\lambda \ln \frac{m}{\lambda} = \lambda \ln \left(1 + \frac{m-\lambda}{\lambda}\right) \leq m - \lambda$. Applying the last inequality to (1), completes the proof. $\qquad \square$

Corollary 1. *Under the notation of the proposition for any $z = (x_1, x_2, \dots, x_m) \in \mathbb{R}^{nm}$ the following inequality holds:*

$$\text{mes}_{n(m-k)} \left(V_1 \times V_2 \times \cdots \times V_m \cap (H_C + z) \right) \leq e^{kn/2} .$$

Proof. Apply the proposition for $V_i - x_i$ instead of V_i. $\qquad \square$

The following result from [G-M1] is an immediate corollary of the Proposition (we refer the reader also to [G-M2] for an alternative proof, which gives a better estimate).

Corollary 2. *Let $V_1, \dots, V_m$, and K be some measurable subsets of $\mathbb{R}^n$ all of them of measure one. Then for any positive t and reals λ_i, $i = 1, \dots, m$, such that $\sum_{i=1}^m \lambda_i^2 = 1$, the following inequality holds:*

$$\text{mes}_{\mathbb{R}^{nm}} \left\{ (x_1, \dots, x_m) \in \prod_{i=1}^m V_i : \sum_{i=1}^m \lambda_i x_i \in tK \right\} \leq (t\sqrt{e})^n .$$

Proof. First we will rewrite the left-hand-term of the inequality as an integral. Let S be a linear map from $\mathbb{R}^d$ to $\mathbb{R}^n$, $d \geq n$ and $B \subset \mathbb{R}^n$, $W \subset \mathbb{R}^d$ be some measurable sets then[1]

$$\text{mes}_{\mathbb{R}^d} \{z \in W : Sz \in B\}$$
$$= \frac{1}{\sqrt{\det(SS^*)}} \int_B \text{mes}_{S^{-1}(\{y\})} \left\{ W \cap S^{-1}(\{y\}) \right\} dy . \quad (2)$$

---

[1] An analogous formula for nonlinear S is known as the Kronrod-Federer coarea formula. In our situation it is quite an elementary fact, which can be obtained by decomposition of $\mathbb{R}^d = \ker S \oplus (\ker S)^{\perp}$, Fubini's theorem and the change of variables formula, applied to the restriction of S on $(\ker S)^{\perp}$.

Now, let $S = S_\Lambda$ be a linear map from $\mathbb{R}^{nm} = \mathbb{R}^n \times \cdots \times \mathbb{R}^n$ to $\mathbb{R}^n$, which assigns to any $(x_1, \ldots, x_m) \in \mathbb{R}^{nm}$ the vector $\sum_{i=1}^m \lambda_i x_i \in \mathbb{R}^n$. It is clear that

$$\det SS^* = \left(\sum_{i=1}^m |\lambda_i|^2 \right)^n = 1 \ .$$

Let us apply (2) for $S = S_\Lambda$, $W = V_1 \times V_2 \times \cdots \times V_m$ and $B = tK$. We get

$$\text{mes}\left\{ (x_1, \ldots, x_m) \in V_1 \times \cdots \times V_m : \sum_{i=1}^m \lambda_i x_i \in tK \right\}$$

$$= \int_{tK} \text{mes}\{ W \cap S_\Lambda^{-1} y \} dy \leq t^n \text{vol} K \max_{y \in tK} \text{mes}\{ W \cap S_\Lambda^{-1} y \} \ .$$

To conclude we use Corollary 1 with $k = 1$ and $C = (\lambda_1, \ldots, \lambda_m)$ by taking into account that $S_\Lambda^{-1}\{y\}$ is a shift of H_C for any $y \in \mathbb{R}^n$. $\square$

Acknowledgement. I would like to express my gratitude to G. Schechtman for his valuable remarks.

References

[B]   Ball, K.: Volume of sections of cubes and related problems. Geometric Aspects of Functional Analysis, Israel Seminar 1987–1988, Springer-Verlag Lecture Notes in Math., **1376**, 251–260 (1989)

[Ba]   Barthe, F.: On a reverse form of the Brascamp–Lieb inequality. Invent. Math., **134** (2), 335–361 (1998)

[B-L]   Brascamp, H.J., Lieb, E.H.: Best constant in Young's inequality, its converse and its generalization to more than three functions. Adv. Math., **20**, 151–173 (1976)

[G-M1] Gluskin, E., Milman V.: Randomizing properties of convex high-dimensional bodies and some geometric inequalities. C.R. Acad. Sci, Paris, Ser. I, **334**, 875–879 (2002)

[G-M2] Gluskin E., Milman V.: Geometric probability and random cotype 2. This volume

[Ya]   Yaglom A.M., Yaglom I.M.: Probability and Information, 1983, 421p

Geometric Probability and Random Cotype 2

E. Gluskin and V. Milman[*]

School of Mathematical Sciences, Tel Aviv University, Tel Aviv 69978, Israel
gluskin@post.tau.ac.il, milman@post.tau.ac.il

1 Introduction

It is a non-trivial task to understand the random properties of a uniform distribution over a convex body, say K, in $\mathbb{R}^n$. To fix terminology, we refer to it as the "geometric probability" (over K), i.e. for any (measurable) set $A \subset \mathbb{R}^n$, the geometric probability of A is $|A \cap K|/|K|$, where $|B|$ is the volume of B. We usually omit the reference to dimension and also write just Prob A if the reference body K is clear.

A very interesting result [ABV], which we cite in the text below, demonstrates what unusual properties of "geometric expectations" we may expect. Some new results in this direction are presented in this paper. In some instances, we compare these results with probabilities coming from some fixed Euclidean structures, e.g. the rotation invariant probability measure $\sigma(x)$ on the unit sphere S^{n-1} of the selected Euclidean norm. Actually, we understand probabilities induced by Euclidean structures much better than geometric probabilities.

Let K be a closed star body in $\mathbb{R}^n$, i.e. for any $0 < t < 1$, $tK \subset K$ and let $0 \in \overset{\circ}{K}$. By analogy with the case of convex centrally symmetric K, we denote by $\| \cdot \|_K$ the gauge function of K, i.e. $\|x\|_K = \inf\{t : \frac{1}{t}x \in K\}$ for $x \in \mathbb{R}^n$. The Lebesgue measure on Euclidean space is denoted by *meas* or by $| \cdot |$. As usual we do not mention the dimension of the space but use the same symbol *meas* to denote the measures of the different Euclidean spaces. We denote by $\mathcal{D}$ the closed unit ball of the standard Euclidean norm on $\mathbb{R}^n$.

The results of this paper were announced in CRAS [GlM].

2 On a Geometric Inequality and the Extremal Properties of Euclidean Balls

The following multi-integral expression was studied in [BoMMP]. Let V_i, $i = 1, \ldots, n$, and K be centrally symmetric convex bodies of volume 1 in $\mathbb{R}^n$, i.e.

---

[*] Supported by the Israel Science Foundation.

$|V_i| = |K| = 1$. Fix $\bar{\lambda} = (\lambda_i \in \mathbb{R})_{i=1}^n$. Consider the norm

$$|||\bar{\lambda}||| = \int_{V_1} \cdots \int_{V_n} \left\| \sum_{i=1}^n \lambda_i x_i \right\|_K dx_1 \cdots dx_n .$$

It was proved ([BoMMP]) that for some universal c,

$$|||\bar{\lambda}||| \geq c\sqrt{n} \left(\prod_{i=1}^n |\lambda_i| \right)^{1/n} .$$

At the same time, it was conjectured that under the cotype condition on the norm $\| \cdot \|_K$ and if all V_i coincide with K, the norm $|||\bar{\lambda}|||$ is equivalent to the ℓ_2-norm (see [BoMMP] for a discussion). In this section we will prove the low ℓ_2-bound.

Theorem 1. *Let V_i, $1 \leq i \leq m$, be measurable sets in $\mathbb{R}^n$ and K be a star body with 0 in its interior. Let $\bar{\lambda} = (\lambda_i)_1^m \in \mathbb{R}^m$. Consider the positively homogeneous function*

$$|||\bar{\lambda}||| := |||\bar{\lambda}|||_{K;\{V_i\}} = \int_{V_1} \cdots \int_{V_m} \left\| \sum_1^m \lambda_i x_i \right\|_K \frac{\prod dx_i}{\prod_1^m |V_i|} . \qquad (1)$$

Then, for a universal constant $c > 0$,

$$|||\bar{\lambda}||| \geq c\sqrt{\Sigma \lambda_i^2 r_i^2} ,$$

where the r_i's denote $r_i = (|V_i|/|K|)^{1/n}$. In particular, if $|V_i| = |K|$ for all $i = 1, \ldots, m$ then

$$|||\bar{\lambda}||| \geq c\sqrt{\Sigma \lambda_i^2} .$$

Here, the universal constant c satisfies $c \geq c(n)\frac{1}{\sqrt{2}}$, for $c(n) \to 1$ when $n \to \infty$.

Proof. We start by using the symmetrization result of Brascamp–Lieb–Luttinger [BLL] to show the following proposition.

Proposition 2. *Let $V_i \subset \mathbb{R}^n$, $i = 1, \ldots, m$, be some measurable sets and $|V_i| = |\mathcal{D}|$ for every i. Let K be a star body with $0 \in \overset{\circ}{K}$ and the same volume condition $|K| = |\mathcal{D}|$. Then for any scalars λ_i, $i = 1, \ldots, m$, and for any positive t the following inequality holds:*

$$meas\left\{ (x_i)_{i=1}^m : x_i \in V_i \text{ for all } i \text{ and } \left\| \sum_1^m \lambda_i x_i \right\|_K < t \right\}$$

$$\leq meas\left\{ (x_i)_{i=1}^m : x_i \in \mathcal{D} \text{ for all } i \text{ and } \left\| \sum_1^m \lambda_i x_i \right\|_\mathcal{D} < t \right\} .$$

Proof. We recall the Brascamp–Lieb–Luttinger inequality ([BLL]) which states that for any measurable functions $f_0, f_1, \ldots, f_m : \mathbb{R}^n \to \mathbb{R}_+$ and for any $(m+1) \times k$ matrix $A = (a_{ji})$ the following inequality holds:

$$\underbrace{\int\limits_{\mathbb{R}^n} \cdots \int\limits_{\mathbb{R}^n}}_{k} \prod_{j=0}^{m} f_j \left(\sum_{i=1}^{k} a_{ji} x_i \right) dx_1 \ldots dx_k$$

$$\leq \int\limits_{\mathbb{R}^n} \cdots \int\limits_{\mathbb{R}^n} \prod_{j=0}^{m} f_j^* \left(\sum_{i=1}^{k} a_{ji} x_i \right) dx_1 \ldots dx_k \; ,$$

where f_j^* is the symmetric decreasing rearrangement of f_j.

For any set $V \subset \mathbb{R}^n$ we write χ_V for its characteristic function. Then for measurable V the function χ_V^* is the characteristic function of the Euclidean ball $r\mathcal{D}$ of such radius r that $|V| = |r\mathcal{D}|$ (just by the definition of the rearrangement). The desired result is obtained now by a direct application of the inequality to the following functions:

$$f_j = \chi_{V_j} \text{ for } j = 1, \ldots, m; \quad f_0 = \chi_{tK}$$

with $k = m$ and

$$a_{ji} = \begin{cases} 1 & \text{if } 1 \leq j = i \leq m, \\ \lambda_i & \text{if } j = 0, \\ 0 & \text{elsewhere.} \end{cases} \qquad \square$$

Returning to the proof of Theorem 1, but still under conditions $|V_i| = |K| = |\mathcal{D}|$, we just restate the previous Proposition 2 for $t = 1$ as an inequality

$$\int\limits_{V_1} \cdots \int\limits_{V_m} \chi_K(\lambda_1 x_1 + \cdots + \lambda_m x_m) dx_1 \cdots dx_m$$

$$\leq \int\limits_{\mathcal{D}} \cdots \int\limits_{\mathcal{D}} \chi_{\mathcal{D}}(\lambda_1 x_1 + \cdots + \lambda_m x_m) dx_1 \cdots dx_m \; . \quad (2)$$

Observe that for every $y \in \mathbb{R}^n$ one has

$$\|y\|_K = \int\limits_0^\infty \left(1 - \chi_K \left(\frac{y}{\lambda} \right) \right) d\lambda \; .$$

Then, the consequence of (2) is

$$|||\bar{\lambda}||| = \int\limits_{V_1} \cdots \int\limits_{V_m} \left\| \sum_1^m \lambda_i x_i \right\|_K \frac{\prod_1^m dx_i}{\prod_1^m |V_i|} \geq \int\limits_{\mathcal{D}} \cdots \int\limits_{\mathcal{D}} \left\| \sum_1^m \lambda_i x_i \right\|_{\mathcal{D}} \frac{\prod_1^m dx_i}{|\mathcal{D}|^m} \; . \quad (3)$$

The last expression can be estimated through the Khinchine inequality

$$
\int_{\mathcal{D}^m} \left\| \sum_{i=1}^m \lambda_i x_i \right\|_D \prod_{i=1}^m \frac{dx_i}{|\mathcal{D}|} = \int_{\mathcal{D}^m} \operatorname*{Ave}_{\varepsilon_i = \pm 1} \left\| \sum_{i=1}^m \varepsilon_i \lambda_i x_i \right\|_D \prod_{i=1}^m \frac{dx_i}{|\mathcal{D}|}
$$

$$
\geq \frac{1}{\sqrt{2}} \int_{\mathcal{D}^m} \sqrt{\sum_{i=1}^m |\lambda_i|^2 \|x_i\|_D^2} \prod_{i=1}^m \frac{dx_i}{|\mathcal{D}|} .
$$

(In the last step we used the Khinchine inequality with a sharp constant, which was found by S. Szarek [S].)

One completes the proof for the case when the volumes of all involved sets are equal to the volume of $\mathcal{D}$ by estimating, in a standard way, the L_1-norm through the L_2-norm (for example as $\|f\|_{L_1} \geq \|f\|_{L_2}(\|f\|_{L_2}/\|f\|_{L_4})^2$). In such a way we get the constant

$$
c = \frac{1}{\sqrt{2}} \left(\frac{n}{n+2} \right)^{3/2} \sqrt{\frac{n+4}{n}} \longrightarrow \frac{1}{\sqrt{2}} \quad \text{(when } n \to \infty) .
$$

The general case follows by trivial homogeneous substitutions, $x_i' = x_i/r_i$, and changing K to $K' = r \cdot K$ such that $|K'| = |\mathcal{D}|$.   □

A Few Remarks Related to Theorem 1

The special case of Theorem 1 with all sets V_i and K being the same centrally symmetric convex body, say K, is the most interesting. Immediately after the problem of computing the symmetric norm (1) in this case was raised, G. Schechtman (unpublished) showed that for K being the unit ℓ_p ball ($1 \leq p < \infty$), $|||\bar{\lambda}|||$ is equivalent to the ℓ_2-norm. Later G. Pisier (again unpublished) proved that, indeed, under cotype condition, this norm is equivalent to the ℓ_2-norm if K has an unconditional basis (which is strange as the problem does not "see" this basis). Later Zinn (again unpublished) showed that it is enough for K to have the GL-property. M. Junge published the proof of this fact in his Ph.D. thesis. [BoMMP] contains the only general results on this problem and besides the lower bound we mentioned before, another lower bound was shown there connecting it with the isotropic constant L_K of the body K. This means, in particular, that any non-trivial upper bound on the norm (1) will also provide a bound on L_K.

We would now like to compute this norm for the case of the unit ball of ℓ_∞^n, i.e. for $K = [-1,1]^n$. The resulting norm $|||\bar{\lambda}|||$ is very close to the ℓ_2-norm, actually $\sqrt{\log n}$-close.

Example. Let $X_K = \ell_\infty^n$, i.e. $K = [-1,1]^n$. Let all $V_i = K$. Then the norm $|||\bar{\lambda}|||_K$ in (1) is uniformly equivalent to $q_n(\bar{\lambda}) = \sum_{1 \leq i < \log n} \lambda_i^* + \sqrt{\log n}(\sum_{i \geq \log n} (\lambda_i^*)^2)^{1/2}$ (where $(\lambda_i^*)_{i \geq 1}$ is the non-increasing rearrangement of the set $\{|\lambda_i|\}_{i\geq 1}$). This means that there are universal constants c_1 and $c_2 > 0$, such that for all integers n

$$c_2 q_n(\bar{\lambda}) \leq |||\bar{\lambda}|||_K \leq c_1 q_n(\bar{\lambda}) . \tag{4}$$

Proof. To estimate $|||\bar{\lambda}|||_K$ from below we denote

$$\xi_j = \sum_{i\geq 1} \lambda_i x_i(j) .$$

Then $|||\bar{\lambda}|||_K = \max_{1\leq j\leq n} |\xi_j|$. Note that coordinates $\{x_i(j)\}_{i\geq 1, j\geq 1}$ are independent, identically and uniformly distributed in $[-1, 1]$ random variables. Then ξ_j are i.i.d. random variables. Let ξ be a random variable with the same distribution. Now we use the following deviation estimate due to Montgomery-Smith[1] [Mo].

Let ξ be as above. Then for some universal numbers a_1 and $a_2 > 0$,

$$e^{-a_1 k} \leq \mathrm{Prob}\left(|\xi| > \sum_{i=1}^{k} \lambda_i^* + \sqrt{k}\Big(\sum_{i\geq k}(\lambda_i^*)^2\Big)^{1/2}\right) \leq e^{-a_2 k} . \tag{5}$$

Denote $q(k, \bar{\lambda}) = \sum_{i=1}^{k} \lambda_i^* + \sqrt{k}(\sum_{i\geq k}(\lambda_i^*)^2)^{1/2}$. Then, for every k,

$$\mathbb{E}\|\Sigma \lambda_i x_i\|_\infty \geq q(k, \bar{\lambda}) \, \mathrm{Prob}\left\{\Big\|\sum_{i\geq k} \lambda_i x_i\Big\|_\infty \geq q(k, \bar{\lambda})\right\}$$

$$= q(k, \bar{\lambda})\left(1 - \mathrm{Prob}\left(\Big\|\sum_{i\geq k} \lambda_i x_i\Big\|_\infty < q(k, \bar{\lambda})\right)\right)$$

$$= q(k, \bar{\lambda})\left(1 - \big[\mathrm{Prob}\left(\xi < q(k, \bar{\lambda})\right)\big]^n\right) .$$

Also

$$\Big[\mathrm{Prob}\left(\xi < q(k, \bar{\lambda})\right)\Big]^n = \left(1 - \mathrm{Prob}\left(\xi \geq q(k, \bar{\lambda})\right)\right)^n$$

$$\leq e^{-n \, \mathrm{Prob}(\xi \geq q(k, \bar{\lambda}))} \leq e^{-n e^{-a_1 k}} ,$$

using (5) in the last step. It remains to choose $k = \lceil \frac{\log n}{a} \rceil$ for a suitable universal constant a to obtain

$$\mathbb{E}\Big\|\sum_{i} \lambda_i x_i\Big\|_\infty \geq \frac{1}{2}\, q\Big(\frac{\log n}{a}, \bar{\lambda}\Big)$$

which is equivalent to the lower bound in (4).

---

[1] Really the Montgomery-Smith result was formulated in the setting of the Rademacher functions, but the same proof works for uniformly distributed in $[-1, 1]$ r.v. without any changes.

To estimate $|||\bar{\lambda}|||_K$ from above, note that for every $p < \infty$

$$\mathbb{E}\|\Sigma\lambda_i x_i\|_\infty \leq \mathbb{E}\|\Sigma\lambda_i x_i\|_p \leq \left(\sum_{j=1}^n \mathbb{E}|\Sigma\lambda_i x_i(j)|^p\right)^{1/p}$$

$$= \left(\sum_{j=1}^n \mathbb{E} \operatorname*{Ave}_{\varepsilon_i=\pm 1} \left|\sum_{i\geq 1}\varepsilon_i\lambda_i x_i(j)\right|^p\right)^{1/p}$$

$$\leq C\left(\sum_{j=1}^{[p]} \lambda_j^* + \sqrt{p}\left(\sum_{j\geq[p+1]}\lambda_j^{*2}\right)^{1/2}\right)n^{1/p},$$

where C is some universal constant. The last inequality is the refinement of Khinchine's inequality (see, e.g. [Hi]). It is enough to take $p = \log n$ in this inequality.

3 Deviation from ℓ_2-Estimate

One may correctly argue that Proposition 2 in the previous section is not complete as it is natural and expected that the measure from above involving the Euclidean norm and vectors from the Euclidean ball will be estimated in terms of $\{\lambda_i\}$ and t. Actually, it is not a trivial computation and it is not easier to deal with the Euclidean ball $\mathcal{D}$ than with the general setting, as the use of the standard concentration phenomenon technique on the product of spheres $\prod S^{n-1}$ will not reveal the complete asymptotic. A better way to analyze it is to use the Brascamp–Lieb inequality as was demonstrated first by K. Ball [B] and extended by F. Barthe [Ba]. This method opens some other interesting connections and is presented in another article in this volume written by the first named author.

We will estimate it by extending an approach from Arias-de-Reyna–Ball–Villa [ABV] and using the multivariable Young inequality with sharp constant, which was found simultaneously by Beckner [Be] and Brascamp and Lieb [BL].

Proposition 3. *Let $K \subset \mathbb{R}^n$ be a star body with $0 \in \overset{\circ}{K}$, and V_i, $i = 1, \ldots, m$, be some measurable subsets of $\mathbb{R}^n$. Let $|K| = |V_i|$ (for any i). Then for any scalars $\lambda_i \in \mathbb{R}$ and any positive $t \leq 1$, the following inequality holds:*

$$\operatorname{Prob}\left\{(x_i)_{i=1}^m : x_i \in V_i \text{ for all } i \text{ and } \left\|\sum_1^m \lambda_i x_i\right\|_K < t\sqrt{\sum_1^m \lambda_i^2}\right\}$$

$$\leq t^n \exp\left[\frac{(1-t^2)}{2}n\right], \quad (6)$$

where Prob *is the geometric probability on* $\prod_{i=1}^m V_i$.

Proof. Without loss of generality suppose that $|K| = 1$. Denote the probability on the left-hand side of (6) by J and $t\sqrt{\Sigma\lambda_i^2} = \tau$. Clearly

$$J = \int_{\mathbb{R}^n} \cdots \int_{\mathbb{R}^n} dx_1 \cdots dx_m \chi_{V_1}(x_1) \cdot \ldots \cdot \chi_{V_m}(x_m) \cdot \chi_{\tau K}\left(\sum_1^m \lambda_i x_i\right) .$$

Introduce a change of the variables $y_k = \sum_{i=1}^k \lambda_i x_i$. Then

$$J = \int_{\mathbb{R}^n} \cdots \int_{\mathbb{R}^n} \chi_{\lambda_1 V_1}(y_1) \chi_{\lambda_2 V_2}(y_2 - y_1) \cdot \ldots \cdot \chi_{\lambda_m V_m}(y_m - y_{m-1}) \cdot$$

$$\cdot \chi_{\tau K}(y_m) \frac{dy_1 \cdot \ldots \cdot dy_m}{\left(\prod_1^m \lambda_i\right)^n} .$$

By Young's inequality, for any numbers p_i, $i = 0, 1, \ldots, m$, satisfying $1 \le p_i \le \infty$ and $\sum_{i=0}^m \frac{1}{p_i} = m$, one has

$$J \le \left(\prod_{i=1}^m \lambda_i\right)^{-n} \left(\prod_{i=0}^m C_i\right)^n \|\chi_{\tau K}\|_{p_0} \cdot \prod_{i=1}^m \|\chi_{\lambda_i V_i}\|_{p_i} , \tag{7}$$

and the (exact) constants are (see [Be] , [BL])

$$C_i = \left[p_i^{1/p_i}\left(1 - \frac{1}{p_i}\right)^{1-1/p_i}\right]^{1/2} .$$

The expression on the right side of (7) may be rewritten as

$$\left(t\sqrt{\sum_1^m \lambda_i^2}\right)^{\frac{n}{p_0}} \prod_{i=1}^m \lambda_i^{n/p_i - n} \prod_0^m p_i^{\frac{n}{2p_i}} \left(1 - \frac{1}{p_i}\right)^{\frac{n}{2}\left(1 - \frac{1}{p_i}\right)} \tag{8}$$

(we put back $\tau = t\sqrt{\Sigma\lambda_i^2}$ and used $|K| = |V_i| = 1$).

To analyze (8) it is convenient to introduce $\varepsilon_i = 1 - \frac{1}{p_i}$, $i = 0, \ldots, m$. Then $\sum_0^n \varepsilon_i = 1$, $0 \le \varepsilon_i \le 1$, and our goal is to select $\{\varepsilon_i\}_{i=0}^m$ satisfying those conditions which will minimize (8). So, putting ε_i instead of p_i, we transform (8) into

$$\left(\sum_1^m \lambda_i^2\right)^{(1-\varepsilon_0)\frac{n}{2}} t^n \left[\frac{1}{t^{2\varepsilon_0}} \frac{\varepsilon_0(1-\varepsilon_0)^{\varepsilon_0}}{1-\varepsilon_0} \prod_{i=1}^m \left(\frac{\varepsilon_i(1-\varepsilon_i)}{\lambda_i^2}\right)^{\varepsilon_i} \frac{1}{1-\varepsilon_i}\right]^{n/2} . \tag{9}$$

Now choose $\varepsilon_0 = t^2$ and $\varepsilon_i = \frac{(1-t^2)\lambda_i^2}{\sum_1^m \lambda_j^2}$, $i = 1, \ldots, m$. (This is probably not the minimizer, but enough for our goal.) With such $\{\varepsilon_i\}_0^m$, the expression (9) is equal to

$$t^n \left[\prod_1^m \frac{1}{(1-\varepsilon_i)^{1-\varepsilon_i}} \right]^{n/2} = t^n e^{\frac{n}{2} \sum_1^m (1-\varepsilon_i) \log \frac{1}{1-\varepsilon_i}} .$$

Because $(1-\varepsilon_i) \log \frac{1}{1-\varepsilon_i} = (1-\varepsilon_i) \log[1 + \frac{\varepsilon_i}{1-\varepsilon_i}] \le \varepsilon_i$ (for $\varepsilon_i \ge 0$) and $\sum_1^m \varepsilon_i = 1 - \varepsilon_0 = 1 - t^2$ we conclude the proof. $\square$

Remark. The right side of the inequality (6) is less than $(t\sqrt{e})^n$ and also less than $\exp(-(1-t)^2 n)$. The first bound is asymptotically sharp when $t \to 0$, while the second one is so for $t \to 1$.

One very interesting example of an application of (6) is the result by Arias-de-Reyna–Ball–Villa [ABV] obtained by selecting $V_1 = V_2 = K$ and $\lambda_1 = -\lambda_2 = 1/\sqrt{2}$.

Another immediate consequence of the Proposition 3 is the following:

Corollary 4. *For any integer m and $\bar{\lambda} = (\lambda_i) \in S^{m-1}$ the geometric probability*

$$\text{Prob}\left\{ (x_i)_{i=1}^m : x_i \in V_i \text{ for all } i \text{ and } \min_{\varepsilon_i = \pm 1} \left\| \sum_1^m \varepsilon_i \lambda_i x_i \right\|_K \ge t \right\} \ge 1 - (t\sqrt{e})^n 2^m$$

(here V_i and K have the same meaning as in Proposition 3 and t is such that $(t\sqrt{e})^n 2^m < 1$).

The following particular case of the corollary looks geometrically interesting. Let all $V_i = K$, $m \le n$ and $\lambda_i = 1/\sqrt{m}$. Then for $t < \frac{1}{2\sqrt{e}}$, with an exponentially close to 1 probability, random independently uniformly on K selected $\{x_i\}_1^m$ satisfy

$$\min_{\pm} \left\| \sum_1^m \pm x_i \right\|_K \ge t\sqrt{m} . \tag{10}$$

Using (6) for t close to 1, but now taking $m \sim c(1 - t^2)n$, we have (10) with t close to 1 which may be interpreted as an almost "orthogonal" behavior of random m vectors in (any) K.

4 Random Cotype 2 Property

Now we apply Proposition 3 to the situation when all $V_i = K$ and K is a centrally symmetric and convex compact body, i.e. $\|x\|_K$ is the usual norm in $\mathbb{R}^n$. So, we consider a normed space $X = (\mathbb{R}^n, \| \cdot \|)$ and K is its unit ball. We recall that X has cotype 2 (with the cotype 2 constant $C_2(X) \le C$) iff for any integer m, for any $\{x_i\}_{i=1}^m \subset X$

$$\text{Ave}_{\varepsilon_i = \pm 1} \left\| \sum_1^m \varepsilon_i x_i \right\| \ge \frac{1}{C} \sqrt{\sum_1^m \|x_i\|^2} .$$

It was observed in [FLM] that it is enough to consider $m \leq n^2$, and it was proved by Tomczak-Jaegermann [T] that it is enough to consider $m = n$ vectors, if we are ready to compromise on a cotype 2 constant up to a factor $\sim \sqrt{2}$.

We say that X has a *random cotype 2* with a constant at most $C > 0$ iff, with a probability above $1 - e^{-an}$ ($a > 0$ is a fixed universal number), the vectors $\{x_i\}_1^n$ iid with respect to the geometric probability on K satisfy for every $\lambda_i \in \mathbb{R}$

$$\operatorname*{Ave}_{\varepsilon_i = \pm 1} \left\| \sum_1^n \varepsilon_i \lambda_i x_i \right\| \geq \frac{1}{C} \sqrt{\sum_1^n |\lambda_i|^2} . \tag{11}$$

Note that we did not put $\|x_i\|$ inside the summation as, with probability exponentially close to 1, $\frac{1}{2} \leq \|x_i\| \leq 1$, and their norms are absorbed in C. For this reason we also added coefficients $\{\lambda_i\}$. Actually, we may be able to think of a weaker notion of a random cotype 2 space, considering (11) only for $\{\lambda_i\}$ from a large measure on the Euclidean sphere S^{n-1}, i.e. with high probability also by $\lambda = \{\lambda_i\}_1^n \in S^{n-1}$. It is immediate from Corollary 4 and Foubini's theorem that any space X has such a weaker property of cotype 2. However, also in the stronger sense it is satisfied by any normed space.

Theorem 5. *There is a universal constant $C > 0$ such that for any n, any n-dimensional space $X = (\mathbb{R}^n, \|\cdot\|)$ has a random cotype 2 with constant at most C.*

Proof. Let K be the unit ball of X. In the proof of this theorem we apply Corollary 4 with all $V_i = K$ and $m = n$. Then, obviously, for any set $\mathcal{A} \subset S^{n-1}$ with $\#\mathcal{A} \leq C_1^n$, we have (11) for iid $x_i \in K$, $i = 1, \ldots, n$, with exponentially close to 1 probability for any $\bar{\lambda} = (\lambda_i)_{i=1}^n \in \mathcal{A}$ for some C which depends on C_1 only.

Now we will choose a special set $\mathcal{A}$ such that inequality (11) for $\bar{\lambda} \in \mathcal{A}$ will imply the same inequality, perhaps with a slightly different constant C, for any $\bar{\lambda} \in \mathbb{R}^n$.

We need the notion of a 1-unconditional norm. We say that a norm $||| \cdot |||$ is 1-unconditional iff for any signs $\varepsilon_i = \pm 1$ and real λ_i

$$|||(\varepsilon_i \lambda_i)||| = |||(\lambda_i)||| .$$

We also say that $x = (x_i)_1^n \leq y = (y_i)_1^n$ iff $x_i \leq y_i$ for every $i = 1, \ldots, n$, and we denote $|x| = (|x_i|)_{i=1}^n$. Note that if $|||x|||$ is 1-unconditional then $|x| \leq |y|$ implies $|||x||| \leq |||y|||$.

Observe also that the norm

$$|||\bar{\lambda}||| \equiv \operatorname*{Ave}_{\varepsilon_i = \pm 1} \left\| \sum_1^n \varepsilon_i \lambda_i x_i \right\| ,$$

which we consider in (11), is 1-unconditional for any selection of $x_i \in K$. We define $\|\bar{\lambda}\|_2 = \sqrt{\Sigma |\lambda_i|^2}$.

Now we define a set $\mathcal{F}$ of all vectors $y = (y_i) \in \mathbb{R}^n$ which satisfy the following two conditions:

a) $\frac{1}{2} \leq \|y\|_2 \leq 1$,

b) for any $i = 1, \ldots, n$ either $y_i = 0$ or $y_i = \sqrt{2^k/4n}$ for some positive integer k.

Claim 1. For any $x \in S^{n-1}$ there is a $y \in \mathcal{F}$ such that $0 \leq y \leq |x|$, i.e. $0 \leq y_i \leq |x_i|$ for every $i = 1, \ldots, n$, where $x = (x_i)_1^n$ and $y = (y_i)_1^n$.

Proof. For every $i = 1, \ldots, n$, either $2nx_i^2 < 1$, and we define in this case $y_i = 0$, or $2nx_i^2 \geq 1$, in which case k_i is defined by the inequalities

$$2^{k_i} \leq 4nx_i^2 < 2^{k_i+1} ,$$

and $y_i = \sqrt{2^{k_i}/4n}$. By the definition, $0 \leq y \leq |x|$, and y is satisfied b). But then $\|y\|_2 \leq \|x\|_2 = 1$ and we should only show that $\|y\|_2 \geq 1/2$. Indeed,

$$\|y\|_2^2 = \Sigma y_i^2 = \frac{1}{4n} \sum_{i:y_i \neq 0} 2^{k_i} > \frac{1}{2} \frac{1}{4n} \sum_{i:y_i \neq 0} 4nx_i^2 = \frac{1}{2}\left(1 - \sum_{i:y_i=0} x_i^2\right), \quad (12)$$

and, by the definition of y_i, the condition $y_i = 0$ implies $x_i^2 < \frac{1}{2n}$, which allows us to continue (12) and to see that $\|y\|_2 \geq \frac{1}{2}$.

The obvious corollary of Claim 1 is

Claim 2. Let $|||x|||$ be a 1-unconditional norm on $\mathbb{R}^n$. Then

$$\inf_{\|x\|_2=1} |||x||| \geq \inf_{y \in \mathcal{F}} |||y||| \geq \frac{1}{2} \inf_{y \in \mathcal{F}} \frac{|||y|||}{\|y\|_2} .$$

Proof. Indeed, for every $x \in S^{n-1}$ find $y \in \mathcal{F}$ such that $0 \leq y \leq |x|$. Then $|||x||| \geq |||y|||$ by 1-unconditionality. This proves the first inequality. The second one follows immediately from property a) of the set $\mathcal{F}$.

To finish the proof of Theorem 5 we take $\mathcal{A} = \{\frac{y}{\|y\|_2}\}_{y \in \mathcal{F}}$. It remains to show that $\#\mathcal{F} \leq C_1^n$ for some universal C_1, which is an easy computation. (Actually, one may show that $C_1 \leq 5$.)   □

5 Spherical Uniform Distribution

We will consider in this section randomized properties of a normed space $X = (\mathbb{R}^n, \|\cdot\|, |\cdot|)$, equipped with a fixed Euclidean norm $|\cdot|$, but now with respect to randomness associated with our Euclidean structure. We need a condition of "non-degeneration" between two norms, the norm $\|\cdot\|$ of X and the Euclidean norm $|\cdot|$. To describe it we introduce two parameters which connect these two norms:

$$b(X) := \max\{\|x\| : |x| = 1\} ,$$

and

$$M(X) := \sqrt{\int_{S^{n-1}} \|x\|^2 d\sigma(x)} ,$$

where $\sigma(x)$ is the probability rotation invariant measure on S^{n-1}. Note that always $b(X) \le M(X)\sqrt{n}$, and for a given $\| \cdot \|$ one may find a Euclidean norm $|\cdot|$ such that, for some $c > 0$,

$$b(X) \le c\sqrt{n/\log n}M(X) , \tag{13}$$

(see, e.g. [MS1]). This is exactly the type of condition we call "non-degeneration". We often write b and M instead of $b(X)$ and $M(X)$. We denote below Π_n to be the permutation group on $\{1, \ldots, n\}$.

Theorem 6. *There is a universal constant $c_0 > 0$ such that, for any space $X = (\mathbb{R}^n, \|\cdot\|, |\cdot|)$ which satisfies the non-degeneration condition (13) with a constant c_0, the following is true:*
consider any set $\{x_i\}_1^n \subset S^{n-1}$; there is an orthogonal operator $u \in O(n)$ such that for any $\lambda_i \in \mathbb{R}$

$$\frac{1}{2}\sqrt{\sum_1^n \lambda_i^2} \cdot M(x) \le \operatorname*{Ave}_{\varepsilon_i = \pm 1; \pi \in \Pi_n} \left\|\sum_1^n \varepsilon_i \lambda_{\pi(i)} u x_i\right\| \le 2\sqrt{\sum_1^n \lambda_i^2 M(x)} . \tag{14}$$

Moreover, the result is true for a large measure of orthogonal operators $u \in O(n)$.

Remarks. 1. Before we start the proof, it should be noted that $\|u x_i\| \sim M(X)$ with a high probability by selections of $u \in O(n)$.

2. Note that we average the norm of the sum not only by signs $\varepsilon_i = \pm 1$, as we did in (11) and Theorem 5, but also by the permutation $\pi(i)$ which creates a symmetric norm

$$\operatorname*{Ave}_{\varepsilon_i = \pm 1; \pi \in \Pi_n} \left\|\sum_1^n \varepsilon_i \lambda_{\pi(i)} u x_i\right\|$$

(depending, of course, on the set $\{x_i\}$ and an operator $u \in O(n)$). Then we guarantee that this is equivalent to the Euclidean norm (with a high probability). This is in contrast with Theorem 5 where we could guarantee only a cotype 2 kind condition, but no information on type 2. If we do not average by Π_n, we can guarantee (14) only for a large measure of $\bar{\lambda} = (\lambda_i) \in S^{n-1}$, but not for any $\bar{\lambda}$.

Proof of Theorem 6. It is known (see [MS2]) that for $T \sim \left(\frac{b}{M}\right)^2$ and random $u_i \in O(n)$ with probability above $1 - e^{-cn}$ (for a universal constant $c > 0$)

$$\frac{1}{\sqrt{2}}|x| \cdot M \le \frac{1}{T} \sum_1^T \|u_i x\| \le 2|x| \cdot M \tag{15}$$

for any $x \in \mathbb{R}^n$. Fix (any) set $\{x_j\}_{j=1}^m \subset S^{n-1}$, any $\lambda_j \in \mathbb{R}$, $j = 1, \ldots, m$, signs $\varepsilon_j = \pm 1$ and $\pi \in \Pi_m$. Then from (15)

$$\frac{1}{\sqrt{2}} M \left| \sum_1^m \varepsilon_j \lambda_{\pi(j)} x_j \right| \le \frac{1}{T} \sum_{i=1}^T \left\| \sum_{j=1}^m \varepsilon_j \lambda_{\pi(j)} u_i x_j \right\| \le 2M \left| \sum_{j=1}^m \varepsilon_j \lambda_{\pi(j)} x_j \right|. \tag{16}$$

Take the average of (16) by $\varepsilon_j = \pm 1$ and $\pi \in \Pi_m$. Then

$$\frac{1}{2} M \sqrt{\sum_1^m |\lambda_j|^2} \le \frac{1}{T} \sum_{i=1}^T \underset{\varepsilon_j=\pm 1, \pi \in \Pi_m}{\mathrm{Ave}} \left\| \sum_{j=1}^m \varepsilon_j \lambda_{\pi(j)} u_i x_j \right\| \le 2M \sqrt{\sum_1^m \lambda_j^2}$$

(we used the exact constant $\sqrt{2}$ in the Khinchine–Kahane inequality between L_1 and L_2 averages; see [LO], [S]). In particular, it means that there exists such $u \in O(n)$ that

$$\underset{\varepsilon_j=\pm 1, \pi \in \Pi_m}{\mathrm{Ave}} \left\| \sum_{j=1}^m \varepsilon_j \lambda_{\pi(j)} u x_j \right\| \le 2M \sqrt{\Sigma \lambda_j^2}.$$

But we are interested in estimating the measure of such $u \in O(n)$. So, fix $\{x_j\}$ and $\{\lambda_j\}$ and let

$$\mathrm{Prob}\left\{ u \in O(n) : \underset{\varepsilon_j=\pm 1, \pi \in \Pi_m}{\mathrm{Ave}} \left\| \sum_1^m \varepsilon_j \lambda_{\pi(j)} u x_j \right\| > 2M \sqrt{\Sigma \lambda_j^2} \right\} = \alpha. \tag{17}$$

Then, for T independent uniformly distributed on $O(n)$ u_i, $i = 1, \ldots, T$, we have

$$\mathrm{Prob}\left\{ u_i \in O(n), i = 1, \ldots, T : \underset{\varepsilon_j=\pm, \pi \in \Pi_m}{\mathrm{Ave}} \left\| \sum_{j=1}^m \varepsilon_j \lambda_{\pi(j)} u_i x_j \right\| \right.$$
$$\left. > 2M \sqrt{\Sigma \lambda_j^2} \text{ for } i = 1, \ldots, T \right\} = \alpha^T$$

and by (15), $\alpha^T < e^{-cn}$ which implies

$$\alpha < e^{-cn/T}.$$

Now use $T \sim (b/M)^2$ and our non-degeneracy condition (13) to derive, for some $\beta > 0$

$$\alpha < \frac{1}{n^\beta}.$$

Moreover, selecting c_0 in the theorem small enough, we may achieve β to be as large as we will need later.

We deal similarly with the upper bound as we did with the lower one in (17). Therefore, with the probability above $1 - 2e^{-cn/T} \geq 1 - \frac{2}{n^\beta}$ we have

$$\operatorname*{Ave}_{\varepsilon_j = \pm 1, \pi \in \Pi_m} \left\| \sum_{j=1}^m \varepsilon_j \lambda_{\pi(j)} u x_j \right\| \sim M \sqrt{\Sigma \lambda_j^2} . \tag{18}$$

Note that the expression on the left side is a 1-unconditional and 1-symmetric norm (the latter means that it does not depend on a permutation of coordinates). This means that the norm is defined by its value for positive λ_j's and is monotonically decayed (i.e. $\lambda_{j_1} \geq \lambda_{j_2}$ for $j_1 \geq j_2$). It is known (see [Go1], [Go2]) that there exists an ε-net of the unit sphere S^{n-1} which is an orbit, with respect to the group generated by the group of permutations and all the changes of signs of the coordinates, of some set $\mathcal{A}$ of cardinality $\#\mathcal{A} \leq c_1 n^{\frac{3}{\varepsilon}} \log \frac{2}{\varepsilon}$. Now the standard argument (see [MS1]) concludes the proof. For every $\bar{\lambda} \in \mathcal{A}$, using (18), we have a subset A of $O(n)$ such that any $u \in A$ satisfies (18). Then we find a subset B of $O(n)$ of the measure $\mu(B) \geq 1 - \frac{2c_1}{n^\beta} m^{3/\varepsilon} \log 2/\varepsilon > 0$ if $\frac{2c_1}{n^\beta} m^{3/\varepsilon} \log 2/\varepsilon < 1$. We achieve this by taking $m = n$, fixing $\varepsilon > 0$ (which plays against the level of precision of (18)) and taking β large enough. Now we find $u \in O(n)$ such that (18) is satisfied for all $\bar{\lambda}$ from $\mathcal{A}$. So (18) is satisfied for all $\bar{\lambda}$ from an ε-net on S^{n-1} and it is extended to any $\bar{\lambda} \in S^{n-1}$. □

Remark. Let us sketch an alternative way to estimate the quantity α from (17). Consider the expression $\operatorname{Ave} \| \Sigma \varepsilon_j \lambda_{\pi(j)} u x_j \|$ as the function on $O(n)$. One easily verifies that it has a Lipschitz constant of at most $b(\Sigma \lambda_j^2)^{1/2}$ and that its mean value is of the order $M(\Sigma \lambda_j^2)^{1/2}$. Now a desired bound for α follows from the measure concentration phenomenon on $O(n)$ (see [GrM]).

6 Can We Check in a "Reasonable Time" that a Normed Space $X = (\mathbb{R}^n, \| \cdot \|)$ Is Very Far from Euclidean?

Reformulating the question in the title of this section, we consider a centrally symmetric compact body K and we want to "check" in a polynomial number of steps (with respect to $\dim K$) whether K is very far from an ellipsoid. Actually, nothing is put precisely in this question: what does "to check" mean? What "steps" are allowed? And how is the body K given? Analyzing some of these questions led us to the problems we studied in this paper. To explain the connection, let us consider an example.

Let X be one of ℓ_p^n spaces for $1 \leq p \leq 2$ (one may think of a similar situation when X is an n-dimensional subspace of $L_1[0,1]$), and B_X be its unit ball. Let us consider the following "game". One player chooses an operator $u \in GL_n$ and $1 \leq p \leq 2$, and builds the normed space with the unit ball

$K_1 = uB_{\ell_p^n}$. Moreover, he may distort it slightly and build a body K such that

$$d(K, K_1) = \inf\{a \cdot b \mid K \subset aK_1, \ K_1 \subset bK\} \leq 10 .$$

Now the body K is fixed.

A second player would like to check if the space X_K is not too far from Euclidean (i.e. K is not too far from an ellipsoid). It would be satisfactory for the second player to verify that, say, for some ellipsoid $\mathcal{E}_0$, $d(K, \mathcal{E}_0) \leq 10$, or, on the contrary, $d(K, \mathcal{E}) > 10$ for any ellipsoid $\mathcal{E}$.

The second player may take any $x \in \mathbb{R}^n$ and the first one will truly tell $\|x\|_K$. This is *a step* of the game. Now one question is well defined: is a polynomial (by dimension n) number of steps enough to decide that K is 10 close to an ellipsoid?

We may put this problem in the framework of probability. Let us call "a step" in the game a random selection of $x \in K$ with respect to the geometric probability over K. Now, is a polynomial number of steps enough to decide the same question with high probability?

Actually, we think that the answer is negative. Moreover, we can start with just two spaces, ℓ_1^n and ℓ_2^n, and we anyway do not see a "polynomial" procedure which recognizes that the first player selected K_1 be $B_{\ell_1^n}$ or slightly distorted $B_{\ell_2^n}$.

The results of this paper show that some natural tests one may perform with a random selection of n independent uniformly distributed on K vectors cannot distinguish any convex body K from a slightly distorted ellipsoid. This is the meaning, say, of Theorem 5 and the random cotype 2 property of any normed space. For this reason, we also conjecture that any normed space has a random type 2 property, but with a constant of order at most $c \log n$. The fact that $\log n$ is necessary follows from the example of the cube (ℓ_∞^n-space) we considered in Section 2. This problem is connected with the slicing problem (see [MP], [BoMMP], or for recent progress in [BoKM]) – one of the central problems of Asymptotic Complexity.

One may also consider different versions of the game, and restrict the selection of operators u to, say, $tu \in O(n)$, for some $t > 0$. In this way the original natural Euclidean structure of the space X is preserved and may be used in the strategy of player B. Theorem 6 shows that it is not really helpful.

Added in Proofs

It is actually true that for *any* c_0 one may find c depending on c_0 only, such that inequalities (14) are satisfied with number c instead of 2 from both sides. The proof is almost the same. However, instead of ε-net $\mathcal{A}$ we should use a similar construction to the construction of the set $\mathcal{F}$ from Section 4 which should be adapted to the case of a symmetric unconditional basis.

References

[ABV]   Arias-de-Reyna, J., Ball, K., Villa, R.: Concentration of the distance in finite-dimensional normed spaces. Mathematika, **45**, no. 2, 245–252 (1998)

[B]   Ball, K.: Volumes of sections of cubes and related problems. Geometric Aspects of Functional Analysis (1987–88), 251–260, Lecture Notes in Math., **1376**. Springer, Berlin (1989)

[Ba]   Barthe, F.: On a reverse form of the Brascamp–Lieb inequality. Invent. Math., **134**, no. 2, 335–361 (1998)

[Be]   Beckner, W.: Inequalities in Fourier analysis. Ann. Math., **102**, 159–182 (1975)

[BL]   Brascamp, H.J., Lieb, E.H.: Best constants in Young's inequality, its converse, and its generalization to more than three functions. Advances in Math., **20**, no. 2, 151–173 (1976)

[BLL]   Brascamp, H.J., Lieb, E.H., Luttinger, J.M.: A general rearrangement inequality for multiple integrals. J. Functional Analysis, **17**, 227–237 (1974)

[BoKM]   Bourgain, J., Klartag, B., Milman,V.: A reduction of the slicing problem to finite volume ratio bodies. C.R. Math. Acad. Sci. Paris, **336**, no. 4, 331–334 (2003)

[BoMMP]   Bourgain, J., Meyer, M., Milman, V., Pajor, A.: On a geometric inequality. Geometric Aspects of Functional Analysis (1986/87), 271–282, Lecture Notes in Math., **1317**. Springer, Berlin (1988)

[FLM]   Figiel, T., Lindenstrauss, J., Milman, V.D.: The dimension of almost spherical sections of convex bodies. Acta Math., **139**, no. 1-2, 53–94 (1977)

[Gl]   Gluskin, E.: On the multivariable version of the Ball's slicing cube theorem. This volume

[GlM]   Gluskin, E., Milman, V.: Randomizing properties of convex high-dimensional bodies and some geometric inequalities. C.R. Math. Acad. Sci. Paris, **334**, no. 10, 875–879 (2002)

[Go1]   Gowers, W.T.: Symmetric block bases of sequences with large average growth. Israel J. Math., **69**, no. 2, 129–151 (1990)

[Go2]   Gowers, W.T.: Symmetric block bases in finite-dimensional normed spaces. Israel J. Math., **68**, no. 2, 193–219 (1989)

[GrM]   Gromov, M., Milman, V.D.: A topological application of the isoperimetric inequality. Amer. J. Math., **105**, no. 4, 843–854 (1983)

[Hi]   Hitczenko, P.: Domination inequality for martingale transforms of Rademacher sequences. Israel J. Math., **84**, 161–178 (1993)

[J]   Junge, M.: Hyperplane conjecture for quotient spaces of L_p. Forum Math., **6**, no. 5, 617–635 (1994)

[LO]   Latala, R., Oleszkiewicz, K.: On the best constant in the Khinchin-Kahane inequality. Studia Math., **109**, no. 1, 101–104 (1994)

[MP]   Milman, V.D., Pajor, A.: Isotropic position and inertia ellipsoids and zonoids of the unit ball of a normed n-dimensional space. Geometric Aspects of Functional Analysis (1987–88), 64–104, Lecture Notes in Math., **1376**. Springer, Berlin (1989)

[MS1]   Milman, V.D., Schechtman, G.: Asymptotic theory of finite-dimensional normed spaces. With an appendix by M. Gromov. Lecture Notes in Mathematics, **1200**. Springer-Verlag, Berlin (1986), viii+156 pp

[MS2]   Milman, V.D., Schechtman, G.: Global versus local asymptotic theories of finite-dimensional normed spaces. Duke Math. J., **90**, no. 1, 73–93 (1997)

[Mo]   Montgomery-Smith, S.J.: The distribution of Rademacher sums. Proc. Amer. Math. Soc., **109**, no. 2, 517–522 (1990)

[S]   Szarek, S.J.: On the best constants in the Khinchin inequality. Studia Math., **58**, no. 2, 197–208 (1976)

[T]   Tomczak-Jaegermann, N.: Computing 2-summing norm with few vectors. Ark. Mat., **17**, no. 2, 273–277 (1979)

Several Remarks Concerning the Local Theory of L_p Spaces

W.B. Johnson[1*] and G. Schechtman[2**]

[1] Department Mathematics, Texas A&M University, College Station, TX, USA
johnson@math.tamu.edu
[2] Department of Mathematics, Weizmann Institute of Science, Rehovot, Israel
gideon.schechtman@weizmann.ac.il

Summary. For the symmetric $X_{p,w}^n$ subspaces of L_p, $p > 2$, we determine the dimension of their approximately Euclidean subspaces and estimate the smallest dimensions of their containing ℓ_p^m spaces. We also show that a diagonal of the canonical basis of ℓ_p, $p > 2$, with an unconditional basic sequence in L_p whose span is complemented, spans a space which is isomorphic to a complemented subspace of L_p.

1 Introduction

One of the main results in [BLM] is that for any n and any $\varepsilon > 0$, every n-dimensional subspace, X, of L_p, $p > 2$, $(1 + \varepsilon)$-embeds into ℓ_p^m for $m \leq c(p, \varepsilon) n^{p/2} \log n$. A result from [BDGJN] implies that, up to the log factor, this is the best possible result for a general subspace of L_p. Actually the "worst" known space is $X = \ell_2^n$. However, unlike the corresponding result for $1 \leq p < 2$ (where the dependence of m on n is linear, up to a logarithmic factor), one may ask for conditions on n-dimensional subspaces of L_p which guarantee a substantially smaller estimate of the dimension of the containing ℓ_p^m space. For a long time the authors speculated that a natural such condition could be that X only contains low-dimensional Euclidean subspaces. More precisely, we were hoping that if k is the largest dimension of a subspace of X (with $X \subset L_p$, $p > 2$) which is 2-isomorphic to a Euclidean space, then X well embeds into ℓ_p^m with $m \leq Ck^{p/2}$.

The main purpose of this note is to show that this hopeful conjecture fails. The examples are symmetric X_p spaces; see the beginning of section 2 for their

* Supported in part by NSF DMS-0200690, US-Israel Binational Science Foundation and Texas Advanced Research Program 010366-0033-20013.
** Supported in part by the Israel Science Foundation and US-Israel Binational Science Foundation; participant, NSF Workshop in Linear Analysis and Probability, Texas A&M University.

definition. We do that by determining, for each $0 < w < 1$, the best (up to universal constants) dimension of an (approximately) Euclidean subspace of $X_{p,w}$ (in Proposition 1), and by giving, for each $0 < w < 1$, a lower bound on the dimension m of an ℓ_p^m containing a well isomorphic copy of $X_{p,w}$ (in Proposition 2). The lower bound on m turns out to be the best possible, up to logarithmic factors, in the range $w > n^{\frac{1}{2}(\frac{1}{p}-\frac{1}{2})}$. In the complementary range $w < n^{\frac{1}{2}(\frac{1}{p}-\frac{1}{2})}$ we give, in Proposition 3, an upper estimate on m which does not match the lower bound and which is most probably not the best one. It is however better than what was previously known.

Section 3 is devoted to another observation regarding subspaces of L_p, $p > 2$; this time complemented subspaces with unconditional basis. We show that given such a space with an unconditional basis $\{x_n\}$, any diagonal of the given unconditional basis with the unit vector basis of ℓ_p, $\{x_n \oplus_p \alpha_n e_n\}$, spans a space which is well isomorphic to a well complemented subspace of L_p (although it may not be complemented in the natural embedding in $L_p \oplus_p \ell_p$). Here and elsewhere in this note we use the terms "well isomorphic" and "well complemented" to indicate that the norms of the operators involved (the isomorphism, its inverse, and the projection) do not depend on the dimensions of the spaces in question or on the weights (α_n here and w below) but only on p.

2 Tight Embeddings of Euclidean Spaces in Symmetric X_p Spaces and of Symmetric X_p Spaces in ℓ_p Spaces

Recall that for each $n \in \mathbb{N}$, $0 \le w \le 1$, and $2 < p < \infty$, $X_{p,w}^n$ is $\mathbb{R}^n$ with the norm

$$\|(x_1, x_2, \ldots, x_n)\|_{X_{p,w}^n} = \max\left\{ \left(\sum_{i=1}^{n} |x_i|^p \right)^{1/p}, \; w\left(\sum_{i=1}^{n} |x_i|^2 \right)^{1/2} \right\}.$$

For w which is $o(1)$ and such that $wn^{\frac{1}{2}-\frac{1}{p}}$ tends to ∞ as $n \to \infty$, we get spaces whose distances from ℓ_2^n and ℓ_p^n tend to ∞ with n. All these spaces are well isomorphic to well complemented subspaces of L_p [Ro] and up to "well isomorphism" they are all the spaces with good symmetric basis which are well isomorphic to well complemented subspaces of L_p [JMST, Theorem 1.1].

Denote by $\{e_i\}$ the canonical basis of $\mathbb{R}^n$ and let $\{g_i\}$ be independent standard Gaussian random variables. Since $\mathbb{E}\|\sum_{i=1}^{n} g_i e_i\|_{X_{p,w}^n} \approx wn^{1/2}$ and $\|\sum_{i=1}^{n} a_i e_i\|_{X_{p,w}^n} \le (\sum_{i=1}^{n} |a_i|^2)^{1/2}$ for all scalars $\{a_i\}$, it follows from the general theory of Euclidean sections of convex bodies (see, for example, [MS]) that ℓ_2^k $(1+\varepsilon)$-embed into $X_{p,w}^n$ as long as $k \le c(\varepsilon)w^2 n$. We shall now show that this estimate on the dimension is the best possible.

Recall that the Khintchine constant B_p, $2 < p < \infty$ is the smallest constant such that

$$\text{Ave}_{\pm}\left|\sum \pm a_i\right|^p \le B_p^p\left(\sum a_i^2\right)^{p/2} \tag{1}$$

for all sequences $\{a_i\}$ of real numbers. The exact value of B_p is known and in particular $B_p/\sqrt{p}$ is bounded away from 0 and ∞. The idea of the proofs of the next two propositions is taken from [BDGJN].

Proposition 1. *Assume ℓ_2^k K-embeds into $X_{p,w}^n$ and $w \ge n^{\frac{1}{p}-\frac{1}{2}}$. Then $k \le (B_p+1)^2 K^2 w^2 n$.*

Proof. If ℓ_2^k K-embeds into $X_{p,w}^n$, then there are numbers $\{\alpha_{i,j}\}_{i=1,j=1}^{k,\ n}$ satisfying

$$\left(\sum_{i=1}^{k} |a_i|^2\right)^{p/2} \le \max\left\{\sum_{j=1}^{n}\left|\sum_{i=1}^{k} a_i\alpha_{i,j}\right|^p, w^p\left(\sum_{j=1}^{n}\left|\sum_{i=1}^{k} a_i\alpha_{i,j}\right|^2\right)^{p/2}\right\} \\ \le K^p\left(\sum_{i=1}^{k} |a_i|^2\right)^{p/2}. \tag{2}$$

For each fixed $1 \le l \le n$, by setting $a_i = \alpha_{i,l}$, we get from the second inequality in (2)

$$K^p\left(\sum_{i=1}^{k} \alpha_{i,l}^2\right)^{p/2} \ge \sum_{j=1}^{n}\left|\sum_{i=1}^{k} \alpha_{i,l}\alpha_{i,j}\right|^p \ge \left(\sum_{i=1}^{k} \alpha_{i,l}^2\right)^p,$$

where the last inequality follows by picking only the $j = l$ term in the sum. We get that for all $1 \le l \le n$,

$$\left(\sum_{i=1}^{k} \alpha_{i,l}^2\right)^{p/2} \le K^p. \tag{3}$$

Using the first inequality in (2) for coefficients $\alpha_i = \pm 1$, averaging over these coefficients, and using Khintchine's inequality (1), we deduce that

$$k^{1/2} \le \text{Ave}_{\pm}\left(\left(\sum_{j=1}^{n}\left|\sum_{i=1}^{k} \pm\alpha_{i,j}\right|^p\right)^{1/p} + w\left(\sum_{j=1}^{n}\left|\sum_{i=1}^{k} \pm\alpha_{i,j}\right|^2\right)^{1/2}\right) \\ \le B_p\left(\sum_{j=1}^{n}\left(\sum_{i=1}^{k} \alpha_{i,j}^2\right)^{p/2}\right)^{1/p} + w\left(\sum_{j=1}^{n}\sum_{i=1}^{k} \alpha_{i,j}^2\right)^{1/2}. \tag{4}$$

Using (3) in (4) we get

$$k^{1/2} \le B_p K n^{1/p} + w K n^{1/2}.$$

Since $w \ge n^{\frac{1}{p}-\frac{1}{2}}$, we get that $k \le (B_p+1)^2 K^2 w^2 n$. □

Next we deal with embeddings of $X_{p,w}^n$ into ℓ_p^m.

Proposition 2. *Assume $X_{p,w}^n$ K-embed into ℓ_p^m. Then $m \ge c\min\{w^{2p}n^{p-1}, n^{p/2}\}$, where the positive constant c depends only on p and K. More precisely, $m \ge (2B_p^p K^p)^{-1}n^{p/2}$ when $w \ge (2B_p^p K^p)^{\frac{1}{2p}}n^{\frac{1}{2}(\frac{1}{p}-\frac{1}{2})}$ and $m \ge (2B_p^p K^p)^{-2}w^{2p}n^{p-1}$ when $w \le (2B_p^p K^p)^{\frac{1}{2p}}n^{\frac{1}{2}(\frac{1}{p}-\frac{1}{2})}$.*

Proof. If $X_{p,w}^n$ K-embeds into ℓ_p^m, then there are $\{\alpha_{i,j}\}_{i=1,j=1}^{n,\ m}$ such that

$$\sum_{j=1}^m \left| \sum_{i=1}^n a_i \alpha_{i,j} \right|^p \le \max \left\{ \sum_{i=1}^n |a_i|^p, w^p \left(\sum_{i=1}^n a_i^2 \right)^{p/2} \right\} \le K^p \sum_{j=1}^m \left| \sum_{i=1}^n a_i \alpha_{i,j} \right|^p \tag{5}$$

for all scalars $\{a_i\}$.

Fix a subset A of $\{1, 2, \ldots, n\}$. It follows from (5) that for every $1 \le l \le n$,

$$\max \left\{ \sum_{i \in A} |\alpha_{i,l}|^p, w^p \left(\sum_{i \in A} \alpha_{i,l}^2 \right)^{p/2} \right\} \ge \sum_{j=1}^m \left| \sum_{i \in A} \alpha_{i,l} \alpha_{i,j} \right|^p \ge \left(\sum_{i \in A} \alpha_{i,l}^2 \right)^p$$

and it follows that

$$\left(\sum_{i \in A} \alpha_{i,l}^2 \right)^{p/2} \le \max \left\{ \left(\sum_{i \in A} |\alpha_{i,l}|^p \right)^{1/2}, w^p \right\}. \tag{6}$$

Assume A is of cardinality k. Letting $a_i = \pm 1$ for $i \in A$ and zero elsewhere and averaging over the signs, we get from (5)

$$\max\{k, w^p k^{p/2}\} \le B_p^p K^p \sum_{j=1}^m \left(\sum_{i \in A} \alpha_{i,j}^2 \right)^{p/2}. \tag{7}$$

Using (6), we get from (7) that

$$\max\{k, w^p k^{p/2}\} \le B_p^p K^p \left(\sum_{j=1}^m \left(\sum_{i \in A} |\alpha_{i,j}|^p \right)^{1/2} + mw^p \right)$$

$$\le B_p^p K^p \left(m^{1/2} \left(\sum_{j=1}^m \sum_{i \in A} |\alpha_{i,j}|^p \right)^{1/2} + mw^p \right).$$

Since, by (5), for all i, $\sum_{j=1}^m |\alpha_{i,j}|^p \le 1$,

$$w^p k^{p/2} \le \max\{k, w^p k^{p/2}\} \le B_p^p K^p (m^{1/2} k^{1/2} + mw^p)$$
$$\le 2 B_p^p K^p \max\{m^{1/2} k^{1/2}, mw^p\} \tag{8}$$

and this holds for all $1 \le k \le n$. Setting $k := n$ in the extreme sides of (8) yields

$$m \ge \min \left\{ (2B_p^p K^p)^{-2} w^{2p} n^{p-1}, (2B_p^p K^p)^{-1} n^{p/2} \right\}.$$

If $w \ge (2B_p^p K^p)^{\frac{1}{2p}} n^{\frac{1}{2}(\frac{1}{p} - \frac{1}{2})}$, we get $m \ge (2B_p^p K^p)^{-1} n^{p/2}$. If $w \le (2B_p^p K^p)^{\frac{1}{2p}} n^{\frac{1}{2}(\frac{1}{p} - \frac{1}{2})}$, we get $m \ge (2B_p^p K^p)^{-2} w^{2p} n^{p-1}$. $\qquad \square$

Remark. For some time we thought the following might be true: Given a subspace X of L_p, $2 < p < \infty$, let k be the largest dimension of a 2-Euclidean subspace of X. Then the smallest m such X 2-embed into ℓ_p^m is at most a constant depending on p times $k^{p/2}$. That is, the smallest dimension of a containing ℓ_p space of a subspace X of L_p depends only on the dimension of the largest Euclidean subspace of X. Proposition 2 shows that this conjecture is wrong: For example, for $w = n^{\frac{1}{2}(\frac{1}{p}-\frac{1}{2})}$, the dimension m of ℓ_p^m which contains a 2-isomorphic copy of $X_{p,w}^n$ is, by Proposition 2, at least of order $m = n^{p/2}$ while, by the discussion preceding Proposition 1, ℓ_2^k $(1+\varepsilon)$-embed into $X_{p,w}^n$ as long as $k \leq c(\varepsilon)n^{\frac{1}{p}+\frac{1}{2}}$ (and $(n^{\frac{1}{p}+\frac{1}{2}})^{\frac{p}{2}} \ll n^{\frac{p}{2}}$).

For a fixed K and $w \geq c_p n^{\frac{1}{2}(\frac{1}{p}-\frac{1}{2})}$, the result of Proposition 2 is best up to a possible $\log n$ factor; it was proved in [BLM] that any n-dimensional subspace of L_p $(1+\varepsilon)$-embeds into ℓ_p^m for $m \leq C(p,\varepsilon)n^{p/2}\log n$. For $w \leq c_p n^{\frac{1}{2}(\frac{1}{p}-\frac{1}{2})}$, it is not clear if the result obtained here is the best possible. In [Sc2] some estimates on the dimension of the containing ℓ_p of an $X_{p,w}^n$ space are given, which in some cases are better than the general estimate of [BLM]. Using the methods of [BLM] one can somewhat improve these results to get the following result, which however still leaves a gap with the lower bound on m in Proposition 2.

Proposition 3. *There is a constant K_p, depending only on $p > 2$, such that, for all $0 < K < \infty$, if $w \leq Kn^{\frac{1}{2}(\frac{1}{p}-\frac{1}{2})}$, then $X_{p,w}^n$ K_p-embeds into ℓ_p^m whenever*

$$m \geq C(K,p)n^{1+\frac{(p-2)(p-1)}{p}}(\log n)w^{2(p-2)}.$$

$C(K,p)$ depends only on its two arguments.

Note that, up to logarithmic factors, the estimate on m is better than $n^{p/2}$.

Sketch of proof. As we said above, the proof is a combination of arguments from [Sc2] and [BLM], neither of which is simple, and it does not seem to give the final answer. We thus only sketch the argument.

We first use a specific embedding of $X_{p,w}^n$ in $L_p(0,1)$ as is given in Proposition 11 of [Sc2]: There is (sign and permutation) exchangeable sequence $\{x_i\}_{i=1}^n$ in L_p such that $\|x_i\|_p = 1$ for all i, $(\sum x_i^2)^{1/2} \equiv wn^{1/2}$ and

$$K_p^{-1}\left\|\sum a_i x_i\right\|_p \leq \max\left\{\left(\sum_{i=1}^n |a_i|^p\right)^{1/p}, w\left(\sum_{i=1}^n |a_i|^2\right)^{1/2}\right\} \leq K_p\left\|\sum a_i x_i\right\|_p.$$

Moreover, the span of the x_i-s is K_p-complemented in L_p. Here and below K_p denotes a constant depending only on p, not necessarily the same in each instance. Lemma 12 in [Sc2] asserts, in particular, that

$$\|x\|_\infty \leq K_p \min\{n^{1/2}, n^{1-1/p}w^2\}\|x\|_p$$

for all x in the span of the x_i-s. (To add to the confusion resulting from the different notations here and in [Sc2], there is a misprint in equation (23) in

[Sc2]: $m^{1/q}$ there should be $m^{1/p}$; see the bottom of the same page in [Sc2].)
Under our assumption on w, this translates to

$$\|x\|_\infty \le K^2 K_p n^{1-1/p} w^2 \|x\|_p.$$

Theorem 13 of [Sc2] and its proof shows that there is some choice of $\bar m$
points of $(0,1)$ with $\bar m \le K_p \min\{n^{1+p/2}, n^p w^{2p}\} \le K_p n^p w^{2p}$ and, letting $\bar x_i$
be the restriction of x_i to these points, considered as elements of $L_p^{\bar m}$ (L_p on
the measure space $\{1, \ldots, \bar m\}$ with the normalized counting measure), we get

$$\frac{1}{2}\left\|\sum a_i x_i\right\|_p \le \left\|\sum a_i \bar x_i\right\|_p \le 2\left\|\sum a_i x_i\right\|_p \qquad (9)$$

and

$$\frac{1}{2}\left(\sum a_i^2\right)^{1/2} \le \left\|\sum a_i w^{-1}\bar x_i\right\|_2 \le 2\left(\sum a_i^2\right)^{1/2} \qquad (10)$$

for all $\{a_i\}_{i=1}^n$. (Moreover, the span of the $\bar x_i$-s is K_p-complemented in $L_p^{\bar m}$.)
Of course we want to improve the bound on $\bar m$, but we need to use the result
above (or something similar) as we shall see shortly.

Note that it is still true that

$$\left(\sum \bar x_i^2\right)^{1/2} \equiv w n^{1/2} \qquad (11)$$

and that

$$\|x\|_\infty \le K^2 K_p n^{1-1/p} w^2 \|x\|_p \qquad (12)$$

for all x in the span of the $\bar x_i$-s.

We would like to apply now the results and techniques of [BLM] and in
particular Theorem 7.3 and its proof. We would like to take advantage of the
improved L_∞ bound in (12) and this leads to a complication since the entropy
bounds in [BLM] are obtained using a "change of density" which, if applied,
may destroy the L_∞ bound we have. We thus would like to see that basically
the same entropy bounds used in the proof of Theorem 7.3 in [BLM] apply in
our situation, without any change of density.

Let us denote by X_r the span of the $\bar x_i$-s in $L_r^{\bar m}$. In the notation of [BLM],
for any $2 < q < \infty$ and any $t > 0$,

$$E(B_{X_p}, B_{X_q}, t) \le E(B_{X_2}, B_{X_q}, t)$$

where $E(U, V, t)$ denotes the minimal number of translates of tV needed to
cover U. Now, by (10), $B_{X_2} \subset 2\{\sum a_i w^{-1}\bar x_i; \sum a_i^2 \le 1\} =: 2C$ so that

$$E(B_{X_p}, B_{X_q}, t) \le E(C, B_{X_q}, t/2).$$

Using now part of the proof of Proposition 4.6 in [BLM] (the part relaying on
Proposition 4.2) we get, for M_{X_q} computed relative to the Euclidean structure
given by C and using (11) above, that

$$M_{X_q} \leq Aq^{1/2} \left\| \left(\sum w^{-2} \bar{x}_i^2 \right)^{1/2} \right\|_{L_q^{\bar{m}}} \leq Aq^{1/2} n^{1/2}$$

for A a universal constant. Using Proposition 4.2 in [BLM], we get as in the proof of Proposition 4.6 there that

$$\log E(B_{X_p}, B_{X_q}, t) \leq A'nq/t^2$$

for some other universal constant A'. Taking $q = \log \bar{m} \leq K_p \log n$ we get

$$\log E(B_{X_p}, B_\infty, t) \leq K_p n (\log n)/t^2, \tag{13}$$

so that we recovered the entropy estimate used in the proof of Theorem 7.3 in [BLM]. We now apply this proof to the space $X = X_p$. We can take $\varepsilon = 1/2$ (there is not much point dealing with small ε since we already have a constant K_p in the original embedding of $X_{p,w}^n$ in L_p). Recall that we have an improved estimate on the L_∞ bound (12) which also implies we can take $l = [\log(K^2 K_p n^{1-1/p} w^2)/\log(3/2)] + 1$ in the beginning of the proof of Theorem 7.3 in [BLM]. Using the notation of [BLM], we get from (13) the estimates

$$\log \bar{\bar{A}}_k, \log \bar{\bar{B}}_k \leq K_p n (\log n)(4/9)^k. \tag{14}$$

Looking now at the end of the proof of Theorem 7.3 in [BLM] and using the parameters above (for ε, l and $\bar{\bar{B}}_k$), we get the right estimate on m (denoted N in [BLM]). $\square$

3 Complemented Subspaces of L_p with Unconditional Bases

In this section we prove

Proposition 4. Let $\{x_n\}$ be a (finite or infinite) C-unconditional basic sequence in L_p, $p > 2$, which is K-complemented. Then for all $\{\alpha_n\} \subset \mathbb{R}$, the span of $\{x_n \oplus \alpha_n e_n\}$ in $L_p \oplus \ell_p$ is K'-isomorphic to a K'-complemented subspace of L_p. K' depends only on K, C, and p. $\{e_n\}$ denotes here the unit vector basis of ℓ_p.

Remark. The case when $\{x_n\}$ is equivalent to the unit vector basis of ℓ_2 is the fundamental result of Rosenthal's [Ro] upon which this entire note rests.

Proof. Denote by $\{h_{n,i}\}_{n=0,i=1}^{\infty, \, 2^n}$ the mean zero L_∞-normalized Haar functions. We first treat the case where $\{x_n\}$ has the special form

$$x_n = \sum_{i=1}^{2^{k_n}} a_{n,i} h_{k_n,i},$$

for some subsequence $\{k_n\} \subset \mathbb{N}$, and the projection onto the span of $\{x_n\}$ has the special form

$$Px = \sum x_n^*(x)x_n$$

where

$$x_n^* = \sum_{i=1}^{2^{k_n}} b_{n,i} h_{k_n,i}.$$

Assume also, as we may, that

$$\|x_n\| = 2^{-k_n/p}\left(\sum_{i=1}^{2^{k_n}} |a_{n,i}|^p\right)^{1/p} = 1.$$

Since the Haar system in its natural order is a monotone basis for L_p (see [LT, p. 3]), we have

$$\|x_n^*\| = 2^{-k_n/q}\left(\sum_{i=1}^{2^{k_n}} |b_{n,i}|^q\right)^{1/q} \leq 2\|P\|.$$

Let $\{e_{n,i}\}_{n=1,i=1}^{\infty,\ n}$ be a rearrangement of the unit vector basis of ℓ_p and put

$$y_n = \alpha_n 2^{-k_n/p} \sum_{i=1}^{2^{k_n}} a_{n,i} e_{k_n,i} \in \ell_p$$

and

$$y_n^* = \alpha_n^{-1} 2^{-k_n/q} \sum_{i=1}^{2^{k_n}} b_{n,i} e_{k_n,i} \in \ell_q.$$

Then

$$y_n^*(y_n) = 1, \quad \|y_n\|_p = \alpha_n, \quad \|y_n^*\|_q \leq \alpha_n^{-1} 2\|P\|$$

so that $\bar{P}(x) = \sum y_n^*(x)y_n$ defines a projection of norm at most $2\|P\|$ from ℓ_p onto the closed span of $\{y_n\}$. Consider the basic sequence

$$z_{n,i} = a_{n,i}(h_{k_n,i} \oplus \alpha_n 2^{-k_n/p} e_{k_n,i}), \quad n = 1, 2, \ldots, \ i = 1, 2, \ldots, 2^{k_n}$$

in $L_p \oplus \ell_p$ and its closed span Z. By [KS] (see [Mü] for an alternative proof), Z is K'-isomorphic to a K'-complemented subspace of L_p, where K' depends only on K.

Put

$$Q = (P, \bar{P}) : Z \to L_p \oplus \ell_p$$

and notice that

$$Qz_{n,i} = 2^{-k_n} a_{n,i} b_{n,i} x_n \oplus \alpha_n^{-1} 2^{-k_n/q} \alpha_n 2^{-k_n/p} a_{n,i} b_{n,i} y_n$$
$$= 2^{-k_n} a_{n,i} b_{n,i}(x_n \oplus y_n).$$

In particular, the range of Q is the closed span of $\{x_n \oplus y_n\}$. Also,

$$Q(x_n \oplus y_n) = Q\sum_{i=1}^{2^{k_n}} z_{n,i} = x_n \oplus y_n \; ;$$

that is, Q is a projection (of norm at most $2\|P\|$) from Z onto the closed span of $\{x_n \oplus y_n\}$, and, since Z is well complemented in $L_p \oplus \ell_p$ and the latter is isomorphic to L_p (with universal constant), it follows that the closed span of $\{x_n \oplus y_n\}$ is well isomorphic to a well complemented subspace of L_p.

The sequence $\{x_n \oplus y_n\}$ is clearly isometrically equivalent to $\{x_n \oplus \alpha_n e_n\}$, so this completes the proof of the special case. In this case the unconditional constant of $\{x_n\}$ is no larger than the unconditional constant of the Haar basis, so that K' depends only on K and p. The reduction of the general case to the special case just treated follows from (the proof in) [Sc1]. This reduction makes the final constant K' also dependent on C. Here is a sketch of this reduction.

Let $Px = \sum x_n^*(x)x_n$ be the given projection. By a standard perturbation argument we may assume that each of the x_n and x_n^* is a linear combination of indicator functions of dyadic intervals in $[0,1]$. It follows that, for some increasing subsequence $\{k_n\} \subset \mathbb{N}$,

$$x_n = \sum_{i=1}^{2^{k_n}} a_{n,i}|h_{k_n,i}| \quad \text{and} \quad x_n^* = \sum_{i=1}^{2^{k_n}} b_{n,i}|h_{k_n,i}|.$$

Put

$$z_n = \sum_{i=1}^{2^{k_n}} a_{n,i}h_{k_n,i} \quad \text{and} \quad z_n^* = \sum_{i=1}^{2^{k_n}} b_{n,i}h_{k_n,i}.$$

The unconditionality of $\{x_n\}$ (and $\{x_n^*\}$) implies that $\{x_n\}$ is equivalent to $\{z_n\}$, $\{x_n^*\}$ is equivalent to $\{z_n^*\}$ and $Qx = \sum z_n^*(x)z_n$ is a bounded projection. The constants involved depend only on K, C and p. $\qquad\square$

References

[BDGJN] Bennett, G., Dor, L.E., Goodman, V., Johnson, W.B., Newman, C.M.: On uncomplemented subspaces of L_p, $1 < p < 2$. Israel J. Math., **26**, no. 2, 178–187 (1977)

[BLM] Bourgain, J., Lindenstrauss, J., Milman, V.: Approximation of zonoids by zonotopes. Acta Math., **162**, no. 1-2, 73–141 (1989)

[FLM] Figiel, T., Lindenstrauss, J., Milman, V.D.: The dimension of almost spherical sections of convex bodies. Acta Math., **139**, no. 1-2, 53–94 (1977)

[JMST] Johnson, W.B., Maurey, B., Schechtman, G., Tzafriri, L.: Symmetric structures in Banach spaces. Mem. Amer. Math. Soc., **19**, no. 217 (1979)

[JS2]   Johnson, W.B., Schechtman, G.: Finite dimensional subspaces of L_p. Handbook of the Geometry of Banach Spaces, vol. I, 837–870, North-Holland, Amsterdam (2001)

[KS]   Kleper, D., Schechtman, G.: Block bases of the Haar system as complemented subspaces of $L_p, 2 < p < \infty$. Proc. Amer. Math. Soc., **131**, no. 2, 433–439(2003)

[LT]   Lindenstrauss, J., Tzafriri, L.: Classical Banach Spaces I: Sequence Spaces. Springer-Verlag, Berlin, Heidelberg, New York (1977)

[MS]   Milman, V.D., Schechtman, G.: Asymptotic theory of finite-dimensional normed spaces. Lecture Notes in Mathematics, **1200**, Springer-Verlag, Berlin (1986)

[Mü]   Müller, P.F.X.: A family of complemented subspaces in VMO and its isomorphic classification. Israel J. Math., **134**, 289–306 (2003)

[Ro]   Rosenthal, H.P.: On the subspaces of L^p ($p > 2$) spanned by sequences of independent random variables. Israel J. Math., **8**, 273–303 (1970)

[Sc1]   Schechtman, G.: A remark on unconditional basic sequences in L_p ($1 < p < \infty$). Israel J. Math., **19**, 220–224 (1974)

[Sc2]   Schechtman, G.: Embedding X_p^m spaces into l_r^n. Geometrical Aspects of Functional Analysis (1985/86), 53–74, Lecture Notes in Math., **1267**, Springer, Berlin (1987)

On John-Type Ellipsoids

B. Klartag[*]

School of Mathematical Sciences, Tel Aviv University, Tel Aviv 69978, Israel
klartagb@post.tau.ac.il

Summary. Given an arbitrary convex symmetric body $K \subset \mathbb{R}^n$, we construct a natural and non-trivial continuous map u_K which associates ellipsoids to ellipsoids, such that the Löwner–John ellipsoid of K is its unique fixed point. A new characterization of the Löwner–John ellipsoid is obtained, and we also gain information regarding the contact points of inscribed ellipsoids with K.

1 Introduction

We work in $\mathbb{R}^n$, yet we choose no canonical scalar product. A centrally-symmetric ellipsoid in $\mathbb{R}^n$ is any set of the form

$$\left\{ \sum_{i=1}^n \lambda_i u_i \ ; \ \sum_i \lambda_i^2 \leq 1, \ u_1, .., u_n \in \mathbb{R}^n \right\}.$$

If $u_1, .., u_n$ are linearly independent, the ellipsoid is non-degenerate. Whenever we mention an "ellipsoid" we mean a centrally-symmetric non-degenerate one. Given an ellipsoid $\mathcal{E} \subset \mathbb{R}^n$, denote by $\langle \cdot, \cdot \rangle_{\mathcal{E}}$ the unique scalar product such that $\mathcal{E} = \{x \in \mathbb{R}^n; \langle x, x \rangle_{\mathcal{E}} \leq 1\}$. There is a group $O(\mathcal{E})$ of linear isometries of $\mathbb{R}^n$ (with respect to the metric induced by $\langle \cdot, \cdot \rangle_{\mathcal{E}}$), and a unique probability measure $\mu_{\mathcal{E}}$ on $\partial \mathcal{E}$ which is invariant under $O(\mathcal{E})$. A body in $\mathbb{R}^n$ is a centrally-symmetric convex set with a non-empty interior. Given a body $K \subset \mathbb{R}^n$, denote by $\| \cdot \|_K$ the unique norm on $\mathbb{R}^n$ such that K is its unit ball:

$$\|x\|_K = \inf\{\lambda > 0; x \in \lambda K\}.$$

Given a body $K \subset \mathbb{R}^n$ and an ellipsoid $\mathcal{E} \subset \mathbb{R}^n$, denote

$$M_{\mathcal{E}}^2(K) = M_{\mathcal{E}}^2(\| \cdot \|_K) = \int_{\partial \mathcal{E}} \|x\|_K^2 d\mu_{\mathcal{E}}(x).$$

[*] Supported in part by the Israel Science Foundation, by the Minkowski Center for Geometry, and by the European Research Training Network "Analysis and Operators".

This quantity is usually referred to as M_2. Let us consider the following parameter:

$$J_K(\mathcal{E}) = \inf_{\mathcal{F} \subset K} M_{\mathcal{E}}(\mathcal{F}) \qquad (1)$$

where the infimum runs over all ellipsoids $\mathcal{F}$ that are contained in K. Since the set of all ellipsoids contained in K (including degenerate ellipsoids) is a compact set with respect to the Hausdorff metric, the infimum is actually attained. In addition, the minimizing ellipsoids must be non-degenerate, since otherwise $J_K(\mathcal{E}) = \infty$ which is impossible for a body K. We are not so much interested in the exact value of $J_K(\mathcal{E})$, as in the ellipsoids where the minimum is obtained.

In Section 2 we prove that there exists a unique ellipsoid for which the minimum in (1) is attained. We shall denote this unique ellipsoid by $u_K(\mathcal{E})$, and we show that the map u_K is continuous. A finite measure ν on $\mathbb{R}^n$ is called $\mathcal{E}$-isotropic if for any $\theta \in \mathbb{R}^n$,

$$\int \langle x, \theta \rangle_{\mathcal{E}}^2 d\nu(x) = L_\nu^2 \langle \theta, \theta \rangle_{\mathcal{E}}$$

where L_ν does not depend on θ. One of the important properties of the map u_K is summarized in the following proposition, to be proved in Section 3.

Proposition 1.1. Let $K \subset \mathbb{R}^n$ be a body, and let $\mathcal{E}$, $\mathcal{F} \subset \mathbb{R}^n$ be ellipsoids such that $\mathcal{F} \subset K$. Then $\mathcal{F} = u_K(\mathcal{E})$ if and only if there exists an $\mathcal{E}$-isotropic measure ν supported on $\partial \mathcal{F} \cap \partial K$.

In particular, given any Euclidean structure (i.e. scalar product) in $\mathbb{R}^n$, there is always a unique ellipsoid contained in K with an isotropic measure supported on its contact points with K. This unexpected fact leads to a connection with the Löwner–John ellipsoid of K, which is the (unique) ellipsoid of maximal volume contained in K. By the characterization of the Löwner–John ellipsoid due to John [J] and Ball [B] (see also [GM]), $u_K(\mathcal{E}) = \mathcal{E}$ if and only if the ellipsoid $\mathcal{E}$ is the Löwner–John ellipsoid of K. Thus, we obtain the following:

Corollary 1.2. Let $K \subset \mathbb{R}^n$ be a body, and let $\mathcal{E} \subset K$ be an ellipsoid such that for any ellipsoid $\mathcal{F} \subset K$,

$$M_{\mathcal{E}}(\mathcal{F}) \geq 1.$$

Then $\mathcal{E}$ is the Löwner–John ellipsoid of K.

As a by-product of our methods, we also obtain an extremality property of the mean width of the Löwner–John ellipsoid (Corollary 5.2). In Section 4 we show that the body K is determined by the map u_K. Further evidence for the naturalness of this map is demonstrated in Section 5, where we discuss optimization problems similar to the optimization problem in (1), and discover connections with the map u_K.

2 Uniqueness

Let D be a minimizing ellipsoid in (1). We will show that it is the only minimizing ellipsoid. We write $|x| = \sqrt{\langle x, x \rangle_D}$. An equivalent definition of $J_K(\mathcal{E})$ is the following:

$$J_K^2(\mathcal{E}) = \min \left\{ \int_{\partial \mathcal{E}} |T^{-1}(x)|^2 d\mu_{\mathcal{E}}(x) \; ; \; \|T : l_2^n \to X_K\| \leq 1 \right\} \quad (2)$$

where $X_K = (\mathbb{R}^n, \|\cdot\|_K)$ is the normed space whose unit ball is K, and where $l_2^n = (\mathbb{R}^n, |\cdot|)$. The definitions are indeed equivalent; $T(D)$ is the ellipsoid from definition (1), as clearly $\|x\|_{T(D)} = |T^{-1}(x)|$. Since D is a minimizing ellipsoid, Id is a minimizing operator in (2). Note that in (2) it is enough to consider linear transformations which are self adjoint and positive definite with respect to $\langle \cdot, \cdot \rangle_D$. Assume on the contrary that T is another minimizer, where $T \neq Id$ is a self adjoint positive definite operator. Let $\{e_1, .., e_n\}$ be an orthogonal basis of eigenvectors of T, and let $\lambda_1, .., \lambda_n > 0$ be the corresponding eigenvalues. Consider the operator $S = \frac{Id+T}{2}$. Then S satisfies the norm condition in (2), and by the strict convexity of the function $x \mapsto \frac{1}{x^2}$ on $(0, \infty)$,

$$\int_{\partial \mathcal{E}} |S^{-1}(x)|^2 d\mu_{\mathcal{E}}(x) = \int_{\partial \mathcal{E}} \sum_{i=1}^n \left(\frac{1}{\frac{1+\lambda_i}{2}} \right)^2 \langle x, e_i \rangle_D^2 d\mu_{\mathcal{E}}(x)$$

$$< \int_{\partial \mathcal{E}} \sum_{i=1}^n \frac{1 + \left(\frac{1}{\lambda_i}\right)^2}{2} \langle x, e_i \rangle_D^2 d\mu_{\mathcal{E}}(x)$$

$$= \frac{\int_{\partial \mathcal{E}} |x|^2 d\mu_{\mathcal{E}}(x) + \int_{\partial \mathcal{E}} |T^{-1}(x)|^2 d\mu_{\mathcal{E}}(x)}{2} = J_K^2(\mathcal{E})$$

since not all the λ_i's equal one, in contradiction to the minimizing property of Id and T. Thus the minimizer is unique, and we may define a map u_K which matches with any ellipsoid $\mathcal{E}$, the unique ellipsoid $u_K(\mathcal{E})$ such that $u_K(\mathcal{E}) \subset K$ and $J_K(\mathcal{E}) = M_{\mathcal{E}}(u_K(\mathcal{E}))$. It is easily verified that for any linear operator T, and $t \neq 0$,

$$u_{TK}(T\mathcal{E}) = T u_K(\mathcal{E}), \quad (3)$$

$$u_K(t\mathcal{E}) = u_K(\mathcal{E}).$$

The second property means that the map u_K is actually defined over the "projective space" of ellipsoids. Moreover, the image of u_K is naturally a "projective ellipsoid" rather than an ellipsoid: If $\mathcal{E}$ and $t\mathcal{E}$ both belong to the image of u_K, then $t = \pm 1$. Nevertheless, we still formally define u_K as a map that matches an ellipsoid to an ellipsoid, and not as a map defined over the "projective space of ellipsoids".

Let us establish the continuity of the map u_K. One can verify that $M_{\mathcal{E}}(\mathcal{F})$ is a continuous function of $\mathcal{E}$ and $\mathcal{F}$ (using an explicit formula as in (4), for example). Fix a body $K \subset \mathbb{R}^n$, and denote by X the compact space of all (possibly degenerate) ellipsoids contained in K. Then $M_{\mathcal{E}}(\mathcal{F}) : X \times X \to [0, \infty]$ is continuous. The map u_K is defined only on a subset of X, the set of non-degenerate ellipsoids. The continuity of u_K follows from the following standard lemma.

Lemma 2.1. *Let X be a compact metric space, and $f : X \times X \to [0, \infty]$ a continuous function. Let $Y \subset X$, and assume that for any $y \in Y$ there exists a unique $g(y) \in X$ such that*

$$\min_{x \in X} f(x, y) = f(g(y), y).$$

Then $g : Y \to X$ is continuous.

Proof. Assume that $y_n \to y$ in Y. The function $\min_{x \in X} f(x, y)$ is continuous, and therefore

$$\min_{x \in X} f(x, y_n) = f(g(y_n), y_n) \overset{n \to \infty}{\longrightarrow} f(g(y), y) = \min_{x \in X} f(x, y).$$

Since $X \times X$ is compact, f is uniformly continuous and

$$\left| f(g(y_n), y) - f(g(y), y) \right|$$
$$\leq \left| f(g(y_n), y) - f(g(y_n), y_n) \right| + \left| f(g(y_n), y_n) - f(g(y), y) \right| \overset{n \to \infty}{\longrightarrow} 0.$$

Therefore, for any convergent subsequence $g(y_{n_k}) \to z$, we must have $f(z, y) = f(g(y), y)$ and by uniqueness $z = g(y)$. Since X is compact, necessarily $g(y_n) \to g(y)$, and g is continuous. $\qquad\square$

3 Extremality Conditions

There are several ways to prove the existence of the isotropic measure announced in Proposition 1.1. One can adapt the variational arguments from [GM], or use the Lagrange multiplier technique due to John [J] (as suggested by O. Guedon). The argument we choose involves duality of linear programming (see e.g. [Ba]). For completeness, we state and sketch the proof of the relevant theorem ($\langle \cdot, \cdot \rangle$ is an arbitrary scalar product in $\mathbb{R}^m$):

Theorem 3.1. *Let $\{u_\alpha\}_{\alpha \in \Omega} \subset \mathbb{R}^m$, $\{b_\alpha\}_{\alpha \in \Omega} \subset \mathbb{R}$ and $c \in \mathbb{R}^m$. Assume that*

$$\langle x^0, c \rangle = \inf \left\{ \langle x, c \rangle; \forall \alpha \in \Omega, \langle x, u_\alpha \rangle \geq b_\alpha \right\}$$

and also $\langle x^0, u_\alpha \rangle \geq b_\alpha$ for any $\alpha \in \Omega$. Then there exist $\lambda_1, .., \lambda_s > 0$ and $u_1, .., u_s \in \Omega' = \{\alpha \in \Omega; \langle x^0, u_\alpha \rangle = b_\alpha\}$ such that

$$c = \sum_{i=1}^{s} \lambda_i u_i.$$

Proof. $K = \{x \in \mathbb{R}^m; \forall \alpha \in \Omega, \langle x, u_\alpha \rangle \geq b_\alpha\}$ is a convex body. x^0 lies on its boundary, and $\{x \in \mathbb{R}^m; \langle x, c \rangle = \langle x^0, c \rangle\}$ is a supporting hyperplane to K at x^0. The vector c is an inner normal vector to K at x^0, hence $-c$ belongs to the cone of outer normal vectors to K at x^0. The crucial observation is that this cone is generated by $-\Omega'$ (e.g. Corollary 8.5 in chapter II of [Ba]), hence

$$c \in \left\{ \sum_{i=1}^s \lambda_i u_i \ ; \ \forall i \ u_i \in \Omega', \lambda_i \geq 0 \right\}. \qquad \square$$

Let $K \subset \mathbb{R}^n$ be a body, and let $\mathcal{E} \subset \mathbb{R}^n$ be an ellipsoid. This ellipsoid induces a scalar product in the space of operators: if $T, S : \mathbb{R}^n \to \mathbb{R}^n$ are linear operators, and $\{e_1, .., e_n\} \subset \mathbb{R}^n$ is any orthogonal basis (with respect to $\langle \cdot, \cdot \rangle_\mathcal{E}$), then

$$\langle T, S \rangle_\mathcal{E} = \sum_{i,j} T_{i,j} S_{i,j}$$

where $T_{i,j} = \langle Te_i, e_j \rangle_\mathcal{E}$ and $S_{i,j} = \langle Se_i, e_j \rangle_\mathcal{E}$ are the entries of the corresponding matrix representations of T and S. This scalar product does not depend on the choice of the orthogonal basis. If $\mathcal{F} = \{x \in \mathbb{R}^n; \langle x, Tx \rangle_\mathcal{E} \leq 1\}$ is another ellipsoid, then

$$M_\mathcal{E}^2(\mathcal{F}) = \int_{\partial\mathcal{E}} \langle x, Tx \rangle_\mathcal{E} d\mu_\mathcal{E}(x) \qquad (4)$$

$$= \sum_{i,j=1}^n T_{i,j} \int_{\partial\mathcal{E}} \langle x, e_i \rangle_\mathcal{E} \langle x, e_j \rangle_\mathcal{E} d\mu_\mathcal{E}(x) = \sum_{i,j=1}^n T_{i,j} \frac{\delta_{i,j}}{n} = \frac{1}{n} \langle T, Id \rangle_\mathcal{E}.$$

The ellipsoid $\mathcal{F} = \{x \in \mathbb{R}^n; \langle x, Tx \rangle_\mathcal{E} \leq 1\}$ is contained in K if and only if for any $x \in \partial K$,

$$\langle x, Tx \rangle_\mathcal{E} = \langle x \otimes x, T \rangle_\mathcal{E} \geq 1$$

where $(x \otimes x)(y) = \langle x, y \rangle_\mathcal{E} x$ is a linear operator. Therefore, the optimization problem (1) is equivalent to the following problem:

$$nJ_K^2(\mathcal{E}) = \min\left\{ \langle T, Id \rangle_\mathcal{E} \ ; \ T \text{ is } \mathcal{E}\text{-positive}, \ \forall x \in \partial K \ \langle x \otimes x, T \rangle_\mathcal{E} \geq 1 \right\}$$

where we say that T is $\mathcal{E}$-positive if it is self adjoint and positive definite with respect to $\langle \cdot, \cdot \rangle_\mathcal{E}$. Actually, the explicit positivity requirement is unnecessary. If K is non-degenerate and $\forall x \in \partial K \ \langle T, x \otimes x \rangle_\mathcal{E} \geq 1$ then T is necessarily positive definite with respect to $\mathcal{E}$. This is a linear optimization problem, in the space $\mathbb{R}^m = \mathbb{R}^{n^2}$. Let T be the unique self adjoint minimizer, and let $\mathcal{F} = \{x \in \mathbb{R}^n; \langle x, Tx \rangle_\mathcal{E} \leq 1\}$ be the corresponding ellipsoid. By Theorem 3.1, there exist $\lambda_1, .., \lambda_s > 0$ and vectors $u_1, .., u_s \in \partial K$ such that

1. For any $1 \leq i \leq s$ we have $\langle u_i \otimes u_i, T \rangle_\mathcal{E} = 1$, i.e. $u_i \in \partial K \cap \partial \mathcal{F}$.
2. $Id = \sum_{i=1}^s \lambda_i u_i \otimes u_i$. Equivalently, for any $\theta \in \mathbb{R}^n$,

$$\sum_{i=1}^s \lambda_i \langle u_i, \theta \rangle_\mathcal{E}^2 = \langle \theta, \theta \rangle_\mathcal{E}^2.$$

Hence we proved the following:

Lemma 3.2. *Let $K \subset \mathbb{R}^n$ be a body and let $\mathcal{E} \subset \mathbb{R}^n$ be an ellipsoid. If $u_K(\mathcal{E}) = \mathcal{F}$, then there exist contact points $u_1, .., u_s \in \partial K \cap \partial \mathcal{F}$ and positive numbers $\lambda_1, .., \lambda_s$ such that for any $\theta \in \mathbb{R}^n$,*

$$\sum_{i=1}^{s} \lambda_i \langle u_i, \theta \rangle_{\mathcal{E}}^2 = \langle \theta, \theta \rangle_{\mathcal{E}}.$$

The following lemma completes the proof of Proposition 1.1.

Lemma 3.3. *Let $K \subset \mathbb{R}^n$ be a body and let $\mathcal{E}, \mathcal{F} \subset \mathbb{R}^n$ be ellipsoids. Assume that $\mathcal{F} \subset K$ and that there exists a measure ν supported on $\partial K \cap \partial \mathcal{F}$ such that for any $\theta \in \mathbb{R}^n$,*

$$\int \langle x, \theta \rangle_{\mathcal{E}}^2 d\nu(x) = \langle \theta, \theta \rangle_{\mathcal{E}}.$$

Then $u_K(\mathcal{E}) = \mathcal{F}$.

Proof. Since $\int x \otimes x d\nu(x) = Id$, for any operator T,

$$\int \langle Tx, x \rangle_{\mathcal{E}} d\nu(x) = \langle T, Id \rangle_{\mathcal{E}} = n \int_{\partial \mathcal{E}} \langle Tx, x \rangle_{\mathcal{E}} d\mu_{\mathcal{E}}(x) \tag{5}$$

where the last equality follows by (4). Let T be such that $\mathcal{F} = \{x \in \mathbb{R}^n; \langle Tx, x \rangle_{\mathcal{E}} \le 1\}$. By (5),

$$\nu(\partial \mathcal{F}) = \int_{\partial \mathcal{F}} \langle Tx, x \rangle_{\mathcal{E}} d\nu(x) = n \int_{\partial \mathcal{E}} \langle Tx, x \rangle_{\mathcal{E}} d\mu_{\mathcal{E}}(x) = n M_{\mathcal{E}}^2(\mathcal{F}).$$

Suppose that $u_K(\mathcal{E}) \ne \mathcal{F}$. Then there exists a linear map $S \ne T$ such that $\langle Sx, x \rangle_{\mathcal{E}} \ge 1$ for all $x \in \partial K$ and such that

$$\int_{\partial \mathcal{E}} \langle Sx, x \rangle_{\mathcal{E}} d\mu_{\mathcal{E}}(x) < M_{\mathcal{E}}^2(\mathcal{F}).$$

Therefore,

$$\nu(\partial K) \le \int_{\partial K} \langle Sx, x \rangle_{\mathcal{E}} d\nu(x) = n \int_{\partial \mathcal{E}} \langle Sx, x \rangle_{\mathcal{E}} d\mu_{\mathcal{E}}(x) < n M_{\mathcal{E}}^2(\mathcal{F})$$

which is a contradiction, since $\nu(\partial K) = \nu(\partial \mathcal{F}) = n M_{\mathcal{E}}^2(\mathcal{F})$. $\qquad \square$

Remark. This proof may be modified to provide an alternative proof of John's theorem. Indeed, instead of minimizing the linear functional $\langle T, Id \rangle$ we need to minimize the concave functional $\det^{1/n}(T)$. The minimizer still belongs to the boundary, and minus of the gradient at this point belongs to the normal cone.

4 Different Bodies Have Different Maps

Lemma 4.1. *Let $K, T \subset \mathbb{R}^n$ be two closed bodies such that $T \not\subset K$. Then there exists an ellipsoid $\mathcal{F} \subset T$ such that $\mathcal{F} \not\subset K$ and n linearly independent vectors $v_1, .., v_n$ such that for any $1 \leq i \leq n$,*

$$v_i \in \partial \mathcal{F} \cap \partial C$$

where $C = \mathrm{conv}(K, \mathcal{F})$ and conv denotes convex hull.

Proof. Let $U \subset T \setminus K$ be an open set whose closure does not intersect K, and let $v_1^*, .., v_n^*$ be linearly independent functionals on $\mathbb{R}^n$ such that for any $y \in K$, $z \in U$,

$$v_i^*(y) < v_i^*(z)$$

for all $1 \leq i \leq n$. Let $\mathcal{F} \subset \mathrm{conv}(U, -U)$ be an ellipsoid that intersects U, and let $v_1, .., v_n \in \partial \mathcal{F}$ be the unique vectors such that for $1 \leq i \leq n$,

$$v_i^*(v_i) = \sup_{v \in \mathcal{F}} v_i^*(v).$$

Then $v_1, .., v_n$ are linearly independent. Also, $v_i^*(v_i) = \sup_{v \in C} v_i^*(v)$ and hence $v_1, .., v_n$ belong to the boundary of $C = \mathrm{conv}(K, \mathcal{F})$. $\square$

Theorem 4.2. *Let $K, T \subset \mathbb{R}^n$ be two closed bodies, such that for any ellipsoid $\mathcal{E} \subset \mathbb{R}^n$ we have $J_K(\mathcal{E}) = J_T(\mathcal{E})$. Then $K = T$.*

Proof. Assume $K \neq T$. Without loss of generality, $T \not\subset K$. Let $\mathcal{F}$ and $v_1, .., v_n$ be the ellipsoid and vectors from Lemma 4.1. Consider the following bodies:

$$L = \mathrm{conv}\{\mathcal{F}, K \cap T\}, \quad C = \mathrm{conv}\{\mathcal{F}, K\}.$$

Then $v_1, .., v_n \in \partial C$ and also $v_1, .., v_n \in \partial L$. Let $\langle \cdot, \cdot \rangle_{\mathcal{E}}$ be the scalar product with respect to which these vectors constitute an orthonormal basis. Then the uniform measure on $\{v_1, .., v_n\}$ is $\mathcal{E}$-isotropic. By Proposition 1.1, $u_L(\mathcal{E}) = u_C(\mathcal{E}) = \mathcal{F}$, and

$$J_L(\mathcal{E}) = M_{\mathcal{E}}\big(u_L(\mathcal{E})\big) = M_{\mathcal{E}}\big(u_C(\mathcal{E})\big) = J_C(\mathcal{E}). \tag{6}$$

Since $K \subset C$, also $u_K(\mathcal{E}) \subset C$. Since $\mathcal{F} = u_C(\mathcal{E}) \not\subset K$, we have $\mathcal{F} \neq u_K(\mathcal{E})$. By the uniqueness of the minimizing ellipsoid for $J_C(\mathcal{E})$,

$$J_C(\mathcal{E}) = M_{\mathcal{E}}(\mathcal{F}) < M_{\mathcal{E}}\big(u_K(\mathcal{E})\big) = J_K(\mathcal{E}). \tag{7}$$

Since $L \subset T$ we have $J_T(\mathcal{E}) \leq J_L(\mathcal{E})$. Combining this with (6) and (7), we get

$$J_T(\mathcal{E}) \leq J_L(\mathcal{E}) = J_C(\mathcal{E}) < J_K(\mathcal{E})$$

and therefore $J_K(\mathcal{E}) \neq J_T(\mathcal{E})$. $\square$

Corollary 4.3. *Let $K, T \subset \mathbb{R}^n$ be two closed bodies such that $u_K = u_T$. Then $K = T$.*

Proof. For any ellipsoid $\mathcal{E} \subset \mathbb{R}^n$,

$$J_K(\mathcal{E}) = M_{\mathcal{E}}\big(u_K(\mathcal{E})\big) = M_{\mathcal{E}}\big(u_T(\mathcal{E})\big) = J_T(\mathcal{E})$$

and the corollary follows from Theorem 4.2. $\square$

5 Various Optimization Problems

Given an ellipsoid $\mathcal{E} \subset \mathbb{R}^n$ and a body $K \subset \mathbb{R}^n$, define

$$K^\circ_{\mathcal{E}} = \left\{ x \in \mathbb{R}^n; \forall x \in K, \ \langle x, y \rangle_{\mathcal{E}} \le 1 \right\}$$

and also $M^*_{\mathcal{E}}(K) = M_{\mathcal{E}} (K^\circ_{\mathcal{E}})$. Consider the following optimization problem:

$$\inf_{K \subset \mathcal{F}} M^*_{\mathcal{E}}(\mathcal{F}) \tag{8}$$

where the infimum runs over all ellipsoids that contain K. Then (8) is simply the dual, equivalent formulation of problem (1) that was discussed above. Indeed, $\mathcal{F}$ is a minimizer in (8) if and only if $u_{K^\circ_{\mathcal{E}}}(\mathcal{E}) = \mathcal{F}^\circ_{\mathcal{E}}$. An apriori different optimization problem is the following:

$$I_K(\mathcal{E}) = \sup_{K \subset \mathcal{F}} M_{\mathcal{E}}(\mathcal{F}) \tag{9}$$

where the supremum runs over all ellipsoids that contain K. The characteristics of this problem are indeed different. For instance, the supremum need not be attained, as shown by the example of a narrow cylinder (in which there is a maximizing sequence of ellipsoids that tends to an infinite cylinder) and need not be unique, as shown by the example of a cube (any ellipsoid whose axes are parallel to the edges of the cube, and that touches the cube - is a maximizer. See also the proof of Corollary 5.2). Nevertheless, we define

$$\bar{u}_K(\mathcal{E}) = \left\{ \mathcal{F} \subset \mathbb{R}^n \ ; \ \mathcal{F} \text{ is an ellipsoid}, \ K \subset \mathcal{F}, \ M_{\mathcal{E}}(\mathcal{F}) = I_K(\mathcal{E}) \right\}.$$

The dual, equivalent formulation of (9) means to maximize M^* among ellipsoids that are contained in K. Apriori, $u_K(\mathcal{E})$ and $\bar{u}_K(\mathcal{E})$ do not seem to be related. It is not clear why there should be a connection between minimizing M and maximizing M^* among inscribed ellipsoids. The following proposition reveals a close relation between the two problems.

Proposition 5.1. *Let* $K \subset \mathbb{R}^n$ *be a body, and let* $\mathcal{E}, \mathcal{F} \subset \mathbb{R}^n$ *be ellipsoids. Then*

$$\mathcal{F} \in \bar{u}_K(\mathcal{E}) \quad \Longleftrightarrow \quad u_{K^\circ_{\mathcal{F}}}(\mathcal{E}) = \mathcal{F}.$$

Proof. $\Longrightarrow$: We write $\mathcal{F} = \{ x \in \mathbb{R}^n; \langle x, Tx \rangle_{\mathcal{E}} \le 1 \}$ for an $\mathcal{E}$-positive operator T. Since $\mathcal{F} \in \bar{u}_K(\mathcal{E})$, the operator T is a maximizer of

$$n I^2_K(\mathcal{E}) = \max \left\{ \langle S, Id \rangle_{\mathcal{E}} \ ; \ S \in L(n), \ \forall x \in \partial K \ 0 \le \langle S, x \otimes x \rangle_{\mathcal{E}} \le 1 \right\}$$

where $L(n)$ is the space of linear operators acting on $\mathbb{R}^n$. Note that the requirement $\langle S, x \otimes x \rangle_{\mathcal{E}} \ge 0$ for any $x \in \partial K$ ensures that S is $\mathcal{E}$-non-negative definite. This is a linear optimization problem. Following the notation of Theorem 3.1, we rephrase our problem as follows:

$$-n I^2_K(\mathcal{E}) = \inf \left\{ \langle S, -Id \rangle_{\mathcal{E}}; \forall x \in \partial K, \ \langle S, x \otimes x \rangle_{\mathcal{E}} \ge 0, \langle S, -x \otimes x \rangle_{\mathcal{E}} \ge -1 \right\}.$$

According to Theorem 3.1, since T is a maximizer, there exist $\lambda_1, .., \lambda_s > 0$ and vectors $u_1, .., u_t, u_{t+1}, .., u_s \in \partial K$ such that

1. For any $1 \leq i \leq t$ we have $\langle T, -u_i \otimes u_i, \rangle_{\mathcal{E}} = -1$, i.e. $u_i \in \partial K \cap \partial \mathcal{F}$.
   For any $t + 1 \leq i \leq s$ we have $\langle T, u_i \otimes u_i \rangle_{\mathcal{E}} = 0$.
2. $Id = \sum_{i=1}^{t} \lambda_i u_i \otimes u_i - \sum_{i=t+1}^{s} \lambda_i u_i \otimes u_i$.

Since we assumed that $\mathcal{F}$ is an ellipsoid, T is $\mathcal{E}$-positive, and it is impossible that $\langle Tu_i, u_i \rangle_{\mathcal{E}} = 0$. Hence, $t = s$ and there exists an $\mathcal{E}$-isotropic measure supported on $\partial K \cap \partial \mathcal{F}$. Since $K \subset \mathcal{F}$, then $\mathcal{F} \subset K_{\mathcal{F}}^{\circ}$ and $\partial K_{\mathcal{F}}^{\circ} \cap \partial \mathcal{F} = \partial K \cap \partial \mathcal{F}$. Therefore, there exists an $\mathcal{E}$-isotropic measure supported on $\partial K_{\mathcal{F}}^{\circ} \cap \partial \mathcal{F}$. Since $\mathcal{F} \subset K_{\mathcal{F}}^{\circ}$ we must have $u_{K_{\mathcal{F}}^{\circ}}(\mathcal{E}) = \mathcal{F}$, according to Proposition 1.1.

$\Longleftarrow$: Since $u_{K_{\mathcal{F}}^{\circ}}(\mathcal{E}) = \mathcal{F}$, then $K \subset \mathcal{F}$ and we can write $Id = \int x \otimes x d\nu(x)$ where $\text{supp}(\nu) \subset \partial K_{\mathcal{F}}^{\circ} \cap \partial \mathcal{F} = \partial K \cap \partial \mathcal{F}$. Reasoning as in Lemma 3.3, $\nu(\partial K) = nM_{\mathcal{E}}^{2}(\mathcal{F})$ and for any admissible operator S,

$$\langle S, Id \rangle_{\mathcal{E}} = \int \langle x, Sx \rangle_{\mathcal{E}} d\nu(x) \leq \nu(\partial K) = nM_{\mathcal{E}}^{2}(\mathcal{F})$$

since $\langle x, Sx \rangle_{\mathcal{E}} \leq 1$ for any $x \in \partial K$. Hence T is a maximizer, and $\mathcal{F} \in \bar{u}_K(\mathcal{E})$. $\square$

If $\mathcal{E}, \mathcal{F} \subset \mathbb{R}^n$ are ellipsoids, then $K_{\mathcal{E}}^{\circ}$ is a linear image of $K_{\mathcal{F}}^{\circ}$. By (3), the map $u_{K_{\mathcal{E}}^{\circ}}$ is completely determined by $u_{K_{\mathcal{F}}^{\circ}}$. Therefore, the family of maps $\{u_{K_{\mathcal{F}}^{\circ}}; \mathcal{F}$ is an ellipsoid$\}$ is determined by a single map $u_{K_{\mathcal{E}}^{\circ}}$, for any ellipsoid $\mathcal{E}$. By Proposition 5.1, this family of maps determines $\bar{u}_K$. Therefore, for any ellipsoid $\mathcal{E}$, the map $u_{K_{\mathcal{E}}^{\circ}}$ completely determines $\bar{u}_K$. An additional consequence of Proposition 5.1 is the following:

Corollary 5.2. *Let $K \subset \mathbb{R}^n$ be a body, and let $\mathcal{E}$ be its Löwner–John ellipsoid. Then for any ellipsoid $\mathcal{F} \subset K$,*

$$M_{\mathcal{E}}^{*}(\mathcal{F}) \leq 1.$$

Equality occurs for $\mathcal{F} = \mathcal{E}$, yet there may be additional cases of equality.

Proof. If $\mathcal{E}$ is the Löwner–John ellipsoid, then $u_K(\mathcal{E}) = \mathcal{E}$. By Proposition 5.1, $\mathcal{E} \in \bar{u}_{K_{\mathcal{E}}^{\circ}}(\mathcal{E})$. Dualizing, we get that $\mathcal{E} \subset K$, and

$$1 = M_{\mathcal{E}}^{*}(\mathcal{E}) = \sup_{\mathcal{F} \subset K} M_{\mathcal{E}}^{*}(\mathcal{F})$$

where the supremum is over all ellipsoids contained in K. This proves the inequality. To obtain the remark about the equality cases, consider the cross-polytope $K = \{x \in \mathbb{R}^n; \sum_{i=1}^{n} |x_i| \leq 1\}$, where $(x_1, .., x_n)$ are the coordinates of x. By symmetry arguments, its Löwner–John ellipsoid is $D = \{x \in \mathbb{R}^n; \sum_i x_i^2 \leq \frac{1}{n}\}$. However, for any ellipsoid of the form

$$\mathcal{E} = \left\{ x \in \mathbb{R}^n; \sum_i \frac{x_i^2}{\lambda_i} \leq 1 \right\}$$

where the λ_i are positive and $\sum_i \lambda_i = 1$, we get that $\mathcal{E} \subset K$, yet $M_D^{*}(\mathcal{E}) = M_D^{*}(D) = 1$. $\square$

Remarks.

1. If $K \subset \mathbb{R}^n$ is smooth and strictly convex, then $\bar{u}_K$ is always a singleton. Indeed, if K is strictly convex and is contained in an infinite cylinder, it is also contained in a subset of that cylinder which is an ellipsoid, hence the supremum is attained. From the proof of Proposition 5.1 it follows that if $\mathcal{F}_1, \mathcal{F}_2$ are maximizers, then there exists an isotropic measure supported on their common contact points with K. Since K is smooth, it has a unique supporting hyperplane at any of these contact points, which is also a common supporting hyperplane of $\mathcal{F}_1$ and $\mathcal{F}_2$. Since these common contact points span $\mathbb{R}^n$, necessarily $\mathcal{F}_1 = \mathcal{F}_2$. Hence, if K is smooth and strictly convex, only John's ellipsoid may cause an equality in Corollary 5.2.

2. If $\mathcal{E}$ is the Löwner–John ellipsoid of K, then for any other ellipsoid $\mathcal{F} \subset K$ we have $M_\mathcal{E}(\mathcal{F}) > M_\mathcal{E}(\mathcal{E}) = 1$. This follows from our methods, yet it also follows immediately from the fact that $\frac{1}{M_\mathcal{E}(\mathcal{F})} \leq \left(\frac{\mathrm{Vol}(\mathcal{F})}{\mathrm{Vol}(\mathcal{E})}\right)^{1/n}$, and from the uniqueness of the Löwner–John ellipsoid.

Acknowledgement. Part of the research was done during my visit to the University of Paris VI, and I am very grateful for their hospitality.

References

[B]   Ball, K.M.: Ellipsoids of maximal volume in convex bodies. Geometriae Dedicata, **41**, 241–250 (1992)

[Ba]   Barvinok, A.: A course in convexity. Amer. Math. Soc., Graduate Studies in Math., **54**, Providence, RI (2002)

[GM]   Giannopoulos, A., Milman, V.D.: Extremal problems and isotropic positions of convex bodies. Israel J. Math., **117**, 29–60 (2000)

[J]   John, F.: Extremum problems with inequalities as subsidiary conditions. In: Courant Anniversary Volume, Interscience, New York, 187–204 (1948)

Isomorphic Random Subspaces and Quotients of Convex and Quasi-Convex Bodies

A.E. Litvak[1], V.D. Milman[2]* and N. Tomczak-Jaegermann[3]**

[1] Department of Mathematical and Statistical Sciences, University of Alberta, Edmonton, Alberta, Canada T6G 2G1 alexandr@math.ualberta.ca
[2] School of Mathematical Sciences, Tel Aviv University, Tel Aviv 69978, Israel milman@post.tau.ac.il
[3] Department of Mathematical and Statistical Sciences, University of Alberta, Edmonton, Alberta, Canada T6G 2G1 nicole@ellpspace.math.ualberta.ca

Summary. We extend the results of [LMT] to the non-symmetric and quasi-convex cases. Namely, we consider a finite-dimensional space endowed with the gauge of either a closed convex body (not necessarily symmetric) or a closed symmetric quasi-convex body. We show that if a generic subspace of some fixed proportional dimension of one such space is isomorphic to a generic quotient of some proportional dimension of another space then for *any* proportion arbitrarily close to 1, the first space has a lot of Euclidean subspaces and the second space has a lot of Euclidean quotients.

0 Introduction

The present paper is a continuation of the study that began in [LMT], and which dealt with the phenomenon suggested by Bourgain and Milman in [BM1]. Recall that they proved that given any two normed spaces X and Y, for a large set of (proportional dimensional) subspaces of X and a large set of quotients of Y, the distance between any two representatives is much less than can be expected in the general case by Gluskin's theorem ([Gl]). In particular, it showed that "random" subspaces and "random" quotients are not of the same nature, and should have some very different properties.

In [LMT] we answered a vaguely posed question from [BM1]. In the present paper we extend the results of [LMT] to the non-symmetric and quasi-convex cases. More precisely, we show that if, for some random structure (described below), a generic subspace of *some* fixed proportional dimension is isomorphic (essentially the same) to a generic quotient of some proportional dimension of another space (with a similarly selected random structure) then for *any* proportion arbitrarily close to 1, the first space has a lot of Euclidean subspaces

* This author is supported in part by the Israel Science Foundation (Grant 142/01).
** This author holds the Canada Research Chair in Geometric Analysis.

and the second space has a lot of Euclidean quotients. So a complete similarity between a generic subspace and a generic quotient implies that most subspaces (respectively, quotients) are Euclidean.

The organization of the paper is as follows. In Sections 1 and 2 we recall some basic notations and some "symmetric" auxiliary results from [LMT]. Next we discuss the non-symmetric case in Sections 3 and 4. For completeness and future references we provide all proofs. Our main theorem in the non-symmetric case is Theorem 4.1. Then in Section 5 we discuss some possible technical improvements of results from [LMT] and dependence of constants on parameters. Although this technical part might not interest a general reader, it will be of use to an expert who wants to advance the subject. Finally, in Section 6, we show that our results hold in the symmetric quasi-convex case as well.

The asymptotic theory of non-symmetric convex bodies was developed in [MPa], [LT], [Ru]. We refer to these papers and references therein for all background not explained here. For quasi-convex bodies many results of the asymptotic theory extend as well ([BBP], [D], [GK], [KPR], [M6]); we shall give more specific references further in the text.

1 Definitions and Notations

We consider $\mathbb{R}^n$ with the standard Euclidean structure and the Euclidean unit ball denoted by B_2. The canonical Euclidean norm on $\mathbb{R}^n$ is denoted by $|\cdot|$, and the corresponding inner product by $\langle \cdot, \cdot \rangle$. We shall also consider other Euclidean structures on $\mathbb{R}^n$, with the unit balls given by ellipsoids.

In this paper by a body we always mean a compact star-shaped set with 0 in its interior. This means that from the very beginning we choose and fix the center of the body in the origin and all definitions below will not allow shifts. We shall call a convex body symmetric if it is centrally symmetric. For a convex body K in $\mathbb{R}^n$ the polar body K^0 is defined by

$$K^0 := \{x \in \mathbb{R}^n \mid \langle x, y \rangle \leq 1 \text{ for every } y \in K\}.$$

Recall that for every subspace E of $\mathbb{R}^n$ the polar (in E) of $K \cap E$ is $P_E K^0$, where P_E is the orthogonal projection onto E.

The n-dimensional volume of a body K in $\mathbb{R}^n$ is denoted by $|K|$. For a body $K \subset \mathbb{R}^n$ we shall occasionally use the notation $\|\cdot\|_K$ for the Minkowski functional of K. The space $(\mathbb{R}^n, \|\cdot\|_K)$ will be also denoted by $(\mathbb{R}^n, K)$. If $L \subset \mathbb{R}^m$ is another body and $T : \mathbb{R}^n \to \mathbb{R}^m$ is a linear operator, by $\|T : K \to L\|$ we shall denote the operator norm of T from $(\mathbb{R}^n, K)$ to $(\mathbb{R}^m, L)$. If $m = n$, the geometric distance between K and L is defined by

$$d_g(K, L) := \inf \{b/a \mid a > 0, b > 0, \ aK \subset L \subset bK\}.$$

If $d_g(K, L) \leq C$ then we say that K and L are C-equivalent. In this paper we define the Banach–Mazur distance between K and L by

$$d(K, L) := \inf \left\{ d_g(K, TL) \right\},$$

where the infimum is taken over all invertible linear operators T from $\mathbb{R}^n$ to $\mathbb{R}^n$. Note again that, compared with the usual definition, we do not minimize over shifts in the infimum above. The Banach–Mazur distance between spaces is the Banach–Mazur distance between their unit balls. If the Banach–Mazur distance between a space and the Euclidean space is bounded by C we say that the space is C-*Euclidean*.

For a real number $a > 0$, by $\lceil a \rceil$ we denote the smallest integer larger than or equal to a.

In this paper we consider ellipsoids centered at 0 only. Given an ellipsoid $\mathcal{E}$ on $\mathbb{R}^n$, by $G_{n,k}^{\mathcal{E}}$ for $1 \le k \le n$ we shall denote the Grassman manifold of k-dimensional linear subspaces of $\mathbb{R}^n$ equipped with the normalized Haar measure $\mu_{n,k}^{\mathcal{E}}$ determined by the Euclidean structure given by $\mathcal{E}$. If $\mathcal{E} = B_2$ we shall write $G_{n,k}$ and $\mu_{n,k}$ instead of $G_{n,k}^{\mathcal{E}}$, and $\mu_{n,k}^{\mathcal{E}}$. We say that some property holds for a *random orthogonal* (in $\mathcal{E}$) projection of rank k whenever the measure of the set of all subspaces $E \in G_{n,k}^{\mathcal{E}}$ for which P_E has the property, is larger than $1 - \exp(ck)$ for some absolute constant $c > 0$.

For a body $K \subset \mathbb{R}^n$ (recall that K contains the origin as an interior point), by $\mathcal{E}_K \supset K$ and $\mathcal{E}_K' \subset K$ we denote ellipsoids of minimal and maximal volume for K respectively. Note that we consider minimal and maximal volume ellipsoids centered at the origin, i.e. we do not allow shifts. We would like also to note that for convex bodies such ellipsoids are uniquely defined. It implies that the minimal volume ellipsoid is unique for any compact star-shaped body $K \subset \mathbb{R}^n$ (since it coincides with the ellipsoid of minimal volume for the convex hull of the body). In general, in the non-convex case, a maximal volume ellipsoid is not necessarily unique, so dealing with non-convex bodies we choose and fix one of them

The volume ratio of K and the outer volume ratio of K, with respect to the origin, are defined by

$$\mathrm{vr}\,(K) := \left(|K| / |\mathcal{E}_K'| \right)^{1/n} \quad \text{and} \quad \mathrm{outvr}\,(K) := \left(|\mathcal{E}_K| / |K| \right)^{1/n}.$$

For a body $K \subset \mathbb{R}^n$, an ellipsoid $\mathcal{E}$ on $\mathbb{R}^n$, and any $0 < \lambda < 1$ we shall consider certain subsets $\mathcal{F}_{\lceil \lambda n \rceil}(K) \subset G_{n, \lceil \lambda n \rceil}^{\mathcal{E}}$ of $\lceil \lambda n \rceil$-dimensional subspaces of $\mathbb{R}^n$. Each element of $\mathcal{F}_{\lceil \lambda n \rceil}(K)$ gives rise to two different normed spaces. Firstly, it can be treated as a *subspace* of the normed space $(\mathbb{R}^n, K)$, in which case we may use a generic notation sK, that is, $sK := (E, K \cap E)$. The set of all these subspaces will be denoted by $\mathcal{F}_{s, \lceil \lambda n \rceil}(K)$. Secondly, every $E \in \mathcal{F}_{\lceil \lambda n \rceil}(K)$ gives rise to a *quotient* space of $(\mathbb{R}^n, K)$, via the orthogonal (in $\mathcal{E}$) projection P_E onto E, and in this case we may use a generic notation of qK; that is, $qK := (E, P_E K)$. The set of all these quotient spaces will be denoted by $\mathcal{F}_{q, \lceil \lambda n \rceil}(K)$. (It should be noted that given a family $\mathcal{F}_k$, the definition of $\mathcal{F}_{s,k}$ does not depend on the ellipsoid $\mathcal{E}$, while the definition of $\mathcal{F}_{q,k}$ depends on this ellipsoid in an essential way.)

2 The Minimal Volume Ellipsoid
of a Symmetric Convex Body

In this section we briefly recall properties of the minimal (resp., maximal) volume ellipsoid associated to a *symmetric* convex body, which we proved in [LMT]. These properties will play an essential role in our constructions. They deal with relations to any other ellipsoid containing (resp., contained in) the same body. These new properties depend on an abstract condition of Dvoretzky–Rogers-type. All results can be dualized in a standard way to the corresponding statements for the maximal volume ellipsoids and their sections.

Let B be a symmetric convex body in $\mathbb{R}^m$ and let $\mathcal{E} \subset \mathbb{R}^m$ be an ellipsoid. Let $\phi : (0,1] \to (0,1]$ be a function. We say that $\mathcal{E}$ has property $(*)$ with respect to B with function ϕ, whenever

$(*)$ for any $1 \le k \le m$ and any projection Q of rank k on $\mathbb{R}^m$ orthogonal with respect to $\mathcal{E}$ we have $\|Q : B \to \mathcal{E}\| \ge \phi(k/m)$.

It is well known that the minimal volume ellipsoid satisfies $(*)$ with the function $\phi(t) = \sqrt{t}$. This is connected to, but simpler than, the Dvoretzky–Rogers Lemma. (Lemma 2.2 below shows that proportional-dimensional projections of the minimal volume ellipsoids satisfy $(*)$ as well.)

Theorem 2.1. *Let $\mathcal{E} \subset \mathbb{R}^m$ and $\mathcal{D} \subset \mathbb{R}^m$ be two ellipsoids, let $B := \mathcal{E} \cap \mathcal{D}$. Let $\phi : (0,1] \to (0,1]$ and set $A := (\prod_{l=1}^{m} \phi(l/m))^{-1/m}$. Assume that $\mathcal{E}$ has property $(*)$ with respect to B with function ϕ. Then*

$$|\mathcal{E}|^{1/m} \le A|\mathcal{D}|^{1/m}. \tag{2.1}$$

A typical situation when this theorem may be used is when a symmetric convex body $\tilde{B} \subset \mathbb{R}^m$ is given, $\mathcal{E} \supset \tilde{B}$ is any ellipsoid satisfying property $(*)$ with respect to $\tilde{B}$ (see e.g., Lemma 2.2 below), and $\mathcal{D} \supset \tilde{B}$ is arbitrary.

As already mentioned, the minimal volume ellipsoid satisfies $(*)$ with $\phi(t) = \sqrt{t}$. A more general class of examples is provided by proportional-dimensional projections of the minimal volume ellipsoids.

Lemma 2.2. *Let $K \subset \mathbb{R}^n$ be a symmetric convex body and let $\mathcal{E}_K$ be the ellipsoid of minimal volume for K. Let P be an arbitrary projection in $\mathbb{R}^n$ with rank $P = m = \alpha n$, for some $0 < \alpha \le 1$. Then $P\mathcal{E}_K$ has property $(*)$ with respect to PK with function $\phi(t) = \sqrt{\alpha t}$.*

For future reference we formulate an important case.

Corollary 2.3. *Let $m \le n = \beta m$ for some $\beta \ge 1$. Let $K \subset \mathbb{R}^n$ be a symmetric convex body and let $\mathcal{E}_K$ be the ellipsoid of minimal volume for K. Let P be an arbitrary projection in $\mathbb{R}^n$ with rank $P = m$. Set $E = P(\mathbb{R}^n)$ and $\mathcal{E} = P(\mathcal{E}_K)$. Let $\mathcal{D} \subset E$ be an arbitrary ellipsoid such that $\mathcal{D} \supset PK$. Then*

$$|\mathcal{E}|^{1/m} \le c\sqrt{\beta}|\mathcal{D}|^{1/m}, \tag{2.2}$$

where $c > 0$ is an absolute constant.

3 Convex Bodies in M-Position

Let us first recall the definition and a few basic facts about M-ellipsoids and M-positions of convex (not necessarily symmetric) bodies.

Let K and L be two sets on $\mathbb{R}^n$. By $N(K, L)$ we denote the covering number, i.e. the minimal number of translations of L needed to cover K.

Let $K \subset \mathbb{R}^n$ be a convex body and let $C > 0$. We say that B_2 is an M-ellipsoid for K with constant C if we have

$$\max \left\{ N(K, B_2), N(B_2, K), N(K^0, B_2), N(B_2, K^0) \right\} \leq \exp(Cn). \qquad (3.1)$$

In this case we shall often say that K is in M-position with constant C. It is a deep theorem first proved in [M3] that there is an absolute constant $C_0 > 0$ such that for every symmetric convex body K in $\mathbb{R}^n$ there exists a linear transformation taking K into M-position with constant C_0. In the non-symmetric case the existence of $C_0 > 0$ such that for any convex body K there exists an affine transformation taking K into M-position was proved in [MPa]. Moreover, the position constructed in that paper determines a choice of the center at the barycenter or at a point close to the barycenter (say $0 \in (K + z)/2$, where z is the barycenter) and, assuming as we do in this paper that the center is the origin, (3.1) holds for $K - K$ and $K \cap (-K)$ as well. (This is a consequence of the fact that bodies $K \cap (-K)$, K and $K - K$ have the same volume (up to c^n), whenever the origin is the barycenter of K, or a point close to the barycenter.) So we may and shall assume, without loss of generality, that in M-position the center is an interior point of K and is fixed at the origin, and K, $K - K$ and $K \cap (-K)$ satisfy (3.1).

Throughout the paper we shall use the notation C_0 for such a constant in (3.1). However, we shall often omit to mention it explicitly and we may just write, for example, that K is in M-position. Still, the reader should always remember that from now on all our absolute constants actually depend on this C_0.

It follows from the definition that if K is in M-position then so is K^0 and that

$$|e^{-C_0} B_2| \leq \min(|K|, |K^0|) \leq \max(|K|, |K^0|) \leq |e^{C_0} B_2|.$$

Without loss of generality we assume from now on that whenever K is in M-position then $|K| = |B_2|$.

In fact the estimates (3.1) are consequences of the conditions $|K| = |B_2|$ and one estimate $N(K, B_2) \leq \exp(C_0'n)$ with $C_0' > 0$ (see Lemma 4.2 of [MS2] for the symmetric case and [MPa] for the general case).

As in [LMT] we shall use that for any two sets in $\mathbb{R}^n$ and every projection P and every subspace E one has

$$N(PK, PL) \leq N(K, L) \quad \text{and} \quad N\big(K \cap E, (L - L) \cap E\big) \leq N(K, L). \qquad (3.2)$$

We now extend a basic functorial construction from [LMT] to the non-symmetric case. For each symmetric convex body K in $\mathbb{R}^n$ in M-position and every $0 < \lambda < 1$ we shall define a certain subset $\mathcal{F}_{\lceil \lambda n \rceil}(K) \subset G_{n, \lceil \lambda n \rceil}$ such that

$$\mu_{n, \lceil \lambda n \rceil}\left(\mathcal{F}_{\lceil \lambda n \rceil}(K)\right) \geq 1 - e^{-c_\lambda n}, \qquad (3.3)$$

where $c_\lambda > 0$ is a function of λ only. In the future we shall refer to a subset satisfying measure estimates of this type as a *random family*.

Given $K \subset \mathbb{R}^n$ as above, recall that the ellipsoids of minimal and maximal volume for K (with respect to the origin) are denoted by $\mathcal{E}_K \supset K$ and $\mathcal{E}'_K \subset K$, respectively. We shall denote the semi-axes of $\mathcal{E}_K$ by $\rho_1 \geq \rho_2 \geq \ldots \geq \rho_n$ and a corresponding orthonormal basis by $\{e_i\}_{i=1}^n$. Similar notation is adopted for $\mathcal{E}'_K$ with the semi-axes $\rho'_1 \geq \rho'_2 \geq \ldots \geq \rho'_n$ and a corresponding orthonormal basis $\{e'_i\}_{i=1}^n$.

Define $\mathcal{F}_{\lceil \lambda n \rceil}(K)$ as the set of all $E \in G_{n, \lceil \lambda n \rceil}$ satisfying

(i) $P_E B_2 \subset C_\lambda P_E K$, where P_E is orthogonal in B_2;
(ii) $K \cap E \subset C_\lambda B_2 \cap E$;
(iii) $|P_E x| \geq b_\lambda |x|$ for every $x \in \text{span}\{e_i\}_{i=1}^m \cup \text{span}\{e'_i\}_{i=n-m+1}^n$, where $m = \lceil \lambda n / 2 \rceil$,

where $C_\lambda > 0$ is an appropriate function on λ, and $b_\lambda = c_0 \sqrt{\lambda}$ with an appropriate absolute constant $c_0 > 0$. Below we keep the notation C_λ, b_λ and c_0 for these constants.

Proposition 3.1. *Let K be a convex body in M-position. Then there exist a choice of C_λ, c_λ and c_0 such that the corresponding family $\mathcal{F}_{\lceil \lambda n \rceil}(K)$ satisfies* (3.3).

For the proof of this proposition in the symmetric case see [LMT]. It was shown in [MPa] that the set of subspaces satisfying conditions (i) and (ii) has measure larger than $1 - 2e^{-n}$. Since condition (iii) is the general fact about orthogonal projections and does not depend on the body (see [LMT]) we obtain Proposition 3.1 for the non-symmetric case as well. We would like to note here that estimates in [MPa] and [LMT] give that C_λ can be chosen such that $C_\lambda \leq c^{1/(1-\lambda)}$, where c is an absolute positive constant. We discuss in Section 5 below how one can improve the dependence of C_λ on λ in the symmetric case.

4 Main Results in the Non-symmetric Case

In this section we extend the main result of [LMT] to the non-symmetric case. The proofs are similar, but a number of modifications is required. Since the arguments are important also in the remaining parts of the paper, we provide the details for completeness.

Theorem 4.1. *Let K and L be two convex bodies in $\mathbb{R}^n$ in M-position, and assume that for some $0 < \lambda < 1$ and some $d > 1$ there is a quotient space*

$qK \in \mathcal{F}_{q,\lceil \lambda n \rceil}(K)$ *and a subspace* $sL \in \mathcal{F}_{s,\lceil \lambda n \rceil}(L)$ *such that the Banach–Mazur distance satisfies*

$$d(qK, sL) \leq d.$$

Then

$$\text{outvr}\,(K) \leq C \quad and \quad \text{vr}\,(L) \leq C,$$

where $C = C(\lambda, d)$ *is a function of* λ *and* d *only.*

The proof of the theorem is based on the following proposition.

Proposition 4.2. *Let* $K \subset \mathbb{R}^n$ *be a convex body in* M*-position. Let* $0 < \lambda < 1$. *Let* $E \in \mathcal{F}_{\lceil \lambda n \rceil}(K)$ *and let* P_E *be the orthogonal projection on* E. *Then*

$$\text{outvr}\,(K) \leq C'_\lambda \big(\text{outvr}\,(P_E K)\big)^2, \tag{4.1}$$

where C'_λ *depends on* λ *(and on constant* C_0 *which defines the* M*-position we use).*

Recall here that our outer volume ratio is in fact the outer volume ratio with respect to the origin.

Remark. As it can be seen from the proof below, the power 2 in the estimate (4.1) can be improved to any $\alpha > 1$.

Proof of Proposition 4.2. Recall that $\mathcal{E}_K \supset K$ is the ellipsoid of minimal volume for K, and we denoted its semi-axes by $\rho_1 \geq \rho_2 \geq \ldots \geq \rho_n$, and the corresponding orthonormal basis by $\{e_i\}_{i=1}^n$. To simplify the notation, set $P := P_E$. Consider the ellipsoid $P\mathcal{E}_K$ in E, and denote its semi-axes by $\rho'_1 \geq \rho'_2 \geq \ldots \geq \rho'_m$ (where $m := \lceil \lambda n \rceil$). (There will be no confusion with the semi-axes of the ellipsoid of maximal volume since we do not consider this ellipsoid in this proof.)

The natural Euclidean structure in E is of course given by $PB_2 = B_2 \cap E$, and by the definition of $\mathcal{F}_m(K)$ we have $PB_2 \subset C_\lambda PK$. On the other hand, clearly, $PK \subset P\mathcal{E}_K$, and hence $\rho'_m \geq C_\lambda^{-1}$.

We first observe that since $E \in \mathcal{F}_m(K)$ then for all $1 \leq j \leq \lceil m/2 \rceil$ we have

$$\rho'_j \leq \rho_j \leq (1/b_\lambda)\rho'_j. \tag{4.2}$$

Indeed, given an ellipsoid $\mathcal{D} \subset \mathbb{R}^m$ with semi-axes $\lambda_1 \geq \lambda_2 \geq \ldots \geq \lambda_m$ one has

$$\lambda_j = \inf_L \sup_{x \in L \cap \mathcal{D}} |x|,$$

where the infimum is taken over all $(m - j + 1)$-dimensional subspaces L. Thus, since $|Px| \leq |x|$ for $x \in \mathbb{R}^n$, we have $\rho'_j \leq \rho_j$ for every $j \leq m$. On the other hand, since $\lceil m/2 \rceil = \lceil \lambda n/2 \rceil$, by the definition of $\mathcal{F}_m(K)$, we have $|Px| \geq b_\lambda |x|$ for every $x \in E_0 := \text{span}\,\{e_i \mid 1 \leq i \leq \lceil m/2 \rceil\}$, which means that the operator

$$P|_{E_0} : (E_0, B_2 \cap E_0) \rightarrow (PE_0, PB_2)$$

is invertible with the norm of the inverse bounded by $1/b_\lambda$. That implies $\rho_j \leq (1/b_\lambda)\rho_j'$.

By (4.2) we get

$$
\left(\frac{|\mathcal{E}_K|}{|K|}\right)^{1/n} = \left(\frac{|\mathcal{E}_K|}{|B_2|}\right)^{1/n} = \left(\prod_{i=1}^{n}\rho_i\right)^{1/n} \leq \left(\prod_{i=1}^{\lceil m/2\rceil}\rho_i\right)^{1/\lceil m/2\rceil}
$$

$$
\leq \frac{C_\lambda}{b_\lambda}\left(\prod_{i=1}^{m}\rho_i'\right)^{2/m} = \frac{C_\lambda}{b_\lambda}\left(\frac{|P\mathcal{E}_K|}{|PB_2|}\right)^{2/m}. \tag{4.3}
$$

Let $\mathcal{D} \supset PK$ be the ellipsoid of minimal volume for PK, so that

$$
|\mathcal{D}|^{1/m} = \text{outvr}\,(PK)\,|PK|^{1/m}. \tag{4.4}
$$

Now let us note that $\mathcal{D} \supset \text{conv}\,\{PK, -PK\}$ and $\mathcal{E}_K$ is also the ellipsoid of minimal volume for $\text{conv}\,\{K, -K\}$. Applying Corollary 2.3 for the symmetric convex body $\text{conv}\,\{PK, -PK\}$ and the ellipsoids $\mathcal{E} = P\mathcal{E}_K \subset E$ and $\mathcal{D} \subset E$ we get, by (2.2),

$$
|P\mathcal{E}_K|^{1/m} \leq \left(c/\sqrt{\lambda}\right)|\mathcal{D}|^{1/m}.
$$

By the definition of M-ellipsoid we have $|PK| \leq \exp(C_0 n)|PB_2|$. Thus we get

$$
\left(\frac{|P\mathcal{E}_K|}{|PB_2|}\right)^{2/m} \leq (c^2/\lambda)\big(\text{outvr}\,(PK)\big)^2\left(\frac{|PK|}{|PB_2|}\right)^{2/m}
$$

$$
\leq (c^2 \exp(2C_0/\lambda)/\lambda)\big(\text{outvr}\,(PK)\big)^2.
$$

Combining this with (4.3) and the formula for b_λ we obtain (4.1) with $C_\lambda' = (c^2/c_0)\,C_\lambda\,\lambda^{-3/2}\,\exp(2C_0/\lambda)$, which completes the proof. □

Now we are in the position to prove Theorem 4.1.

Proof of Theorem 4.1. By the definition of M-position (3.1) and by (3.2), we have

$$
|PK| \leq \exp(C_0 n)\,|PB_2|
$$

for every projection P. Thus, by the definitions of $\mathcal{F}_{\lceil \lambda n\rceil}(K)$ we have that every quotient $qK \in \mathcal{F}_{q,\lambda n}(K)$ admits an estimate for the volume ratio,

$$
\text{vr}\,(qK) \leq \left(|P_E K|/|C_\lambda^{-1}P_E B_2|\right)^{1/\lceil \lambda n\rceil} \leq C_\lambda \exp(C_0/\lambda).
$$

By duality every subspace $sL \in \mathcal{F}_{s,\lceil \lambda n\rceil}(L)$ admits an estimate for the outer volume ratio, $\text{outvr}\,(sL) \leq a_\lambda := c\,C_\lambda \exp(C_0/\lambda)$, where c is an absolute constant. Now, let qK and sL satisfy the hypothesis of the theorem; then

$$
\text{outvr}\,(qK) \leq d\,\text{outvr}\,(sL) \leq a_\lambda d.
$$

By Proposition 4.2 we obtain

$$\mathrm{outvr}\,(K) \le C_\lambda'\,(a_\lambda d)^2.$$

The estimate for $\mathrm{vr}\,(L)$ follows by duality.   □

Remark. Instead of using the duality argument we could use entropy estimates, since, as we discussed above, they hold for bodies $K - K$, $K \cap (-K)$, $L - L$, $L \cap (-L)$ as well.

Remark. The dependence on d in $C(\lambda, d)$ can be improved by using the remark after Proposition 4.2 and a modification of the family $\mathcal{F}_{\lceil \lambda n \rceil}(K)$. We then obtain that for every $\alpha > 1$, $C(\lambda, d) \le C_{\lambda,\alpha} d^\alpha$, where $C_{\lambda,\alpha}$ depends on λ and α only.

Setting $K = B_2$ in Theorem 4.1 we get an interesting corollary.

Corollary 4.3. *Let L be a convex body in $\mathbb{R}^n$ in M-position. If for some $0 < \lambda < 1$ and some $d > 1$ a random $\lceil \lambda n \rceil$ section sL of L is d-Euclidean, then $\mathrm{vr}\,(L) \le C$, where $C = C(\lambda, d)$ is a function of λ and d only.*

This corollary was proved in [MS2] in the case when $L = -L$ and the Euclidean distance was replaced by the geometric distance to the ball B_2. In this case it is shown by combining Theorems 3.1' and 2.2 in [MS2], that for any $0 < \xi < 1$, a random section of L is C-equivalent to B_2.

Theorem 4.1 has the following standard consequence about the existence of a large family of Euclidean quotients and subspaces.

Corollary 4.4. *Let K and L be two convex bodies in $\mathbb{R}^n$ in M-position, and assume that for some $0 < \lambda < 1$ and some $d > 1$ there is a quotient space $qK \in \mathcal{F}_{q,\lceil \lambda n \rceil}(K)$ and a subspace $sL \in \mathcal{F}_{s,\lceil \lambda n \rceil}(L)$ such that the Banach–Mazur distance satisfies*

$$d(qK, sL) \le d.$$

Then for every $0 < \xi < 1$ a random orthogonal (in $\mathcal{E}_K$) projection of K of rank $\lceil \xi n \rceil$ is $\bar{C}$-Euclidean and a random (in $\mathcal{E}_L'$) $\lceil \xi n \rceil$-dimensional section of L is $\bar{C}$-Euclidean, where $\bar{C} = \bar{C}(\lambda, \xi, d)$ is a function of λ, ξ and d only.

This corollary follows from the well-known properties of spaces with finite volume ratio and their duals ([Sz1], see also Chapter 6 of [Pi], and note that the proof does not require the symmetry).

As we have just seen, the closeness of spaces qK and sL in Theorem 4.1 and Corollary 4.4 implies the existence of many Euclidean quotients and subspaces, of an arbitrary proportional dimension, for K and L, respectively. However, one may ask whether spaces qK and sL themselves are isomorphic to Euclidean as well. Surprisingly, the answer in general is *no* even in the symmetric case. An example in [LMT] shows that for some symmetric convex bodies K and L one may select M-ellipsoids in such a way that random quotients of K and subspaces of L are far from Euclidean, while being close together. At the same time we believe that it might be true that for a judiciously selected M-ellipsoid, the hypothesis of our theorem indeed implies that qK and sL are Euclidean, with a high probability.

Similarly to [LMT], we now pass to a discussion of the global form of the results of the first part of this section. Although we always have an analogy between local and global results, there is no abstract argument proving this. In the present context the global result is much easier.

Instead of working with random families of subspaces of $\mathbb{R}^n$ we will work with random families of orthogonal operators. Let $O(n)$ denote the group of orthogonal operators on $\mathbb{R}^n$ and let ν denote the normalized Haar measure on $O(n)$. We consider a convex body $K \subset \mathbb{R}^n$ in M-position and recall that we always assume (as we may) that $K - K$ and $K \cap (-K)$ satisfy (3.1). Define $\mathcal{H}(K)$ as the set of all operators $U \in O(n)$ satisfying

(i) $cB_2 \subset K + UK$,
(ii) $c(K \cap UK) \subset B_2$

for some absolute constant $c > 0$. These two conditions are the global form of the conditions (i) and (ii) in the definition of the random family $\mathcal{F}_{[\lambda n]}(K)$. It can be shown that there exists a choice of $c > 0$ such that

$$\nu\big(\mathcal{H}(K)\big) \geq 1 - e^{-c_1 n},$$

where $c_1 > 0$ is an absolute constant (see [M5] for the symmetric case and [MPa] (and also Theorem 2′ of [LMP]) for the non-symmetric case).

Note that $(K + UK)^0$ is 2-equivalent to $K^0 \cap (U^*)^{-1} K^0$ and $(K \cap UK)^0$ is 2-equivalent to $K^0 + (U^*)^{-1} K^0$. Thus, since $(U^*)^{-1} = U$ for $U \in O(n)$, we obtain that $\mathcal{H}(K) = \mathcal{H}(K^0)$, possibly replacing the constant c in the definition by $c/2$.

The following theorem is the global version of Theorem 4.1.

Theorem 4.5. *Let K and L be two convex bodies in $\mathbb{R}^n$ in M-position. Assume that there are operators $U \in \mathcal{H}(K)$, $V \in \mathcal{H}(L)$, and some $d > 1$ such that*

$$d(K_0, L_0) \leq d,$$

where $K_0 = K + UK$ and $L_0 = L \cap VL$. Then

$$\text{outvr}\,(K) \leq (C_1/c) \exp(4C_0)d \quad and \quad \text{vr}\,(L) \leq (C_1/c) \exp(4C_0)d,$$

where C_1 is an absolute constant and c is from the definition of the families $\mathcal{H}(K)$ and $\mathcal{H}(L)$.

In particular we obtain that for some linear operators u and v (not necessarily orthogonal) $K + uK$ and $L \cap vL$ are (uniformly) equivalent to some ellipsoids.

Proof. By the definition of the set $\mathcal{H}(K)$ we have $cB_2 \subset K_0$. On the other hand, by the definition of M-ellipsoid and covering numbers we obtain that $K + UK$ can be covered by $\exp(2C_0 n)$ translations of $2B_2$. That implies

$$\mathrm{vr}\,(K_0) \le (|K_0|/|cB_2|)^{1/n} \le (2/c)\exp(2C_0).$$

To find the upper bound for the outer volume ratio of L_0 we are going to use duality. Indeed, $cB_2 \subset L_0^0 = \mathrm{conv}\,(L_0, VL_0) \subset L^0 + VL_0$. Thus repeating the proof above and using Bourgain–Milman's inverse Santaló inequality ([BM2]) we obtain

$$\mathrm{outvr}\,(L_0) \le (|(1/c)B_2|/|L_0|)^{1/n} \le C_1(|L_0^0|/|cB_2|)^{1/n} \le (2C_1/c)\exp(2C_0),$$

where $C_1 > 0$ is an absolute constant. Since $d(K_0, L_0) \le d$, then K_0 has outer volume ratio bounded by $(2C_1 d/c)\exp(2C_0)$. Let $\mathcal{E}$ be the minimal volume ellipsoid for K_0. Then $K \subset K_0 \subset \mathcal{E}$ and

$$\left(\frac{|\mathcal{E}|}{|K|}\right)^{1/n} = \left(\frac{|\mathcal{E}|}{|K_0|}\right)^{1/n} \left(\frac{|K_0|}{|B_2|}\right)^{1/n} \left(\frac{|B_2|}{|K|}\right)^{1/n} \le (4C_1/c)\,\exp(2C_0)\,d,$$

which implies boundedness of $\mathrm{outvr}\,(K)$. The result for $\mathrm{vr}\,L$ follows by a similar argument. $\qquad\square$

Remark. It is clear from the proof that the theorem can be generalized to the case of many orthogonal operators. Namely, let K, L, U, and V be as in the theorem. Assume further that U_1, ..., U_k and V_1, ..., V_m are *arbitrary* orthogonal operators on $\mathbb{R}^n$. Let $K_0 = K + UK + U_1 K + ... + U_k K$ and $L_0 = L \cap VL \cap V_1 L \cap ... \cap V_m L$. Then if $d(K_0, L_0) \le d$ then

$$\mathrm{outvr}\,(K) \le Cd \quad \text{and} \quad \mathrm{vr}\,(L) \le Cd,$$

where C is a function of k, m, c and C_0 only.

5 Technical Remarks and Improvements

In this section we discuss possible improvements of our results and dependence of functions and constant on parameters.

First we would like to note that in the symmetric case, choosing a stronger definition of an M-ellipsoid, the dependence of C_λ on λ in the definition of our random family $\mathcal{F}_{\lceil \lambda n \rceil}(K)$ can be improved to a polynomial dependence on $1/(1 - \lambda)$.

Indeed, for every $\varepsilon > 0$ there exists an ellipsoid D such that

$$\max\{N(K, tD), N(D, tK), N(K^0, tD^0), N(D^0, tK^0)\}$$
$$\le \exp\left(n(C_\varepsilon/t)^{\alpha(\varepsilon)}\right) \quad (5.1)$$

for $\alpha(\varepsilon) = (\varepsilon + 1/2)^{-1}$ and some C_ε depending on ε only (such an M-ellipsoid was constructed by Pisier, see e.g. Corollary 7.16 of [Pi]). If we take such estimates as a definition of M-ellipsoid then repeating the proof of Theorem 3 in [LMP] (cf. the proof of Proposition 7′ of [MPi]), one can obtain $C_\lambda \le f(\varepsilon)(1 - \lambda)^{-(1+\varepsilon)}$. Moreover, if we use the definition of M-ellipsoid given by Theorem 7.13 of [Pi] then, applying Theorem 3.2 of [LT], we immediately observe that $C_\lambda \le C\varepsilon^{-3/2}(1 - \lambda)^{-(\varepsilon+1/2)}$.

Now we pass to the discussion related to Theorem 4.1 and Proposition 4.2 and their consequences.

We start with Proposition 4.2. It says in qualitative terms that for a body K in M-position, if $\mathrm{outvr}\,(P_E K)$ is small for a random projection P_E, then $\mathrm{outvr}\,(K)$ is small. As was noted after the statement of this proposition the power 2 in estimate (4.1) can be improved to any $\alpha > 1$. This follows from the proof. Namely, fix $\alpha > 1$; then for any $0 < \lambda < 1$ we can define a family $\mathcal{F}^{(\alpha)}_{\lceil \lambda n \rceil}(K)$ satisfying an appropriate measure estimate depending additionally on α such that for any $qK \in \mathcal{F}^{(\alpha)}_{q,\lceil \lambda n \rceil}(K)$ we have $\mathrm{outvr}\,(K) \le C'_{\lambda,\alpha}(\mathrm{outvr}\,(qK))^\alpha$, where $C'_{\lambda,\alpha}$ depends on λ and α only.

Using the same idea as in the proof of Proposition 4.2 we can show that in the situation from the proposition one can get the inclusion $QK \subset C''QB_2$, for a certain projection Q of an arbitrarily large proportional dimension. In fact we get more.

Proposition 5.1. *Let $K \subset \mathbb{R}^n$ be a convex body in M-position. Let $0 < \lambda < 1$. Let $E \in \mathcal{F}_{\lceil \lambda n \rceil}(K)$ and let P_E be the orthogonal projection on E. Then for every $0 < \xi < 1$ there exists a projection Q with rank $Q \ge \xi n$, orthogonal in the standard Euclidean structure and in $\mathcal{E}_K$, such that $Q\mathcal{E}_K \subset C''QB_2$ where*

$$C'' = C(\lambda, \beta)\bigl(\mathrm{outvr}\,(P_E K)\bigr)^{2\beta},$$

with some $\beta \le 1 + \lambda/2(1 - \xi)$, and $C(\lambda, \beta)$ depending only on λ, β (and, of course, on c_0 and C_0). In particular, $QK \subset C''QB_2$.

Remark. As is seen from the proof below one can take

$$C(\lambda, \beta) = C_\lambda^{-1}\left(\frac{c}{c_0}\frac{C_\lambda}{\lambda^{3/2}}\exp(2C_0/\lambda)\right)^\beta,$$

where $c > 0$ is an absolute constant and $\beta = \max\{1, \lceil \lambda n/2 \rceil / \lceil (1 - \xi)n \rceil\}$.

Proof. We use the same setting and notations as in the proof of Proposition 4.2.

Let us note that the proof of the Proposition also gives

$$\left(\prod_{i=1}^{\lceil m/2 \rceil} \rho_i\right)^{1/\lceil m/2 \rceil} \le A := \frac{c^2 C_\lambda}{c_0 \lambda^{3/2}}\exp(2C_0/\lambda)\bigl(\mathrm{outvr}\,(PK)\bigr)^2,$$

and $\rho_{m/2} \ge 1/C_\lambda$.

Since for every $1 \le k \le \lceil m/2 \rceil$ we have

$$\prod_{i=1}^{\lceil m/2 \rceil} \rho_i \ge \rho_k^k \cdot \rho_{\lceil m/2 \rceil}^{\lceil m/2 \rceil - k},$$

we obtain $\rho_k \le C_\lambda^{-1}(C_\lambda A)^{\lceil m/2 \rceil / k}$, for every $1 \le k \le \lceil m/2 \rceil$.

Now let $0 < \xi < 1$ and let $k_0 = \lceil (1 - \xi)n \rceil$. If $k_0 \ge \lceil m/2 \rceil$ then we have $\rho_{k_0} \le \rho_{\lceil m/2 \rceil} \le A$. If $k_0 < \lceil m/2 \rceil$ then $\rho_{k_0} \le C_\lambda^{-1}(C_\lambda A)^{\lceil m/2 \rceil / k_0}$. Let Q be the orthogonal projection onto span $\{e_i \mid k_0 \le i \le n\}$, that is, Q "kills" the $k_0 - 1$ largest semi-axes of $\mathcal{E}_K$. Then we obtain

$$Q\mathcal{E}_K \subset C_\lambda^{-1}(C_\lambda A)^\beta Q B_2,$$

where $\beta = \lceil m/2 \rceil / \lceil (1 - \xi)n \rceil$ if $\lceil (1 - \xi)n \rceil < \lceil m/2 \rceil$ and $\beta = 1$ otherwise. Since $\lceil \lambda n/2 \rceil = \lceil m/2 \rceil$ we obtain the result. $\qquad \square$

We turn now to Theorem 4.1.

First let us comment on the dependence of C on d in this theorem. The proof we presented gives C of the form $C = C(\lambda, d) \le C_\lambda'' d^2$, where C_λ'' depends on λ only. Using Remark after Proposition 4.2 and the discussion before Proposition 5.1 we can obtain that for every $\alpha > 1$, this dependence can be improved to $C_{\lambda,\alpha}'' d^\alpha$, where $C_{\lambda,\alpha}''$ depends on λ and α only.

Let us also note that, as can be seen from the proof, the theorem holds even for bodies of different dimensions. Namely, the following theorem holds.

Theorem 5.2. *Let $K \subset \mathbb{R}^n$ and $L \subset \mathbb{R}^m$ be two convex bodies in M-position. Let $1 \le k < \min(n, m)$ and assume that for some quotient space $qK \in \mathcal{F}_{q,k}(K)$ and some subspace $sL \in \mathcal{F}_{s,k}(L)$ one has $d(qK, sL) \le d$. Then $\mathrm{outvr}(K) \le C$ and $\mathrm{vr}(L) \le C$, where $C = C(k/n, k/m, d)$ depends only on k/n, k/m and d.*

Since Corollary 4.4 follows immediately from Theorem 4.1 we can also extend Corollary 4.4 to the case of different dimensions of K and L. Namely, we have the following corollary of Theorem 5.2.

Corollary 5.3. *Let $K \subset \mathbb{R}^n$ and $L \subset \mathbb{R}^m$ be two convex bodies in M-position. Let $1 \le k < \min(n, m)$ and assume that for some quotient space $qK \in \mathcal{F}_{q,k}(K)$ and some subspace $sL \in \mathcal{F}_{s,k}(L)$ one has $d(qK, sL) \le d$. Then for every $0 < \xi < 1$ a random orthogonal (in $\mathcal{E}_K$) projection of K of rank $\lceil \xi n \rceil$ is $\bar{C}$-Euclidean and a random (in $\mathcal{E}_L'$) $\lceil \xi m \rceil$-dimensional section of L is $\bar{C}$-Euclidean, where $\bar{C} = \bar{C}(k/n, k/m, \xi, d)$ is a function of k/n, k/m, ξ and d only.*

Remarks. **1.** In the situation of Corollaries 4.4 and 5.3, it follows from the proof that a random projection QK of K is C-equivalent to $Q\mathcal{E}_K$. Similarly, a random section $L \cap E$ of L is C-equivalent to $\mathcal{E}_L' \cap E$.

2. In the situation of Theorems 4.1 and 5.2 we also have that for every $0 < \xi < 1$, there exists a projection Q with rank $P \geq \xi n$, orthogonal in the standard Euclidean structure and in $\mathcal{E}_K$, such that $Q\mathcal{E}_K \subset C'QB_2$. Similarly, there exists a subspace $F \subset \mathbb{R}^n$ with $\dim F \geq \xi n$ such that $\mathcal{E}'_L \cap F \supset (1/C')B_2 \cap F$. This follows from Proposition 5.1 above.

6 The Symmetric Quasi-Convex Case

Here we discuss briefly the symmetric quasi-convex cases.

Let $C \geq 1$. Recall that a body K is called C-quasi-convex if $\alpha K + (1 - \alpha)K \subset CK$ for every $\alpha \in [0,1]$. Given $p \in (0,1]$ we say that a body K is p-convex if $\alpha K + (1 - \alpha^p)^{1/p}K \subset K$ for every $\alpha \in [0,1]$. Note that if $C = 1$ or $p = 1$ then the definitions above give a convex body. Clearly every p-convex symmetric body is C-quasi-convex with $2C = 2^{1/p}$. By the Aoki–Rolewicz theorem (see [KPR], [Ko], [Ro]) every C-quasi-convex body is $2C$-equivalent to some p-convex body, where p satisfies $2C = 2^{1/p}$. Hence it is enough to restrict ourselves to a p-convex case. Note that the gauge of a p-convex body satisfies $\|x + y\|^p \leq \|x\|^p + \|y\|^p$. It should be also noted that since $K^0 = (\text{conv } K)^0$, duality arguments usually lead to very weak results for p-convex bodies. Thus to extend proofs which originally used duality, we need to find direct arguments.

We say that a p-convex body K is in M-position if there exists $q > 0$ and C_p, depending on p and q only, such that

$$\max\left\{N(B_2, tK), N(K, tB_2)\right\} \leq \exp\left(n(C_p/t)^q\right), \tag{6.1}$$

for every $t \geq 1$. The existence of M-position for every symmetric p-convex body and every $q < (2p)/(2 - p)$ was proved in [BBP] (see also [L] for the dependence of C_p on p and q). We can assume, as we did before, that a body K in M-position satisfies $|K| = |B_2|$. We will now show that all theorems of Section 3 can be extended to the p-convex setting. Of course all constants appearing in the statements will depend on p.

It was shown in [LMP] that if a p-convex body satisfies entropy estimates (6.1) then the set of $\lceil \lambda n \rceil$-dimensional subspaces E of $\mathbb{R}^n$ satisfying

$$P_E B_2 \subset (1 - \lambda)^{-1-1/p} (c/p)^{c/p} P_E K$$

and

$$K \cap E \subset (1 - \lambda)^{-1-1/p} (c/p)^{c/p} B_2 \cap E$$

for some absolute constant $c > 0$, has measure exponentially close to 1. This corresponds to the conditions (i) and (ii) of the definition of random family. (Actually only the existence of such a subspace was stated explicitly, but the measure estimates follow from the proof.) This allows us to define, for every symmetric p-convex body K, a random family $\mathcal{F}_{\lceil \lambda n \rceil}(K)$ in the same way as we did before. Note in passing that a maximal volume ellipsoid might not be unique in the p-convex case, so before defining the random family we fix one of the maximal volume ellipsoids for the rest of the argument.

Since the arguments in Section 3 use duality several times, we need to introduce a new property $(**)$, which plays the role of the dual property to $(*)$. Recall that property $(*)$ is connected to the Dvoretzky–Rogers lemma. This lemma was extended to the quasi-convex case by Dilworth, who provided also many applications of it and other properties of a maximal volume ellipsoid in the quasi-convex case ([D]). We say that an ellipsoid $\mathcal{E} \in \mathbb{R}^m$ has property $(**)$ with respect to a p-convex body $K \in \mathbb{R}^m$ with constant $a \geq 1$ if

$(**)$ for any $1 \leq k \leq m$ and any k-dimensional subspace $E \subset \mathbb{R}^m$ we have

$$\|i_{|E} : \mathcal{E} \cap E \to K\| \geq (1/a)\,(k/m)^{1/p-1/2},$$

where $i_{|E}$ is the identity map on E.

Note that for simplicity we restrict ourselves to the case of function $\phi(t) = t^{1/p-1/2}$.

Theorem 2.1 is stated for ellipsoids, so we can dualize it. We then get

Theorem 6.1. *Let $\mathcal{E} \subset \mathbb{R}^m$ and $\mathcal{D} \subset \mathbb{R}^m$ be two ellipsoids, let $B := \mathrm{conv}\,(\mathcal{E}, \mathcal{D})$ and assume that $\mathcal{E}$ has property $(**)$ with constant $a \geq 1$ with respect to B. Then*

$$|\mathcal{D}|^{1/m} \leq c^{1/p} a |\mathcal{E}|^{1/m},$$

where $c > 0$ is an absolute constant.

The following lemma is the dual version of Lemma 2.2.

Lemma 6.2. *Let $p \in (0,1)$, K be a symmetric p-convex body and $\mathcal{E}$ be an ellipsoid of maximal volume for K. Let F be an αn-dimensional subspace of $\mathbb{R}^n$. Then $\mathcal{E} \cap F$ has property $(**)$ with respect to $K \cap F$ with constant $a = (e/\alpha)^{1/p-1/2}$*

We postpone the proof of this lemma until the end of this section.

Now we will show that Proposition 4.2 as well as its dual form are valid in the p-convex case. Namely, the following statement holds.

Proposition 6.3. *Let $p \in (0,1)$, $K \subset \mathbb{R}^n$ be a symmetric p-convex body in M-position. Let $0 < \lambda < 1$. Let $E \in \mathcal{F}_{\lceil \lambda n \rceil}(K)$ and let P_E be the orthogonal projection on E. Then*

$$\mathrm{outvr}\,(K) \leq C\big(\mathrm{outvr}\,(P_E K)\big)^2 \quad and \quad \mathrm{vr}\,(K) \leq C\big(\mathrm{vr}\,(K \cap E)\big)^2,$$

where C depends on λ and p only.

Remark. Of course the power 2 can be improved the same way as in the convex case.

Proof. The case outvr (K) repeats the proof of Proposition 4.2, using the observation that the ellipsoid of minimal volume for a p-convex symmetric body K is the ellipsoid of minimal volume for conv K as well.

To prove the estimate for vr (K) we observe that the same ideas (with obvious modifications) as were used in the "convex" proof before (4.4) imply for $E \subset \mathcal{F}_{\lceil \lambda n \rceil}$

$$(|K|/|\mathcal{E}'_K|)^{1/n} \le C(\lambda, p) \, b_\lambda^{-1} \, (|B_2 \cap E|/|\mathcal{E}'_K \cap E|)^{2/m},$$

where $\mathcal{E}'_K$ is the ellipsoid of maximal volume for K which was fixed before the definition of the random family. Now let $\mathcal{D} \subset K \cap E$ be an ellipsoid of maximal volume for $K \cap E$ and let $\mathcal{E} = \mathcal{E}'_K \cap E$. Then, by Lemma 6.2, the ellipsoid $\mathcal{E}$ has property $(**)$ with respect to $K \cap E$ with constant $a = (e/\lambda)^{1/p-1/2}$. Since K is p-convex, then conv $(\mathcal{D}, \mathcal{E}) \subset 2^{-1+1/p}K$. Hence, $\mathcal{E}$ has property $(**)$ with respect to conv $(\mathcal{D}, \mathcal{E})$ with constant $a = (2e/\lambda)^{1/p-1/2}$. Therefore, by Theorem 6.1 we obtain $|\mathcal{D}|^{1/m} \le c_p a |\mathcal{E}|^{1/m}$, which implies

$$(|B_2 \cap E|/|\mathcal{E}|)^{2/m} \le c_p^2 a^2 \, (|B_2 \cap E|/|\mathcal{D}|)^{2/m}$$
$$= c_p^2 a^2 \big(\mathrm{vr}\,(K \cap E) \big)^2 \, (|B_2 \cap E|/|K \cap E|)^{2/m}.$$

The proposition now follows by covering estimates and (3.2) (note that $K = -K$ and $K - K \subset 2^{1/p}K$). $\qquad\qquad\qquad\qquad\qquad\qquad\qquad\qquad\quad$ $\square$

Now we can prove our main result, Theorem 4.1, for symmetric p-convex bodies.

Theorem 6.4. *Let $p \in (0,1)$, K and L be two symmetric p-convex bodies in $\mathbb{R}^n$ in M-position, and assume that for some $0 < \lambda < 1$ and some $d > 1$ there is a quotient space $qK \in \mathcal{F}_{q,\lceil \lambda n \rceil}(K)$ and a subspace $sL \in \mathcal{F}_{s,\lceil \lambda n \rceil}(L)$ such that the Banach–Mazur distance satisfies*

$$d(qK, sL) \le d.$$

Then

$$\mathrm{outvr}\,(K) \le C \quad and \quad \mathrm{vr}\,(L) \le C,$$

where $C = C(\lambda, d, p)$ is a function of λ, d and p only.

As in the convex case the proof is based on the previous proposition.

Proof. By the definition of M-position (6.1) and by (3.2), we have

$$|PK| \le \exp\,(C_0 n) \, |PB_2| \quad and \quad |B_2 \cap E| \le \exp\,(C_0 n) \, |2^{1/p}L \cap E|$$

for every projection P and every subspace E (we used again that $L - L \subset 2^{1/p}L$). Thus, by the definitions of $\mathcal{F}_{\lceil \lambda n \rceil}(K)$ we have that every quotient $qK \in \mathcal{F}_{q,\lambda n}(K)$ admits an estimate for the volume ratio,

$$\mathrm{vr}\,(qK) \le \left(|P_E K|/|C_\lambda^{-1} P_E B_2|\right)^{1/\lceil \lambda n \rceil} \le C_\lambda \exp\left(C_0/\lambda\right).$$

Similarly, every subspace $sL \in \mathcal{F}_{s,\lceil \lambda n \rceil}(L)$ admits an estimate for the outer volume ratio, $\mathrm{outvr}\,(sL) \le a_\lambda := 2^{1/p} C_\lambda \exp\left(C_0/\lambda\right)$. Now, let qK and sL satisfy the hypothesis of the theorem; then

$$\mathrm{outvr}\,(qK) \le d\,\mathrm{outvr}\,(sL) \le a_\lambda d \quad \text{and} \quad \mathrm{vr}\,(sL) \le \mathrm{vr}\,(qK) \le a_\lambda d.$$

By Proposition 6.3 we obtain

$$\mathrm{outvr}\,(K) \le C_\lambda'\,(a_\lambda d)^2 \quad \text{and} \quad \mathrm{vr}\,(L) \le C_\lambda'\,(a_\lambda d)^2. \qquad \square$$

Corollary 4.4 in the p-convex case follows from Theorem 6.4. Indeed, it is not difficult to check that the volume ratio argument works for p-convex bodies as well (with constants depending on p). If a p-convex symmetric body K has bounded outer volume ratio then so does conv K. Thus, by the corresponding "convex" result P conv $K = $ conv PK is C-Euclidean. One can show that if the convex hull of a quasi-convex body is isomorphic to a Euclidean ball then so is the body itself (see e.g. Lemma 3 and Remark that follows in [GK] or proof of Theorem 13 in [KT]). It implies that PK is $C(p)$-Euclidean.

To prove the global results let us note first that the existence of orthogonal operator U satisfying conditions (i) and (ii) of Section 4 was proved in [M6] (see also Lemmas 6.3 and 6.8 of [LMS] and Theorem $2'$ of [LMP]). Moreover, it follows from this proof that the measure of the set of such operators is exponentially close to 1. This means that we can define the family of operators $\mathcal{H}(K)$ the same way as before. The only place in the proof of Theorem 4.5, where convexity was used is the estimate for outvr L_0. The necessary estimate in the p-convex case follows from Lemma 6.4 of [LMS].

Finally, to prove Lemma 6.2, we prove the following lemma which obviously implies it.

Lemma 6.5. *Let K be a symmetric p-convex body and $\mathcal{E}$ be an ellipsoid of maximal volume for K. Let E be a k-dimensional subspace of $\mathbb{R}^n$. Then*

$$\|i_{|E} : \mathcal{E} \cap E \to K\| \ge (k/e\,n)^{1/p-1/2}.$$

Proof. Without loss of generality we assume that $\mathcal{E} = B_2$ and $k < n$. Denote $B_2^k = B_2 \cap \mathbb{R}^k$. Clearly it is enough to show that

$$\|i_{|\mathbb{R}^k} : B_2^k \to K\| \le 1/A$$

implies $A \le (en/k)^{\frac{1}{p}-\frac{1}{2}}$. In other words, $AB_2^k \subset K$ implies $A \le (en/k)^{\frac{1}{p}-\frac{1}{2}}$.

Assume $AB_2^k \subset K$. Set $1/r = 1/p - 1/2$. Let $b = A(k/n)^{1/r}$ and $a = (1 - k/n)^{-1/r} > 1$. Consider the ellipsoid

$$\mathcal{D} = \left\{ x \in \mathbb{R}^n \mid \sum_{i=1}^{k} x_i^2/b^2 + \sum_{i=k+1}^{n} x_i^2 a^2 \le 1 \right\}.$$

We shall show that

$$\mathcal{D} \subset L := \bigcup_{\alpha^p + \beta^p \leq 1} \left(\alpha B_2 + A\beta B_2^k\right).$$

Since $B_2 \subset K$, $AB_2^k \subset K$, by p-convexity of K, it will imply that $\mathcal{D} \subset K$.

Let $x \in \mathcal{D}$. Set $\gamma_1 = (\sum_{i=1}^k x_i^2)^{1/2}$ and $\gamma_2 = (\sum_{i=k+1}^n x_i^2)^{1/2}$.

Put $\alpha = \gamma_2$ and $\beta = \gamma_1/A$. Then we have $x = \alpha y + \beta z$ where $y = (1/\gamma_2)(0, ..., 0, x_{k+1}, ..., x_n) \in B_2$ and $z = (A/\gamma_1)(x_1, ..., x_k) \in AB_2^k$. Furthermore,

$$\alpha^p + \beta^p = \left(\frac{1}{a}(\gamma_2 a)\right)^p + \left(\frac{b}{A}(\gamma_1/b)\right)^p.$$

Note that since $x \in \mathcal{D}$ then

$$(\gamma_2 a)^2 + (\gamma_1/b)^2 \leq 1.$$

So now, using Hölder's inequality for $1/p = 1/r + 1/2$ we get that $\alpha^p + \beta^p$ is less than or equal to

$$\left(\frac{1}{a}\right)^r + \left(\frac{b}{A}\right)^r = 1,$$

by definitions of a and b.

Thus $\mathcal{D} \subset K$. Now we compare the volumes of $\mathcal{D}$ and B_2 (recall that B_2 is an ellipsoid of maximal volume for K).

$$|\mathcal{D}|/|B_2| = b^k a^{k-n} = (A^r k/n)^{k/r} (1 - k/n)^{(n-k)/r} \geq (A^r k/n)^{k/r} e^{-k/r} > 1$$

if $A^r > en/k$. This contradicts the maximality of the volume of B_2 and hence completes the proof.   □

Remark. In the convex case $p = 1$ a similar argument gives the best possible result

$$\|i_{|E} : \mathcal{E} \cap E \to K\| \geq \sqrt{k/n}.$$

Indeed, with the same notations as above, assume that $A > \sqrt{k/n}$. Consider the ellipsoid $\mathcal{D}$, defined as before, for parameters $b = A\sqrt{k/n}$ and $a = \sqrt{(A^2 - 1)/(A^2 - b^2)}$. Let $L := \text{conv}\,(B_2, AB_2^k) \subset K$. The direct computations shows that $\mathcal{D}^0 \supset L^0 = B_2 \cap (1/A)B_2^k$. Thus $\mathcal{D} \subset K$ and

$$(|\mathcal{D}|/|B_2|)^2 = (b^k a^{k-n})^2 = (A^2 k/n)^k \left(\frac{1 - k/n}{1 - 1/A^2}\right)^{n-k} > 1,$$

since $A > \sqrt{n/k}$.

References

[BBP]   Bastero, J., Bernués, J., Peña, J.: An extension of Milman's reverse Brunn-Minkowski inequality. Geom. Funct. Anal., **5**, no. 3, 572–581 (1995)

[BM1]   Bourgain, J., Milman, V.D.: Distances between normed spaces, their subspaces and quotient spaces. Integral Equations Operator Theory, **9**, 31–46 (1986)

[BM2]   Bourgain, J., Milman, V.D.: New volume ratio properties for symmetric bodies in $\mathbb{R}^n$. Invent. Math., **88**, no. 2, 319–340 (1987)

[D]   Dilworth, S.J.: The dimension of Euclidean subspaces of quasi-normed spaces. Math. Proc. Camb. Phil. Soc., **97**, 311–320 (1985)

[GM]   Giannopoulos, A., Milman, V.D.: Euclidean structure in finite-dimensional normed spaces. In: Handbook of the Geometry of Banach Spaces, Vol. I. North Holland, Amsterdam, 707–779 (2001)

[Gl]   Gluskin, E.D.: The diameter of Minkowski compactum roughly equals to n. Funct. Anal. Appl., **15**, 57–58 (1981) (English translation)

[G]   Gordon, Y.: On Milman's inequality and random subspaces which escape through a mesh in $\mathbb{R}$. In: Geometric Aspects of Functional Analysis (1986/87), 84–106, Lecture Notes in Math., **1317**. Springer, Berlin-New York (1988)

[GK]   Gordon, Y., Kalton, N.J.: Local structure theory for quasi-normed spaces. Bull. Sci. Math., **118**, 441–453 (1994)

[KPR]   Kalton, N.J., Peck, N.T., Roberts, J.W.: An F-space sampler. London Mathematical Society Lecture Note Series, **89**. Cambridge University Press, Cambridge–New-York (1984)

[KT]   Kalton, N.J., Tam, S.: Factorization theorems for quasi-normed spaces. Houston J. Math., **19**, 301–317 (1993)

[Ko]   König, H.: Eigenvalue distribution of compact operators. Operator Theory: Advances and Applications, **16**. Birkhäuser Verlag, Basel-Boston, Mass. (1986)

[L]   Litvak, A.E.: On the constant in the reverse Brunn-Minkowski inequality for p-convex balls. Convex Geometric Analysis (Berkeley, CA, 1996), 129–137, Math. Sci. Res. Inst. Publ., **34**. Cambridge Univ. Press, Cambridge (1999)

[LMP]   Litvak, A.E., Milman, V.D., Pajor, A.: The covering numbers and "low M^*-estimate" for quasi-convex bodies. Proc. Amer. Math. Soc., **127**, no. 5, 1499–1507 (1999)

[LMS]   Litvak, A.E., Milman, V.D., Schechtman, G.: Averages of norms and quasi-norms. Math. Ann., **312**, 95–124 (1998)

[LMT]   Litvak, A.E, Milman, V.D., Tomczak-Jaegermann, N.: When random proportional subspaces are also random quotients. Preprint

[LT]   Litvak, A.E., Tomczak-Jaegermann, N.: Random aspects of high-dimensional convex bodies. GAFA, Lecture Notes in Math., **1745**, 169–190, Springer-Verlag (2000)

[MT]   Mankiewicz, P., Tomczak-Jaegermann, N.: Geometry of families of random projections of symmetric convex bodies. Geometric and Functional Analysis, **11**, 1282–1326 (2001)

[M1]   Milman, V.D.: Almost Euclidean quotient spaces of subspaces of a finite dimensional normed space. Proc. Amer. Math. Soc., **94**, 445–449 (1985)

[M2]   Milman, V.D.: Geometrical inequalities and mixed volumes in the local theory of Banach spaces. Asterisque, **131**, 373–400 (1985)

[M3]   Milman, V.D.: An inverse form of the Brunn-Minkowski inequality, with applications to the local theory of normed spaces. C.R. Acad. Sci. Paris Sr. I Math., **302**, no. 1, 25–28 (1986)

[M4]   Milman, V.D.: Isomorphic symmetrizations and geometric inequalities. Geometric Aspects of Functional Analysis (1986/87), 107–131, Lecture Notes in Math., **1317**. Springer, Berlin (1988)

[M5]   Milman, V.D.: Some applications of duality relations. Geometric Aspects of Functional Analysis (1989–90), 13–40, Lecture Notes in Math., **1469**. Springer, Berlin (1991)

[M6]   Milman, V.D.: Isomorphic Euclidean regularization of quasi-norms in $\mathbb{R}$. C.R. Acad. Sci. Paris, **321**, 879–884 (1996)

[MPa]   Milman, V.D., Pajor, A.: Entropy and asymptotic geometry of non-symmetric convex bodies. Adv. Math., **152**, no. 2, 314–335 (2000); see also Entropy methods in asymptotic convex geometry. C.R. Acad. Sci. Paris, Ser. I Math., **329**, no. 4, 303–308 (1999)

[MPi]   Milman, V.D., Pisier, G.: Banach spaces with a weak cotype 2 property. Israel J. Math., **54**, no. 2, 139–158 (1986)

[MS1]   Milman, V.D., Schechtman, G.: Asymptotic theory of finite-dimensional normed spaces. With an appendix by M. Gromov. Lecture Notes in Mathematics, **1200**. Springer-Verlag, Berlin (1986)

[MS2]   Milman, V.D., Schechtman, G.: Global versus local asymptotic theories of finite-dimensional normed spaces. Duke Math. J., **90**, no. 1, 73–93 (1997)

[PT]   Pajor, A., Tomczak-Jaegermann, N.: Subspaces of small codimension of finite-dimensional Banach spaces. Proc. Amer. Math. Soc., **97**, no. 4, 637–642 (1986)

[Pi]   Pisier, G.: The Volume of Convex Bodies and Banach Space Geometry. Cambridge University Press, Cambridge (1989)

[Ro]   Rolewicz, S.: Metric linear spaces. Monografie Matematyczne, Tom. 56. [Mathematical Monographs, Vol. 56] PWN-Polish Scientific Publishers, Warsaw (1972)

[Ru]   Rudelson, M.: Distances between non–symmetric convex bodies and the MM^*-estimate. Positivity, **4**, no. 2, 161–178 (2000)

[Sz1]   Szarek, S.J.: On Kasin's almost Euclidean orthogonal decomposition of ℓ_1. Bull. Acad. Polon. Sci., **26**, 691–694 (1978)

[Sz2]   Szarek, S.J.: Spaces with large distance to ℓ_∞^n and random matrices. Amer. J. Math., **112**, 899–942 (1990)

[T]   Tomczak-Jaegermann, N.: Banach–Mazur distances and finite-dimensional operator ideals. Pitman Monographs and Surveys in Pure and Applied Mathematics, **38**. Longman Scientific & Technical, Harlow; co-published in the United States with John Wiley & Sons, Inc., New York (1989)

Almost Euclidean Subspaces of Real ℓ_p^n with p an Even Integer

Yu. I. Lyubich

Department of Mathematics, Technion, Haifa 32000, Israel
lyubich@techunix.technion.ac.il

Summary. An asymptotic lower bound for the distance between ℓ_2^m and m-dimensional subspaces of ℓ_p^n is obtained for $p \in 2\mathbf{N}$, $p \to \infty$ and $n \le \binom{m+p/2-1}{m-1}$. In the case $m = 2$, $p/2 - n = o(\sqrt{p})$, this bound is exact up to a factor asymptotically lying between 1 and 2.

1 Introduction

In the family of n-dimensional real Banach spaces ℓ_p^n, $1 \le p \le \infty$, only the spaces with even integer p may have an Euclidean subspace of a given dimension $m \ge 2$, see [L]. Moreover, for any $p \in 2\mathbf{N}$ and for any m such a subspace does exist for $n \ge N(m,p)$, where

$$\binom{m + \frac{p}{2} - 1}{m - 1} \le N(m,p) \le \binom{m + p - 1}{m - 1}, \tag{1.1}$$

see [LV], [R]. Earlier the existence result with the upper bound which exceeds (1.1) by 1 was obtained in [M]. In fact, even the upper bound (1.1) is never attained [DJP]. In contrast, the lower bound (1.1) can be attained. For example, $N(2,p) = \frac{p}{2} + 1$, $N(3,4) = 6$, etc., see [LV], [R]. However, $N(3,6) = 11$ according to [R], while the lower bound (1.1) equals 10 in this case.

For $n < N(m,p)$ it is interesting to consider "almost Euclidean" subspaces, see [LV] for $(m,p,n) = (3,6,10)$ and $(3,8,15)$. On the other hand, the famous Dvoretzky Theorem [Dv] asserts that for every $\varepsilon > 0$ and every m there exists an ε-Euclidean m-dimensional subspace E of every n-dimensional Banach space X as long as $n \ge \nu(m,\varepsilon)$. Recall that a Banach space E is said to be ε-Euclidean if there exists an inner product $(\ ,\ )$ in E such that

$$\sqrt{(x,x)} \le \|x\| \le (1+\varepsilon)\sqrt{(x,x)}, \quad x \in E. \tag{1.2}$$

The value $1 + \varepsilon$ with the minimal possible ε is nothing but the Banach-Mazur distance $\mathrm{dist}\,(l_2^m, E)$.

It is known that

$$\nu(m,\varepsilon) \le e^{\frac{cm}{\varepsilon^2}} \tag{1.3}$$

with $\varepsilon < \frac{1}{2}$ and an absolute constant $c > 0$, see [G], [S]. Dvoretzky's phenomenon in the family ℓ_p^n takes place for $n \ge \nu_p(m,\varepsilon)$ with $\nu_p(m,\varepsilon) \le \nu(m,\varepsilon)$. For a fixed $p > 2$ and $m \to \infty$ the dimension $\nu_p(m,\varepsilon)$ is asymptotically $\le C(\varepsilon)m^{p/2}$ [F]. The order $p/2$ is exact in this setting since, according to [BDGJN], $\mathrm{dist}\,(l_2^m, l_p^n) \to \infty$ if $n = o(m^{p/2})$.

It was proven in [BoL] that if p is not an even integer then

$$c(m,p)\left(\frac{1}{\varepsilon}\right)^{\frac{2m-2}{m+2p}} \le \nu_p(m,\varepsilon) \le \tilde{c}(m,p)\left(\frac{1}{\varepsilon}\right)^{\frac{2m-2}{m+2p}} |\log \varepsilon|^{m+2} \tag{1.4}$$

with some coefficients $c(p,m) > 0$ and $\tilde{c}(p,m) < \infty$. The situation $p \in 2\mathbf{N}$ and "small" n, i.e. $n < N(m,p)$, is not considered in [BoL]. In the present paper we investigate this complementary case. More precisely, for an ε-Euclidean m-dimensional subspaces of ℓ_p^n with $p \in 2\mathbf{N}$, $p \ge 4$, and with

$$m \le n < \binom{m + \frac{p}{2} - 1}{m - 1} \tag{1.5}$$

we estimate the minimal possible $\varepsilon = \varepsilon(m,p,n)$ from below. Thus, we obtain a lower bound for the Banach-Mazur distance from ℓ_2^m to the m-dimensional subspaces of ℓ_p^n. This yields an explicit asymptotic inequality which depends on m and p only, $p \to \infty$. (Note that the dependence on p in (1.4) is implicit.) For $m = 2$ the lower bound can be sharpened and, on the other hand, there is an upper bound which is asymptotically double the lower one.

Our technique is an "approximative" version of the method based on the relations between isometric embeddings and cubature formulas, cf. [LV], [R]. We combine it with the use of spherical harmonics which is typical for investigations concerning Dvoretzky's Theorem. All information on the spherical harmonics we need can be found in [Mu].

Beyond (1.5), i.e. for

$$\binom{m + \frac{p}{2} - 1}{m - 1} \le n < N(m,p),$$

our technique does not work and the problem of an estimate of $\mathrm{dist}\,(\ell_2^m, E)$ $(E \subset \ell_p^n)$ from below remains open.

2 A General Scheme

Let us assume that (1.2) is valid on an m-dimensional subspace $E \subset \ell_p^n$ provided with an inner product $(\ ,\ )$. Let $\{e_k\}_1^n$ be the canonical basis of ℓ_p^n and let $x \in E$. In the decomposition

$$x = \sum_{k=1}^{n} \xi_k e_k$$

the coordinates ξ_k are linear functionals of x. Therefore, $\xi_k = (x, u_k)$ where u_k are uniquely determined vectors from E, $1 \le k \le n$. The inequality (1.2) can be rewritten as

$$(x, x)^{\frac{p}{2}} \le \sum_{k=1}^{n} (x, u_k)^p \le (1 + \varepsilon)^p (x, x)^{\frac{p}{2}},$$

i.e.

$$\left| \sum_{k=1}^{n} (x, u_k)^p - \left(1 + \frac{\delta}{2} \right) (x, x)^{\frac{p}{2}} \right| \le \frac{\delta}{2} (x, x)^{\frac{p}{2}} \tag{2.1}$$

where

$$\delta = (1 + \varepsilon)^p - 1, \quad \varepsilon = (1 + \delta)^{\frac{1}{p}} - 1. \tag{2.2}$$

Further, similarly to the isometric case ($\varepsilon = 0$), we use Hilbert's formula

$$(x, x)^{\frac{p}{2}} = \gamma_{m,p} \int (x, y)^p \, d\sigma(y) \tag{2.3}$$

where σ is the normalized Lebesgue measure on the Euclidean unit sphere $S \subset E$ and

$$\gamma_{m,p} = \frac{\Gamma(\frac{p+m}{2})\Gamma(\frac{1}{2})}{\Gamma(\frac{m}{2})\Gamma(\frac{p+1}{2})} = \prod_{k=1}^{\frac{p}{2}} \frac{m + p - 2k}{p - 2k + 1}. \tag{2.4}$$

On the other hand,

$$\sum_{k=1}^{n} (x, u_k)^p = \int (x, y)^p \, d\rho(y)$$

where ρ is the measure on S concentrated at those points $\hat{u}_k = u_k / \|u_k\|$ which correspond to $u_k \ne 0$, $\rho(\{u_k\}) = \|u_k\|^p$. Setting

$$\tau = \rho - \left(1 + \frac{\delta}{2} \right) \gamma_{m,p} \sigma \tag{2.5}$$

we rewrite (2.1) as

$$\left| \int_S (x, y)^p \, d\tau(y) \right| \le \frac{\delta}{2} (x, x)^{\frac{p}{2}}. \tag{2.6}$$

Let $V(m, p)$ be the linear space of all complex-valued forms (homogeneous polynomials) of degree p on E. Its dimension is

$$d = \dim V(m, p) = \binom{m + p - 1}{m - 1} \tag{2.7}$$

since the canonical basis in $V(m, p)$ consists of monomials of degree p in m variables. The standard inner product in $V(m, p)$ is

$$\langle \varphi, \psi \rangle = \int \varphi \bar{\psi} \, d\sigma.$$

From now on the letters $x, y, z, \ldots$ denote points on S, and the functions from $V(m, p)$ are restricted to S if otherwise not noted. Such a restriction is not substantial because of homogeneity.

Lemma 2.1. (cf. [LV], [R]) *The family of functions $\varphi_x(y) = (x, y)^p$ generates the linear space $V(m, p)$.*

Proof. Let $\{v_k\}_1^m$ be an orthonormal basis in E and let

$$x = \sum_{k=1}^{m} \alpha_k v_k, \quad y = \sum_{k=1}^{m} \beta_k v_k.$$

Then

$$(x, y)^p = \left(\sum_{k=1}^{m} \alpha_k \beta_k \right)^p = \sum_I \frac{p!}{I!} \alpha^I \beta^I \tag{2.8}$$

where $\alpha = (\alpha_1 \ldots, \alpha_m)$, $I = (i_1, \ldots, i_m)$ with $i_k \geq 0$, $\sum i_k = m$, and, as usual, $\alpha^I = \alpha_1^{i_1} \cdot \ldots \cdot \alpha_m^{i_m}$, $I! = i_1! \ldots i_m!$. Suppose that a function $\psi \in V(m, p)$ is orthogonal to all φ_x, i.e.

$$\int (x, y)^p \, \overline{\psi(y)} \, d\sigma(y) = 0.$$

Then by (2.8)

$$\sum_I \frac{p!}{I!} \alpha^I \int \beta^I \bar{\psi} \, d\sigma = 0. \tag{2.9}$$

Since the left hand side of (2.9) is a homogeneous polynomial of α, we have

$$\int \beta^I \bar{\psi} \, d\sigma = 0$$

for all of I. Hence, $\psi = 0$ since the monomials β^I form a basis in $V(m, p)$. □

Now let us introduce an orthonormal basis $\{s_i\}_0^{d-1}$ in $V(m, p)$ and consider the expansion

$$(x, y)^p = \sum_{i,k=0}^{d-1} \bar{\lambda}_{ik} s_i(x) s_k(y) \tag{2.10}$$

with

$$\lambda_{ik} = \iint (x, y)^p \, s_i(x) s_k(y) \, d\sigma(x) \, d\sigma(y). \tag{2.11}$$

Lemma 2.2. *The matrix* $\Lambda = (\lambda_{ik})$ *is invertible.*

Proof. Suppose that

$$\sum_{k=0}^{d-1} \lambda_{ik}\eta_k = 0, \quad 0 \le i \le d-1$$

with some η_k, $0 \le k \le d-1$. By substitution of λ_{ik} from (2.11) we obtain

$$\int \left(\int (x,y)^p \sum_{k=0}^{d-1} \eta_k s_k(y)\, d\sigma(y) \right) s_i(x)\, d\sigma(x) = 0, \quad 0 \le i \le d-1.$$

Hence,

$$\int (x,y)^p \sum_{k=0}^{d-1} \eta_k s_k(y)\, d\sigma(y) = 0.$$

By Lemma 2.1

$$\sum_{k=0}^{d-1} \eta_k s_k(y) = 0$$

that yields $\eta_k = 0$, $0 \le k \le d-1$. $\qquad\square$

In order to derive a lower bound for ε (or, equivalently, for δ, according to (2.2)) we note that (1.5) means

$$m \le n < \dim V\left(m, \frac{p}{2}\right).$$

Therefore, the system of linear homogeneous equations

$$\theta(\hat{u}_k) = 0, \quad 1 \le k \le n, \tag{2.12}$$

has a nontrivial solution $\theta \in V(m, \frac{p}{2})$. The function $\psi = |\theta|^2$ belongs to $V(m,p)$ and $\psi \ge 0$. In addition, one can assume that

$$\int \psi\, d\sigma = 1. \tag{2.13}$$

Then (2.5) yields

$$\int \psi\, d\tau = -\left(1 + \frac{\delta}{2}\right)\gamma_{m,p} \tag{2.14}$$

since $\psi|\mathrm{supp}\rho = 0$. The right hand side of (2.14) can be expanded as follows.

Lemma 2.3. *The formula*

$$\left(1 + \frac{\delta}{2}\right)\gamma_{m,p} = -\sum_{k,i=0}^{d-1} \mu_{ki} \int \psi(z)s_k(z)\, d\sigma(z) \int \overline{s_i(x)}\, d\sigma(x) \int (x,y)^p\, d\tau(y),$$

$$\tag{2.15}$$

holds.

Proof. By decomposition of ψ for the orthonormal basis $\{s_k\}_0^{d-1}$ the equality (2.14) can be rewritten as

$$\left(1 + \frac{\delta}{2}\right)\gamma_{m,p} = -\sum_{k=0}^{d-1} \int s_k(y)\,d\tau(y)\int \psi(z)\overline{s_k(z)}\,d\sigma(z). \tag{2.16}$$

In order to determine the integrals

$$\int s_k(y)\,d\tau(y), \quad 0 \le k \le d-1, \tag{2.17}$$

we return to (2.10) and get

$$\int (x,y)^p\,d\tau(y) = \sum_{i,k=0}^{d-1} \bar\lambda_{ik}s_i(x)\int s_k(y)\,d\tau(y).$$

Further,

$$\iint (x,y)^p\,\overline{s_i(x)}\,d\sigma(x)\,d\tau(y) = \sum_{k=0}^{d-1} \bar\lambda_{ik}\int s_k(y)\,d\tau(y), \quad 0 \le i \le d-1.$$

This is a system of linear equations with the unknowns (2.17). Lemma 2.2 allows us to solve this system. Namely, if (μ_{ki}) are the entries of the inverse matrix $\bar\Lambda^{-1}$ then

$$\int s_k(y)\,d\tau(y) = \sum_{i=0}^{d-1} \mu_{ki}\iint (x,y)^p\,\overline{s_i(x)}\,d\sigma(x)\,d\tau(y), \quad 0 \le k \le d-1. \tag{2.18}$$

The result follows by substitution from (2.18) into (2.16). □

The inequality (2.6) allows us to estimate the right hand side of formula (2.15). Since ψ is non-negative and satisfies (2.13) we obtain

$$\left(1 + \frac{\delta}{2}\right)\gamma_{m,p} \le \frac{\delta}{2}\sum_{k,i=0}^{d-1} |\mu_{ki}|M_k L_i \tag{2.19}$$

where

$$M_k = \max\{|s_k(z)| : z \in S\}, \quad L_i = \int |s_i(x)|\,d\sigma(x). \tag{2.20}$$

By (2.2) the inequality (2.19) results in

Theorem 2.4. *Let $p \in 2\mathbf{N}$ and let (1.5) be fulfilled. For any m-dimensional subspace $E \subset \ell_p^n$ the inequality*

$$\operatorname{dist}(\ell_2^m, E) \ge \left(\frac{1+\chi}{1-\chi}\right)^{\frac{1}{p}} \tag{2.21}$$

holds with $\chi = \chi_{m,p}$ *where*

$$\chi_{m,p} = \frac{\gamma_{m,p}}{\displaystyle\sum_{k,i=0}^{d-1} |\mu_{ki}| M_k L_i} < 1. \tag{2.22}$$

Certainly, the quantity $\chi_{m,p}$ depends on the choice of an orthonormal basis $\{s_k\}_0^{d-1}$ in $V(m,p)$. In addition to the inequality (2.22) we have

$$\chi_{m,p} \geq \frac{\gamma_{m,p}}{\displaystyle\sum_{k,i=0}^{d-1} |\mu_{ki}| M_k M_i} \tag{2.23}$$

since $L_i \leq M_i$ for all i.

3 Asymptotic Lower Bound for dist $\left(\ell_2^m, E\right)$

Here we apply out general scheme to the following asymptotical situation.

Theorem 3.1. *Let* $p \in 2\mathbf{N}$ *and let* (1.5) *be fulfilled. For any* m-*dimensional subspace* $E \subset \ell_p^n$ *the inequality*

$$\operatorname{dist}\left(\ell_2^m, E\right) \geq 1 + \omega_{m,p} \tag{3.1}$$

where asymptotically

$$\omega_{m,p} \sim \frac{c_m}{2^p p^{\frac{3m-5}{2}}}, \quad (p \to \infty) \tag{3.2}$$

with

$$c_m = \frac{(m-2)!^{\frac{3}{2}} \sqrt{\pi}}{3 \cdot 2^{\frac{m-2}{2}}}. \tag{3.3}$$

Proof. The standard orthonormal basis in $V(m,p)$ consists of the spherical harmonics $Y_{kl}(x)$, $k = 0, 2, \ldots, p$, $1 \leq l \leq d_{mk}$, where

$$d_{mk} = \binom{m+k-1}{m-1} - \binom{m+k-3}{m-1} = \binom{m+k-3}{m-3}\frac{2k+m-2}{m-2}. \tag{3.4}$$

Then the Funk–Hecke formula

$$\int (x,y)^p \, Y_{kl}(y) \, d\sigma(y) = \lambda_{m,p,k} Y_{kl}(x) \tag{3.5}$$

is valid with

$$\lambda_{m,p,k} = \frac{\Gamma(\frac{m}{2})}{\Gamma(\frac{1}{2})\Gamma(\frac{m-1}{2})} \int_{-1}^{1} t^p P_k^{\frac{m-2}{2}}(t)(1-t^2)^{\frac{m-3}{2}} \, dt. \tag{3.6}$$

Here

$$P_k^{\frac{m-2}{2}}(t) = \frac{(-1)^k}{2^k} \prod_{j=0}^{k-1}\left(\frac{m-1}{2}+j\right)^{-1}(1-t^2)^{-\frac{m-3}{2}}\frac{d^k}{dt^k}(1-t^2)^{\frac{m-3}{2}} \qquad (3.7)$$

is a Gegenbauer polynomial normalized as

$$P_k^{\frac{m-2}{2}}(1) = 1.$$

By substitution from (3.7) into (3.6) we obtain

$$\lambda_{m,p,k} = \frac{(-1)^k \Gamma(\frac{m}{2})}{2^k \Gamma(\frac{1}{2})\Gamma(\frac{m-1}{2}+k)}\int_{-1}^{1} t^p \frac{d^k}{dt^k}(1-t^2)^{\frac{m-3}{2}}\, dt$$

and then integration by parts yields

$$\lambda_{m,p,k} = \frac{\Gamma(\frac{m}{2})}{2^k \Gamma(\frac{1}{2})\Gamma(\frac{m-1}{2}+k)}\prod_{j=0}^{k-1}(p-j)\int_{-1}^{1} t^{p-k}(1-t^2)^{\frac{m-3}{2}+k}\, dt.$$

The latter integral is equal to

$$2\int_{-1}^{1} t^{p-k}(1-t^2)^{\frac{m-3}{2}+k}\, dt = \int_0^1 s^{\frac{p-k-1}{2}}(1-s)^{\frac{m-3}{2}+k}ds$$

$$= \frac{\Gamma(\frac{p-k+1}{2})\Gamma(\frac{m-1}{2}+k)}{\Gamma(\frac{m+p+k}{2})}$$

since $p-k$ is even. Therefore,

$$\lambda_{m,p,k} = \frac{\Gamma(\frac{m}{2})\Gamma(\frac{p-k+1}{2})}{2^k \Gamma(\frac{1}{2})\Gamma(\frac{m+p+k}{2})}\prod_{j=0}^{k-1}(p-j)$$

$$= \frac{\Gamma(\frac{m}{2})}{2^k \Gamma(\frac{m+p+k}{2})}\prod_{j=0}^{k-1}\left(\frac{p-k-1}{2}-j\right)\prod_{j=0}^{k-1}(p-j).$$

Finally,

$$\lambda_{m,p,k} = \frac{\Gamma(\frac{m}{2})p!}{2^p \Gamma(\frac{m+p+k}{2})\left(\frac{p-k}{2}\right)!}. \qquad (3.8)$$

According to (2.11) and (3.5)

$$\Lambda = \mathrm{diag}(\lambda_{m,p,0}; \underbrace{\lambda_{m,p,2},\ldots,\lambda_{m,p,2}}_{d_{m,2}};\ldots;\underbrace{\lambda_{m,p,p},\ldots,\lambda_{m,p,p}}_{d_{m,p}}) \qquad (3.9)$$

$(d_{m,0} = 1$ by (3.4)).

Now we estimate the quantity $\chi_{m,p}$ defined by (2.22). This is the fraction with the numerator $\gamma_{m,p}$ and the denominator

$$K_{m,p} = \sum_{j=0}^{p/2} \lambda_{m,p,2j}^{-1} \sum_{l=1}^{d_{m,2j}} M_{2j,l} L_{2j,l} \tag{3.10}$$

where

$$M_{2j,l} = \max\{|Y_{2j,l}(x)| : x \in S\}, \quad L_{2j,l} = \int |Y_{2j,l}(x)| \, d\sigma(x). \tag{3.11}$$

Since $Y_{2j,l}$ are L_2-normalized, we have

$$M_{2j,l} \le \sqrt{d_{m,2j}}, \quad L_{2j,l} \le 1. \tag{3.12}$$

(The former bound (3.12) follows from the Schwartz inequality; for the latter see [BoL, p.14].) Hence,

$$K_{m,p} \le \sum_{j=0}^{p/2} \lambda_{m,p,2j}^{-1} d_{m,2j}^{\frac{3}{2}}. \tag{3.13}$$

The formulae (3.4) and (3.8) show that $d_{m,k}$ increases with k but $\lambda_{m,p,k}$ decreases. Hence,

$$d_{m,2j} \le d_{mp}, \quad \lambda_{m,p,2j}^{-1} \le \lambda_{m,p,p}^{-1}.$$

Moreover, if $j \le \frac{p}{2} - 1$ then

$$\frac{\lambda_{m,p,2j}^{-1}}{\lambda_{m,p,p}^{-1}} \le \frac{\lambda_{m,p,p-2}^{-1}}{\lambda_{m,p,p}^{-1}} = \frac{\Gamma(\frac{m}{2} + p - 1)}{\Gamma(\frac{m}{2} + p)} = \frac{1}{\frac{m}{2} + p - 1} \le \frac{1}{p}.$$

Therefore,

$$K_{m,p} \le \frac{3}{2} \lambda_{m,p,p}^{-1} (d_{mp})^{\frac{3}{2}} \le \frac{3}{2} \cdot \frac{2^p \Gamma(\frac{m}{2} + p)(m + p - 3)!^{\frac{3}{2}}(2p + m - 2)^{\frac{3}{2}}}{\Gamma(\frac{m}{2})(m - 2)!^{\frac{3}{2}} p!^{\frac{5}{2}}}. \tag{3.14}$$

For $p \to \infty$ the expression on the right hand side of (3.14) asymptotically equals

$$\frac{3\sqrt{2}}{\Gamma(\frac{m}{2})(m - 2)!^{\frac{3}{2}}} \cdot 2^p \cdot p^{2m-4}. \tag{3.15}$$

In addition,

$$\gamma_{m,p} \sim \frac{\sqrt{\pi}}{\Gamma(\frac{m}{2})} \left(\frac{p}{2}\right)^{\frac{m-1}{2}}, \tag{3.16}$$

according to (2.4). Combining (2.21), (2.22) and (3.14)-(3.16) we get what we need. □

4 The 2-Dimensional Case

This case merits special attention because the lower bound (3.1) can be sharpened for $m = 2$ so that the final result turns out to be very close to an upper bound. A preliminary statement of this kind was announced in [LS].

It is convenient to consider a 2-dimensional Euclidean subspace $E \subset \ell_p^n$ as the complex plane $\mathbf{C}$ provided with the standard inner product

$$(z, w) = \operatorname{Re}(z\bar{w}) = \frac{z\bar{w} + \bar{z}w}{2}. \tag{4.1}$$

Thus,

$$S = \{z : |z| = 1\}, \quad d\sigma(z) = \frac{dz}{2\pi i z}.$$

An orthonormal basis in $V(2, p)$ is

$$s_k(z) = z^{p-k}\bar{z}^k, \quad 0 \le k \le p.$$

For $z \in S$ we have $\bar{z} = z^{-1}$ so,

$$s_k(z) = z^{p-2k}, \quad 0 \le k \le p. \tag{4.2}$$

Accordingly,

$$(z, w)^p = \frac{1}{2^p} \sum_{k=0}^p \binom{p}{k} s_k(z)\overline{s_k(w)}. \tag{4.3}$$

Comparing (4.3) to (2.10) we see that the matrix Λ turns out to be diagonal with

$$\lambda_{kk} = \frac{1}{2^p}\binom{p}{k}, \quad 0 \le k \le p.$$

Then, the diagonal entries of $\bar{\Lambda}^{-1}$ are

$$\mu_{kk} = 2^p \binom{p}{k}^{-1}, \quad 0 \le k \le p.$$

Now we could directly refer to Theorem 2.4 with $m = 2$ and, accordingly, $2 \le n \le \frac{p}{2}$. However, it is better to return to the formula (2.15) in order to estimate its right hand side more carefully.

We have

$$\left(1 + \frac{\delta}{2}\right)\gamma_{2,p} =$$

$$- 2^p \sum_{k=0}^p \binom{p}{k}^{-1} \int \psi(z)\overline{s_k(z)}\, d\sigma(z) \int \overline{s_k(u)}\, d\sigma(u) \int (u, v)^p\, d\tau(v), \tag{4.4}$$

with a function ψ satisfying (2.13) and (2.14). Since

$$\left| \int (u,v)^p \, d\tau(v) \right| \le \frac{\delta}{2},$$

(4.5)

it follows from (4.4) that

$$(2+\delta)\gamma_{2,p} \le 2^p \delta \sum_{k=0}^{p} \binom{p}{k}^{-1} |\psi_k|$$

(4.6)

where ψ_k are the Fourier coefficients of ψ,

$$\psi_k = \int \psi(z) \overline{s_k(z)} \, d\sigma(z), \quad 0 \le k \le p,$$

(4.7)

so that

$$\psi(z) = \sum_{k=0}^{p} \psi_k z^{p-2k}, \quad z \in S.$$

(4.8)

We can take $\psi = |\theta|^2$ where

$$\theta(z) = \sum_{l=0}^{n} \theta_l z^{n-2l}$$

(4.9)

since $n \le p/2$ and, on the other hand, the interpolation condition (2.12) can be satisfied by some $\theta \ne 0$ of the form (4.9). By normalization,

$$\int |\theta|^2 \, d\sigma = 1,$$

(4.10)

so that (2.13) is fulfilled and then (2.14) is also true. Thus, our starting formula (4.4) is valid in spite of a difference between (4.9) and the choice of θ in the general scheme.

For $\psi = |\theta|^2$ the decompositions (4.8) and (4.9) can be compared. This yields $\psi_k = 0$ if $|k - \frac{p}{2}| > n$ and $\psi_{\frac{p}{2}-n} = \theta_0 \bar{\theta}_n$, $\psi_{\frac{p}{2}+n} = \theta_n \bar{\theta}_0$. Since $|\theta_0|^2 + |\theta_n|^2 \le 1$, we have $|\psi_{\frac{p}{2}\pm n}| \le 1/2$. In any case, $|\psi_k| \le 1$ because of (2.13), (4.7) and (4.2). Hence, the sum on the right hand side of (4.6) can be replaced by

$$\binom{p}{p/2-n}^{-1} + 2 \sum_{s=1}^{n-1} \binom{p}{p/2-s}^{-1} + \binom{p}{p/2}^{-1}$$

taking the symmetry of the binomial coefficients into account. Since

$$\gamma_{2,p} = 2^p \binom{p}{p/2}^{-1}$$

(4.11)

according to (2.4), we have the inequality

$$2\binom{p}{p/2}^{-1} \leq \delta\left\{\binom{p}{p/2-n}^{-1} + 2\sum_{s=1}^{n-1}\binom{p}{p/2-s}^{-1}\right\},$$

i.e.

$$\delta\left\{\binom{p}{r}^{-1} + 2\sum_{j=r+1}^{p/2-1}\binom{p}{j}\right\} \geq 2\binom{p}{p/2}^{-1}$$

where $r = p/2 - n$. It is easy to prove that

$$\sum_{j=r+1}^{p/2-1}\binom{p}{j}^{-1} \leq \frac{(r+1)(r+2)}{p-r}\binom{p}{r}^{-1}.$$

As a result, we obtain

Theorem 4.1. *Let $p \in 2\mathbf{N}$ and let $3 \leq n \leq p/2$. Then for any Euclidean plane $E \subset \ell_p^n$ the inequality*

$$\operatorname{dist}\left(\ell_2^2, E\right) \geq (1 + \delta_{p,n})^{\frac{1}{p}} \tag{4.12}$$

holds with

$$\delta_{p,n} = 2\left(1 + \frac{(r+1)(r+2)}{p-r}\right)^{-1}\binom{p}{r}\binom{p}{p/2}^{-1}, \quad r = n - \frac{p}{2}. \tag{4.13}$$

Asymptotically for $p \to \infty$ and $r = o(\sqrt{p})$ we have

$$\delta_{p,n} \sim 2\binom{p}{r}\binom{p}{p/2}^{-1} \sim \sqrt{2\pi p}\cdot\frac{p^r}{r!}\cdot\frac{1}{2^p} \tag{4.14}$$

and, accordingly,

$$\min_E \operatorname{dist}\left(\ell_2^2, E\right) = 1 + \theta_{p,n}\sqrt{\frac{2\pi}{p}}\cdot\frac{p^r}{r!}\cdot\frac{1}{2^p}, \quad \liminf \theta_{p,n} \geq 1. \tag{4.15}$$

Indeed, $\delta_{p,n} \to 0$, hence,

$$(1 + \delta_{p,n})^{\frac{1}{p}} - 1 \sim \frac{1}{p}\delta_{p,n}.$$

The lower bound (4.14) is very close to the following upper bound.

Theorem 4.2. *For $2 \leq n \leq \frac{p}{2}$ there exists an ε-Euclidean plane $E \subset \ell_p^n$ with $\varepsilon = \varepsilon_{p,n}$ where*

$$\varepsilon_{p,n} = \left(\frac{1 + \kappa_{p,n}}{1 - \kappa_{p,n}}\right)^{\frac{1}{p}} - 1, \quad \kappa_{p,n} = 2\left[\frac{p}{2n}\right]\binom{p}{r}\binom{p}{p/2}^{-1} \tag{4.16}$$

where $[\cdot]$ means the integral part.

Proof. Consider the **R**-linear embedding $f_{p,n} : \mathbf{C} \to \ell_p^n$ defined as

$$fz = \sum_{k=0}^{n-1} (z, \omega^k) e_{k+1}, \quad z \in \mathbf{C}, \tag{4.17}$$

where $\omega = e^{\frac{2\pi i}{p}}$ and $\{e_k\}_1^n$ is the canonical basis of ℓ_p^n. Then for $|z| = 1$ we have

$$\|fz\|^p = \sum_{k=0}^{n-1} (z, \omega^k)^p = \frac{1}{2^p} \sum_{k=0}^{n-1} (z\omega^{-k} + z^{-1}\omega^k)^p$$

$$= \frac{1}{2^p} \sum_{l=0}^{p} \binom{p}{l} z^{p-2l} \sum_{k=0}^{n-1} \omega^{(2l-p)k}.$$

The interior sum vanishes for $l \not\equiv p/2 \pmod{n}$, otherwise it equals n. Therefore,

$$A(\alpha) \equiv \|f_{p,n}(e^{i\alpha})\|^p$$

$$= \frac{n}{2^p} \left(\binom{p}{p/2} + 2\sum \left\{ \binom{p}{p/2 - nj} \cos 2nj\alpha : 1 \le j \le \left[\frac{p}{2n}\right] \right\} \right).$$

The plane $E = \mathrm{Im} f_{p,n} \subset \ell_p^n$ is $\tilde{\varepsilon}$-Euclidean where

$$\tilde{\varepsilon} = \left(\frac{\max A(\alpha)}{\min A(\alpha)} \right)^{\frac{1}{p}} - 1$$

and the inner product in E is transferred from $\mathbf{C}$ by $f_{p,n}$. Obviously,

$$\max A(\alpha) = \frac{n}{2^p} \left(\binom{p}{p/2} + 2\sum \left\{ \binom{p}{p/2 - n} : 1 \le j \le \left[\frac{p}{2n}\right] \right\} \right)$$

and

$$\min A(\alpha) \ge \frac{n}{2^p} \left(\binom{p}{p/2} - 2\sum \left\{ \binom{p}{p/2 - nj} : 1 \le j \le \left[\frac{p}{2n}\right] \right\} \right).$$

It remains to note that

$$\sum \left\{ \binom{p}{p/2 - nj} : 1 \le j \le \left[\frac{p}{2n}\right] \right\} \le \left[\frac{p}{2n}\right] \binom{p}{r}. \qquad \square$$

Corollary 4.3. *The factor $\theta_{p,n}$ in the asymptotic formula (4.15) satisfies*

$$\limsup \theta_{p,n} \le 2 \tag{4.18}$$

as $p \to \infty$, $r = o(\sqrt{p})$.

Remark 4.4. For $m = 2$ the right hand sides of (3.2) and (4.15) with $r = 0$ differ by a constant factor only. In (4.15) we have the coefficient $\sqrt{2\pi}$ instead of $c_2 = \frac{1}{3}\sqrt{\pi}$ in (3.2), see (3.3). An interesting problem is to introduce a parameter like r into a lower bound for the case $m \geq 3$.

The family of planes $E_{p,n}$ remains asymptotically Euclidean in a substantially wider range of r. For instance, one can derive the following result from the estimates above.

Theorem 4.5. *If* $p \to \infty$ *and* $n \geq \frac{1}{2}\sqrt{p \ln p}$ *then* dist $\left(\ell_2^2, E_{p,n}\right) \to 1$.

Acknowledgement. The author is grateful to G. Schechtman for pointing out the paper [BDGJN].

References

[BDGJN] Benette, A., Dor, L.E., Goodman, V., Johnson, W.B., Newman, C.M.: On uncomplemented subspaces of L_p, $1 < p < 2$. Israel J. Math., **26**, no. 2, 178–187 (1977)

[BoL] Bourgain, J., Lindenstrauss, J.: Almost Euclidean sections in spaces with a symmetric bases. Springer Lecture Notes in Math., **1376**, 278–288 (1989)

[DJP] Delbaen, F., Jarchow, H., Pelczynski, A.: Subspaces of L_p isometric to subspaces of ℓ_p. Positivity, **2**, 339-367 (1998)

[Dv] Dvoretzky, A.: Some results on convex bodies and Banach spaces. Proc. Symp. on Linear Spaces, Jerusalem, 123–160 (1961)

[F] Figiel, T., Lindenstrauss, J., Milman, V.D.: The dimension of almost spherical sections of convex bodies. Acta Math., **129**, 53–94 (1979)

[G] Gordon, Y.: Some inequalities for Gaussian processes and applications. Israel Journal of Mathematics, **50**, no. 4, 265–289 (1985)

[L] Lyubich, Yu.I.: On the boundary spectrum of contractions in Minkowsky spaces. Sib. Math. J., **11**, 271–279 (1970)

[LS] Lyubich, Yu.I., Shatalova, O.A.: Almost Euclidean planes in ℓ_p^n. Func. Anal. and its Appl., **32**, no. 1, 59–61 (1998)

[LV] Lyubich, Yu.I., Vaserstein, L.N.: Isometric embeddings between classical Banach spaces, cubature formulas and spherical designs. Geom. Dedicata, **47**, 327–362 (1993)

[M] Milman, V.D.: A few observations on the connections between local theory and some other fields. Springer Lecture Notes in Math., **1317**, 283–289 (1988)

[Mu] Müller, C.: Spherical harmonics. Lecture Notes in Math., **17**, 283–289, (1966)

[R] Reznick, B.: Sums of even powers of real linear forms. Memoirs AMS, **96** (1992)

[S] Schechtman, G.: A remark concerning the dependence on ε in Dvoretzky's theorem. Springer Lecture Notes in Math., **1376**, 278–288 (1989)

Geometric Parameters in Learning Theory

S. Mendelson

Research School of Information Sciences and Engineering, Australian National University, Canberra, ACT 0200, Australia shahar.mendelson@anu.edu.au

Contents

1 Introduction

In these notes we present some mathematics aspects of Statistical Learning Theory. By no means is this a complete survey which captures the variety of interesting mathematical problems encountered in Machine Learning, but rather, it focuses on one particular problem, called the *sample complexity* problem, which has a geometric flavor. Relevant manuscripts which offer a much broader exposition of Machine Learning (though not necessarily from the mathematical viewpoint) are, for example [AnB, DGL, DL, Lu, Va, Vi]. We also refer the reader to [BiM, BaBM] for results of a similar flavor to the ones we present here.

The surprising fact we wish to stress is that many topics in Machine Learning (and in particular, the sample complexity problem), which have been viewed as problems in Computer Sciences and Nonparametric Statistics, are connected to natural questions arising in the local theory of normed spaces. Our hope is that this article would encourage mathematicians to investigate these seemingly "applied" problems, which are, in fact, linked to interesting theoretical questions.

The starting point of our discussion is the formulation of a *learning problem*. Consider a space Ω endowed with an unknown probability measure μ, let G be a given class of real valued functions defined on Ω, and set T to be a real valued function. The aim of the learner is to approximate T in some sense using a function in G. Throughout these notes we assume that each $g \in G$ and T map Ω into $[0, 1]$. G is called the *base class* and T is referred to as the *target concept*.

The notion of approximation investigated here is with respect to the $L_p(\mu)$ norm (which cannot be computed because the underlying probability measure is unknown). In other words, given $\varepsilon > 0$, the learner attempts to construct a function $g_0 \in G$ which satisfies that

$$\|g_0 - T\|_{L_p(\mu)}^p \leq \inf_{g \in G} \|g - T\|_{L_p(\mu)}^p + \varepsilon. \tag{1.1}$$

The data the learner receives in order to construct the approximating function is a sample $s_n = \left(X_i, T(X_i)\right)_{i=1}^n$, where $(X_i)_{i=1}^n$ are independent random variables distributed according to μ.

The construction is based on the *learning rule*, which is a mapping that assigns to every sample s_n some $A_{s_n} \in G$. The effectiveness of the learning rule is measured by "how much data" it requires to produce an almost optimal function in the sense of (1.1).

The learning rule we focus on is the *empirical loss minimization*. For the sake of simplicity, assume that the $L_p(\mu)$ distance between T and G is attained at a point denoted by $P_G T$, which is an element in the metric projection of T onto G. Define a new function class which depends on G and T in the following manner; for every $g \in G$, let $\ell_p(g) = |g - T|^p - |P_G T - T|^p$ and set $\mathcal{L}_p(G, T) = \{\ell_p(g) | g \in G\}$. The class $\mathcal{L}_p(G, T)$ is called the *p-loss class* associated with G and T. Since

$$\mathbb{E}_\mu \ell_p(g) = \|g - T\|^p_{L_p(\mu)} - \inf_{h \in G} \|h - T\|^p_{L_p(\mu)},$$

it follows that given $\varepsilon > 0$, the learner's aim is to find some $g \in G$ for which $\mathbb{E}_\mu \ell_p(g) < \varepsilon$. Recall that this has to be obtained without apriori knowledge on the measure μ, and thus, with no knowledge on the particular L_p structure one is interested in.

The empirical loss minimization algorithm assigns to every sample a function in G which is "almost optimal" on the data. For every sample $(X_1, ..., X_n)$ and $\varepsilon > 0$, let $g^* \in G$ be any function which satisfies that

$$\frac{1}{n} \sum_{i=1}^n \left(g^*(X_i) - T(X_i)\right)^p \le \inf_{g \in G} \frac{1}{n} \sum_{i=1}^n \left(g(X_i) - T(X_i)\right)^p + \frac{\varepsilon}{2}. \qquad (1.2)$$

Thus, any g^* is an "almost minimizer" of the *empirical L_p distance* between members of G and the target T.

From the definition of the loss class it is evident that if μ_n is the empirical measure $n^{-1} \sum_{i=1}^n \delta_{X_i}$, then $\mathbb{E}_{\mu_n} \ell_p(g^*) \le \varepsilon/2$, since the second term in every loss function is the same - $|T - P_G T|^p$, and thus the infimum is determined only by the first term $|g - T|^p$. Hence,

$$\mathbb{E}_{\mu_n} \ell_p(g^*) \le \inf_{f \in \mathcal{L}_p(G,T)} \mathbb{E}_{\mu_n} f + \frac{\varepsilon}{2} \le \frac{\varepsilon}{2},$$

because $\inf_{f \in \mathcal{L}(G,T)} \mathbb{E}_{\mu_n} f \le 0$, simply by looking at $f = \ell_p(P_G T)$. Therefore, the empirical minimization algorithm is effective if, for any $\varepsilon > 0$, the fact that $\mathbb{E}_{\mu_n} \ell_p(g) < \varepsilon/2$ implies that $\mathbb{E}_\mu \ell_p(g) < \varepsilon$.

Formally, we tackle the following

Question 1.1. Given a class of functions G, what are the conditions which ensure that for every $\varepsilon > 0$, $0 < \delta < 1$, there exists n_0 such that for every $n \ge n_0$, every target T and any probability measure μ, the following holds. If $(X_i)_{i=1}^n$ are independent and distributed according to μ and g^* is a function which satisfies (1.2), then $Pr\{\mathbb{E}_\mu \ell_p(g^*) \ge \varepsilon\} \le \delta$?

Note that we require an estimate which holds uniformly with respect to all probability measures and all possible targets, because both are unknown. In what follows we shall also be interested in results which are measure dependent.

The reason for the requirement that $\mathbb{E}_\mu \ell_p(g^*)$ is "small" only with high probability, is because it is too much to hope for that any g^* which is almost optimal empirically will be almost optimal with respect to the original L_p structure. In fact, one can encounter arbitrarily large samples which give misleading information on the behavior of the target. The hope is that an affirmative answer can be established with a relatively high probability as the size of the sample increases. The quantitative tradeoff between the desired accuracy ε, the high probability required and the size of the sample is called the *sample complexity problem*.

There are several learning scenarios which are common and which were investigated in the literature. All of them are variations on the theme presented above. The first and simplest case is called proper learning, in which the target concept T belongs to the class. Improper function learning is when T is a function which does not necessarily belong to the given class. A more general case is called *agnostic learning* in which both the target and μ are represented by a probability measure ν on $\Omega \times [0,1]$. In this case, the given data points are $(X_i, Y_i)_{i=1}^n \subset \Omega \times [0,1]$, where $(X_i, Y_i)_{i=1}^n$ are independent random variables distributed according to ν, and the goal is to minimize $\mathbb{E}_\nu |g(x) - y|^p$. The reason for the more general setup is mainly to handle the possibility of noisy observations, which is the standard situation in real-world problems. The analysis we present holds in the agnostic setup with essentially the same proofs. We present the results in the improper case only for the sake of simplicity.

Historically, Learning Theory originated from pattern recognition problems (sometimes referred to as "classification problems"), in which both the base class and the target are assumed to be binary valued, or more generally, to have a finite range. The main tools used in the analysis of these problems were combinatorial in nature (e.g. the Vapnik–Chervonenkis dimension [VaC]).

In the late 80s the interest in real-valued problems increased, but even then the approach to the sample complexity problem was based on generalized versions of the same tools used in the binary case. In recent years, due to influences from Mathematical Statistics, more modern tools (such as Rademacher and gaussian processes) were introduced in the context of the sample complexity problem. The goal of these notes is to survey these developments from a geometric viewpoint.

In the analysis, we estimate the sample complexity of learning problems using geometric parameters associated with the given class, but we do not attempt to estimate these parameters for particular classes. A more extensive survey would have to contain such estimates, which exist in some cases (see, for example, [AnB]) but are unknown in many others.

Let us mention some easy observations regarding Question 1.1. Note that any attempt to approximate T with respect to any measure other than the one according to which the sampling is made will not be successful. As an extreme example, if one has two probability measures which are supported on disjoint subsets of Ω, any data obtained by sampling according to one measure will be meaningless when computing distances with respect to the other. Thus, among the various L_p structures possible, the notion of $L_p(\mu)$ approximation is the best one can hope for.

Another observation is that if the class G is "too large" it would be impossible to construct any worthwhile approximating function using empirical data. As an example, assume that G consists of all the measurable functions $g : [0,1] \to [0,1]$, that $T : [0,1] \to [0,1]$ and that μ is the Lebesgue measure on $[0,1]$. Clearly, there are functions which agree with T completely on any finite sample, but are very far apart with respect to the L_p norm.

The remark above demonstrates a general phenomenon; if G is too large, then for an arbitrarily large sample there are still "too many" very different functions in the class which behave in a similar way to (or even coincide with) the target on the sample. Thus, if one wants an effective learning scheme, additional empirical data (i.e. a larger sample) should decrease the number of class members which are "close" to the target on the sample. Therefore, a question which emerges is which is the right notion of the "size" of a class?

The approach we present is an outcome of this line of reasoning. Firstly, the easiest notion of size is the ability to approximate means by empirical means. To that end, assume that one can ensure that when the sample size is "large enough", then for *any* probability measure μ there is a set of "high probability" on which empirical means of members of $\mathcal{L}_p$ are uniformly close to their actual means (that is, if $n > n_0$, with high probability every $f \in \mathcal{L}_p$ satisfies that $|\mathbb{E}_\mu f - \mathbb{E}_{\mu_n} f| < \varepsilon/2$). In particular, since $\mathbb{E}_{\mu_n} \ell_p(g^*) < \varepsilon/2$ then $\mathbb{E}_\mu \ell_p(g^*) < \varepsilon$ as required. This leads to the definition of *uniform Glivenko–Cantelli* classes.

Definition 1.2. *Let F be a class of functions. We say that F is a uniform Glivenko–Cantelli class (uGC class) if for every $\varepsilon > 0$,*

$$\lim_{n \to \infty} \sup_\mu Pr\left\{ \sup_{f \in F} \left| \mathbb{E}_\mu f - \frac{1}{n} \sum_{i=1}^n f(X_i) \right| \geq \varepsilon \right\} = 0,$$

where $(X_i)_{i=1}^\infty$ are independent random variables distributed according to μ.

The uGC condition is simply a uniform version of the law of large numbers, where the uniformity is with respect to all members of F and all probability measures μ.

This definition is natural in the context of Learning Theory because the supremum is taken with respect to all probability measures, and this is essential since the learner does not have apriori information on the probability measure according to which the data is sampled.

Let us mention that if F is uncountable, there is a measurability issue that needs to be addressed, and which will be completely ignored in these notes. It can be resolved by imposing mild conditions on F (see [Du3] for more details).

The definition of the uGC condition has a quantified version. For every $0 < \varepsilon, \delta < 1$, let $S_F(\varepsilon, \delta)$ be the first integer n_0 such that for every $n \geq n_0$ and any probability measure μ,

$$Pr\left\{ \sup_{f \in F} |\mathbb{E}_\mu f - \mathbb{E}_{\mu_n} f| \geq \varepsilon \right\} \leq \delta, \tag{1.3}$$

where μ_n is the random empirical measure $n^{-1} \sum_{i=1}^n \delta_{X_i}$.

$S_F(\varepsilon, \delta)$ is called the *Glivenko–Cantelli sample complexity* of the class F with accuracy ε and confidence δ.

One can show [ABCH] that if G consists of uniformly bounded functions then it can be used to approximate *any* target T in the agnostic version of

(1.1) (and thus uniformly with respect to the measure) if and only if G is a uniform Glivenko–Cantelli class. Therefore, uniform Glivenko–Cantelli classes are the natural base classes that should be investigated, and from the qualitative viewpoint, the uGC property for the base class captures the correct notion of size. However, from the quantitative point of view, this result is not satisfactory. Firstly, exact estimates on the uGC sample complexity are essential, and should depend on a more fine-grained notion of size. Moreover, the uniform Glivenko–Cantelli condition is not necessarily the optimal approach to the sample complexity problem. Indeed, the uGC condition provides the ability to control the deviation between empirical means and the actual mean of *every* function within the class. This is a very strong property, and is only a (loose!) condition which suffices to ensure that g^* is a "good approximation" of T. In fact, it is enough to show that empirical means are close to the actual means for functions which are produced by the learning algorithm, i.e., functions for which $\mathbb{E}_{\mu_n} f < \varepsilon$.

Formally, for every $\varepsilon > 0$, one would like to estimate

$$\sup_T \sup_\mu Pr\{\exists f \in \mathcal{L}_p(G,T),\ \mathbb{E}_{\mu_n} f < \varepsilon, \mathbb{E}_\mu f \geq 2\varepsilon\}, \qquad (1.4)$$

and define the *learning sample complexity* as the first integer n_0 such that for every $n \geq n_0$ the term in (1.4) is smaller than δ. For such a value of n, there is a set of large probability on which any function which is an "almost minimizer" of the empirical loss will be an "almost minimizer" of the actual loss, regardless of the underlying probability measure. Clearly, this would imply that the empirical minimization algorithm was successful.

Let us remark that an estimate on

$$Pr\{\exists f \in F,\ \mathbb{E}_{\mu_n} f < \varepsilon,\ \mathbb{E}_\mu f \geq 2\varepsilon\} \qquad (1.5)$$

has a geometric interpretation. Indeed, fix a probability measure μ and assume that G is a $\sqrt{\varepsilon}$ separated class in $L_2(\mu)$. For every $(x_1, ..., x_n)$, let μ_n be the empirical measure supported on the set, and let $\tilde{G} = \{(g(x_1), ..., g(x_n)) \mid g \in G\}$ be the projection of G onto the coordinates $(x_1, ..., x_n)$, endowed with the $L_2(\mu_n)$ norm. It is natural to ask when such a coordinate projection of G is also separated in $L_2(\mu_n)$ at a scale proportional to $\sqrt{\varepsilon}$. This is exactly determined by (1.5) - if $F = \{(g - h)^2 \mid g \neq h,\ g, h \in G\}$, then (1.5) is the probability that for a random set $(X_1, ..., X_n)$ there is some $f \in F$ for which $n^{-1} \sum_{i=1}^n f(x_i) < \varepsilon$, although one knows that $\mathbb{E}_\mu f \geq 2\varepsilon$.

The analysis of the sample complexity problem is presented as follows. In the next section we present an outline of the classical and modern probabilistic approaches used to estimate the Glivenko–Cantelli sample complexity. These approaches lead to the introduction of various notions of size, such as the uniform entropy, the combinatorial dimensions and the Rademacher averages, all of which are investigated in section 3. In section 4 we explore the parameters that govern the learning sample complexity needed to improve the bounds obtained via the Glivenko–Cantelli sample complexity. In particular, we show that the learning sample complexity of an arbitrary class can be reduced to "Glivenko–Cantelli" tail estimates for a class of functions with "small variance", an observation which leads to sharper complexity estimates. We also present a slightly different approach to the sample complexity problem, known as *error bounds* and compare the empirical and actual structures endowed on the class by μ_n and μ respectively using the notion of *isomorphic coordinate projections*, leading to even sharper bounds. Finally, we present a detailed example of a family of particularly interesting function classes, called *Kernel classes*.

We end this introduction with some notation. We denote the class in question by F, which should be thought of as a loss class associated with a base class G and a target T, although all the results we present hold for general classes of functions which are uniformly bounded. All the constants we use are denoted by c, C or K. Their value may change from line to line, or even within the same line. C_p denoted a constant which depends only on p. We write $a \sim b$ if there are absolute constants c and C such that $cb \leq a \leq Cb$. Finally, if X is a Banach space, we denote its unit ball by B_X or $B(X)$.

2 Glivenko–Cantelli Classes and Learnability

In this section we investigate two approaches to the problem of bounding the tails of the random variable $\sup_{f \in F} |\mathbb{E}_\mu f - n^{-1} \sum_{i=1}^n f(X_i)|$. The first approach was the basis to the best known sample complexity estimates up to the mid 90s. We then present a more modern result which is based on a sharper concentration inequality.

2.1 The Classical Approach

Our starting point is a symmetrization argument which originated in the works of Kahane [K] and Hoffman–Jørgensen [Ho1, Ho2] (see also [GiZ1]).

Theorem 2.1. *Let* $F \subset B(L_\infty(\Omega))$. *Then, for any* $\varepsilon > 0$ *and every* $n \geq 8/\varepsilon^2$,

$$Pr\left\{ \sup_{f \in F} \left| \mathbb{E}_\mu f - \frac{1}{n} \sum_{i=1}^n f(X_i) \right| > \varepsilon \right\} \leq 4 Pr\left\{ \sup_{f \in F} \left| \sum_{i=1}^n \varepsilon_i f(X_i) \right| > \frac{n\varepsilon}{4} \right\}, \quad (2.1)$$

where $(\varepsilon_i)_{i=1}^n$ *are independent Rademacher random variables (that is, symmetric and* $\{-1, 1\}$*-valued), and the probability on the right-hand side is with respect to the product measure.*

Having the symmetrization idea in mind, the course of action one can take is clear: restricted to any sample $(X_1, ..., X_n)$, F is projected onto a random set $V \subset \mathbb{R}^n$. To control the possibly infinite set V, one uses the notion of covering.

If (Y, d) is a metric space and $F \subset Y$ then for every $\varepsilon > 0$, $N(\varepsilon, F, d)$ is the minimal number of open balls (with respect to the metric d) needed to cover F, and the set of the centers of the balls is called an ε-cover of F. For obvious reasons, we are interested in metrics endowed by samples. Given a sample $(X_1, ..., X_n)$ let μ_n be the empirical measure supported on that sample, and for $1 \le p < \infty$, denote by $N(\varepsilon, F, L_p(\mu_n))$ the covering numbers of F at scale ε with respect to the $L_p(\mu_n)$ norm.

To bound the right-hand side of (2.1), fix a sample $(X_1, ..., X_n)$ which corresponds to fixing a coordinate projection, and set $\mu_n = n^{-1} \sum_{i=1}^n \delta_{X_i}$. It is clear that given a fixed projection one can replace F with an $\varepsilon/8$ cover of F with respect to the $L_1(\mu_n)$ norm. Bounding the probability of the supremum over the cover by the sum of probabilities, all that remains is to estimate $Pr\{|\sum_{i=1}^n \varepsilon_i v_i| \ge t\}$ for a fixed vector $(v_i)_{i=1}^n \in \mathbb{R}^n$, which is the coordinate projection of an element of the cover of F with respect to the $L_1(\mu_n)$ norm. By Hoeffding's inequality [VW],

$$Pr\left\{\left|\sum_{i=1}^n \varepsilon_i v_i\right| > t\right\} \le 2e^{-t^2/2\|v\|_2},$$

where $\|v\|_2 = \left(\sum_{i=1}^n v_i^2\right)^{1/2}$.

Finally, taking the expectation with respect to μ, the following is evident:

Theorem 2.2. *Let $F \subset B(L_\infty(\Omega))$ and set μ to be a probability measure on Ω. Let $(X_i)_{i=1}^\infty$ be independent random variables distributed according to μ. For every $\varepsilon > 0$ and any $n \ge 8/\varepsilon^2$,*

$$Pr\left\{\sup_{f \in F}\left|\mathbb{E}_\mu f - \frac{1}{n}\sum_{i=1}^n f(X_i)\right| > \varepsilon\right\} \le 8\mathbb{E}_\mu\left[N\left(\varepsilon/8, F, L_1(\mu_n)\right)\right]e^{-\frac{n\varepsilon^2}{128}},$$

where μ_n is the (random) empirical measure supported on $(X_1, ..., X_n)$.

If one is interested in measure independent bounds, it is possible to replace $\mathbb{E}_\mu N(\varepsilon, F, L_1(\mu_n))$ with the *uniform covering numbers*, which is the first notion of size we investigate.

Definition 2.3. *For every class F, $1 \le p \le \infty$ and $\varepsilon > 0$, let*

$$N_p(\varepsilon, F, n) = \sup_{\nu_n} N(\varepsilon, F, L_p(\nu_n)),$$

and

$$N_p(\varepsilon, F) = \sup_n \sup_{\nu_n} N(\varepsilon, F, L_p(\nu_n)),$$

where ν_n is an average of n point-mass measures.

The logarithm of the covering numbers is called the metric entropy of the set. $\log N_p(\varepsilon, F)$ is the uniform L_p entropy of F.

Corollary 2.4. *If* $F \subset B(L_\infty(\Omega))$ *then for any probability measure* μ,

$$Pr\left\{\sup_{f\in F}\left|\mathbb{E}_\mu f - \frac{1}{n}\sum_{i=1}^{n}f(X_i)\right| > \varepsilon\right\} \leq 8N_1(\varepsilon/8, F, n)e^{-\frac{n\varepsilon^2}{128}}.$$

In particular, if F *is such that* $\log N_1(\varepsilon, F) = O(\varepsilon^{-p})$ *for some* $0 < p < \infty$, *then the uGC sample complexity satisfies that* $S_F(\varepsilon, \delta) = O(\varepsilon^{-(2+p)})$ *up to logarithmic factors in* $1/\varepsilon$ *and* $1/\delta$.

As an example, consider $F = \mathcal{L}_q(G, T)$, and observe that the entropy of F is controlled by that of G. Indeed, since each $g \in G$ maps Ω into $[0,1]$, then for every $g_1, g_2 \in G$ and every $\omega \in \Omega$, $|\ell_q(g_1) - \ell_q(g_2)|(\omega) \leq q|g_1 - g_2|(\omega)$. In particular, if $\log N_1(\varepsilon, G) = O(\varepsilon^{-p})$, then for every target T, $S_{\mathcal{L}_q(G,T)}(\varepsilon, \delta) \leq C_{p,q}\varepsilon^{-(2+p)}$ up to logarithmic factors in $1/\varepsilon$ and $1/\delta$.

Although these tail bounds will be shown to be loose, the uniform entropy does characterize uniform Glivenko–Cantelli classes. This line of investigation - connecting the metric entropy to the uniform Glivenko–Cantelli property originated in the work of Vapnik and Chervonenkis [VaC], and then extended by various authors.

Theorem 2.5. [DuGZ] *A class* $F \subset B(L_\infty(\Omega))$ *is a uniform Glivenko–Cantelli class if and only if the following holds. There is some* $1 \leq p \leq \infty$ *such that for every* $\varepsilon > 0$

$$\lim_{n\to\infty} \frac{\log N_p(\varepsilon, F, n)}{n} = 0.$$

The fact that uniform entropy can be used to characterize uniform Glivenko–Cantelli classes (and thus, classes in which any target can be approximated) demonstrates its significance. Unfortunately, in most cases it is not easy to estimate the uniform entropy directly. It turns out that one can control the uniform entropy of the class using combinatorial dimensions which measure the "richness" of the class (such as the VC dimension) and those are sometimes easier to estimate (see [AnB] for some examples). These *uniform measures of complexity* will be discussed in the next section.

It is clear that the main source of possible looseness in the uGC sample complexity bounds is due to the weak concentration result used in the proof. In order to obtain sharper bounds one has to find better ways of controlling the supremum of the random variable $\sup_{f\in F}|\mathbb{E}_\mu f - n^{-1}\sum_{i=1}^{n}f(X_i)|$.

2.2 Talagrand's Inequality for Empirical Processes

Let us begin by recalling Bernstein's inequality in a form analogous to Talagrand's inequality below [M, VW].

Theorem 2.6. *Let* μ *be a probability measure on* Ω *and set* $X_1, ..., X_n$ *to be independent random variables distributed according to* μ. *Given a function* $f : \Omega \to \mathbb{R}$, *set* $Z = \sum_{i=1}^{n}f(X_i)$, *let* $b = \|f\|_\infty$ *and put* $u = n\mathbb{E}_\mu f^2$. *Then,*

$$Pr\{|Z - \mathbb{E}_\mu Z| \geq x\} \leq 2e^{-\frac{x^2}{2(u+bx/3)}}.$$

This exponential inequality is sharper than Hoeffding's inequality when one has additional information on the variance of the random variable Z. It has been a long standing question whether a similar result can be obtained when Z is replaced by $\sup_{f\in F}|\sum_{i=1}^n (f(X_i) - \mathbb{E}_\mu f)|$. Such an inequality was first established by Talagrand [Ta2] and later modified by Ledoux [L], Massart [M], Rio [R] and Bousquet [Bou] who currently has the best estimates on the constants appearing in the inequality. The following is Bousquet's version of Talagrand's inequality.

Theorem 2.7. [Bou] *Let F be a class of functions defined on a probability space (Ω, μ) such that $\sup_{f\in F}\|f\|_\infty \leq 1$. Let $(X_i)_{i=1}^n$ be independent random variables distributed according to μ, put $\sigma^2 \geq \sup_{f\in F} \mathrm{Var}[f(X_1)]$ and set $Z = \sup_{f\in F}|\sum_{i=1}^n (f(X_i) - \mathbb{E}_\mu f)|$. Then, for every $x > 0$,*

$$Pr\{Z \geq \mathbb{E}Z + x\} \leq \exp\left(-vh\left(\frac{x}{v}\right)\right),$$

where $v = n\sigma^2 + 2\mathbb{E}Z$ and $h(x) = (1+x)\log(1+x) - x$. Moreover, for every $x > 0$,

$$Pr\left\{Z \geq \mathbb{E}Z + \sqrt{2xv} + \frac{x}{3}\right\} \leq e^{-x}. \tag{2.2}$$

Hence, if F consists of functions which are bounded by 1, then by selecting $x = n\varepsilon^2/4$ it follows that with probability larger than $1 - e^{-\frac{n\varepsilon^2}{4}}$,

$$\sup_{f\in F}\left|\mathbb{E}_\mu f - \frac{1}{n}\sum_{i=1}^n f(X_i)\right| \leq 2\mathbb{E}\sup_{f\in F}\left|\mathbb{E}_\mu f - \frac{1}{n}\sum_{i=1}^n f(X_i)\right| + \frac{3\varepsilon}{4}.$$

The dominating term in (2.2) is the expectation of the random variable Z, and this expectation $\mathbb{E}Z$ can be estimated by a symmetrization argument.

Definition 2.8. *Let μ be a probability measure on Ω and set F to be a class of uniformly bounded functions on Ω. For every integer n, let*

$$R_n(F, \mu) = \frac{1}{\sqrt{n}}\mathbb{E}\sup_{f\in F}\left|\sum_{i=1}^n \varepsilon_i f(X_i)\right| = \mathbb{E}_\mu \mathbb{E}_\varepsilon \frac{1}{\sqrt{n}}\sup_{f\in F}\left|\sum_{i=1}^n \varepsilon_i f(X_i)\right|,$$

where $(X_i)_{i=1}^n$ are independent random variables distributed according to μ and $(\varepsilon_i)_{i=1}^n$ are independent Rademacher random variables (which are also independent of $X_1, ..., X_n$). $R_n(F, \mu)$ are the (global) Rademacher averages associated with the class F and the measure μ.

Proposition 2.9. *Let μ be a probability measure and set $F \subset B(L_\infty(\Omega))$. Denote*

$$H = \sup_{f \in F} \left| \mathbb{E}_\mu f - \frac{1}{n} \sum_{i=1}^{n} f(X_i) \right| = \frac{Z}{n},$$

where $(X_i)_{i=1}^n$ are independent random variables distributed according to μ. Then,

$$\mathbb{E}_\mu H \le 2 \frac{R_n(F,\mu)}{\sqrt{n}} \le 4 \mathbb{E}_\mu H + 2 \left| \sup_{f \in F} \mathbb{E}_\mu f \right| \cdot \mathbb{E}_\varepsilon \left| \frac{1}{n} \sum_{i=1}^{n} \varepsilon_i \right| = 4 \mathbb{E}_\mu H + O\left(\frac{1}{\sqrt{n}} \right).$$

Combining this with Talagrand's inequality yields sharp estimates on the uGC sample complexity.

Theorem 2.10. *Let μ be a probability measure on Ω, set $F \subset B\big(L_\infty(\Omega)\big)$ and put $\sigma^2 = \sup_{f \in F} \mathrm{var}(f(X_1))$. There is an absolute constant $C \ge 1$ such that for every $x > 0$, there is a set of probability larger than $1 - e^{-x}$ on which*

$$\sup_{f \in F} \left| \mathbb{E}_\mu f - \frac{1}{n} \sum_{i=1}^{n} f(X_i) \right| \le \frac{4 R_n(F,\mu)}{\sqrt{n}} + C \left(\sigma \sqrt{\frac{x}{n}} + \frac{x}{n} \right). \qquad (2.3)$$

In particular, there is an absolute constant C such that if

$$n \ge \frac{C}{\varepsilon^2} \max \left\{ R_n^2(F,\mu), \log \frac{1}{\delta} \right\},$$

then $Pr \left\{ \sup_{f \in F} \left| \mathbb{E}_\mu f - \frac{1}{n} \sum_{i=1}^{n} f(X_i) \right| \ge \varepsilon \right\} \le \delta$.

We will show in the following sections that although the Glivenko–Cantelli sample complexity is governed by the global Rademacher averages, the learning sample complexity can be controlled by several local versions, one of which is

$$\frac{1}{\sqrt{n}} \mathbb{E}_{\mu \times \varepsilon} \sup_{f \in F_r} \left| \sum_{i=1}^{n} \varepsilon_i f(X_i) \right|,$$

where $F_r = \{h = \lambda f \mid f \in F, \ 0 \le \lambda \le 1, \ \mathbb{E}_\mu h^2 \le r\}$. In other words, F_r is the intersection of the *star-shaped hull* of F and 0 with an $L_2(\mu)$ ball of radius $\sqrt{r}$.

It is evident that the averages associated with every fixed sample $(X_i)_{i=1}^n$ can be written as $\mathbb{E}_\varepsilon \| \sum_{i=1}^{n} \varepsilon_i e_i \|_{(F/s_n)^\circ}$, where

$$F/s_n = \left\{ \frac{1}{\sqrt{n}} \sum_{i=1}^{n} f(X_i) e_i \mid f \in F \right\} \subset \ell_2^n$$

is the natural coordinate projection of F onto ℓ_2^n endowed with the Euclidean structure of $L_2(\mu_n)$, and $(F/s_n)^\circ$ is the polar of F/s_n. This can be interpreted as a decoupling of $R_n(F,\mu)$: the "richness" associated with every sample is determined by the geometry of the coordinate projection F/s_n, and the complexity of the underlying measure μ is the "average size" (with respect to μ) of individual coordinate projections.

Sometimes it is useful to use the gaussian averages instead of the Rademacher ones, i.e,

$$\mathbb{E}_\mu \mathbb{E}_g \left\| \sum_{i=1}^n g_i e_i \right\|_{(F/s_n)^\circ} = \mathbb{E}_\mu \frac{1}{\sqrt{n}} \mathbb{E}_g \sup_{f \in F} \left| \sum_{i=1}^n g_i f(X_i) \right|,$$

where $(g_i)_{i=1}^n$ are independent standard gaussian random variables. The gaussian average associated with a fixed coordinate projection is simply the ℓ-norm of the polar body of the projected class.

Recall that for any index set T, the expectation of the supremum of the gaussian process indexed by T dominates the expectation of the supremum of the Rademacher process (up to an absolute multiplicative constant), and thus the gaussian averages may be used instead of the Rademacher ones to obtain upper bounds.

The reason for the normalizing factor of $1/\sqrt{n}$ in the definition of the random averages (instead of the more transparent $1/n$) becomes clearer in the gaussian case. If $T \subset \ell_2$, the isonormal process indexed by T is the gaussian process which has the covariance structure endowed by the inner product in ℓ_2, i.e., for every $t_1, t_2 \in T$, $\mathbb{E} X_{t_1} X_{t_2} = \langle t_1, t_2 \rangle$. Under the $1/\sqrt{n}$ normalization, the gaussian process indexed by F/s_n is the isonormal process indexed by F when considered as a subset of $L_2(\mu_n)$, since the covariance structure of the process is endowed by the inner product in $L_2(\mu_n)$. As such, all the usual tools of gaussian processes, and in particular, the Dudley–Sudakov inequalities may be applied, and the underlying metric entropy is the natural empirical L_2 entropy.

Note that Theorem 2.10 is measure dependent. Hence, to obtain uniform estimates one would require measure independent estimates on the Rademacher averages $R_n(F, \mu)$.

3 Uniform Measures of Complexity

As stated above, the sample complexity problem has two determining factors. The first is the probability measure according to which the samples are selected and the other is the geometry of coordinate projections of F. The probability measure determines which samples, or equivalently - which projections are encountered frequently and thus influence the "size" of the class with respect to that measure. To analyze the worst case scenario (which would hold for all probability measures) one has to estimate the size of the largest coordinate projection of F.

In this section we investigate various geometric parameters of the projected sets which are uniform with respect to the measure. In section 5 we present measure dependent bounds which can be computed using the given sample. The notions of size we focus on are the random averages, the combinatorial dimension and the metric entropy.

3.1 Metric Entropy and the Combinatorial Dimension

The interest in the metric entropy is two-fold. Firstly, the "classical" approach shows that the uniform metric entropy $N_p(\varepsilon, F)$ is essential in obtaining the (loose) complexity bound in Theorem 2.2. Secondly, the Rademacher/gaussian averages associated with the class (which lead to sharper bounds) can be estimated via Dudley's entropy integral. Indeed, for every empirical measure μ_n supported on the sample $s_n = (X_1, ..., X_n)$,

$$\frac{1}{\sqrt{n}} \mathbb{E}_g \sup_{f \in F} \left| \sum_{i=1}^{n} g_i f(X_i) \right| \leq C \int_0^1 \sqrt{\log N(\varepsilon, F, L_2(\mu_n))} d\varepsilon,$$

where C is an absolute constant.

Below, we present estimates on the uniform L_2 entropy of a class based on its *combinatorial dimension*.

Binary Valued Classes

Let F be a class of $\{0, 1\}$-valued functions. One possible way of measuring how "rich" a class is, is by identifying the largest dimension of a combinatorial cube that can be found in a coordinate projection of F. This combinatorial dimension was introduced by Vapnik and Chervonenkis and is known as the VC dimension.

Definition 3.1. *Let F be a class of $\{0, 1\}$-valued functions on a space Ω. We say that F shatters $\{x_1, ..., x_n\} \subset \Omega$, if for every $I \subset \{1, ..., n\}$ there is a function $f_I \in F$ for which $f_I(x_i) = 1$ if $i \in I$ and $f_I(x_i) = 0$ if $i \notin I$. Let*

$$\mathrm{vc}(F, \Omega) = \sup \left\{ |A| \mid A \subset \Omega, \ A \text{ is shattered by } F \right\}.$$

$\mathrm{vc}(F, \Omega)$ is called the VC dimension of F, but when the underlying space is clear we denote the Vapnik–Chervonenkis dimension by $\mathrm{vc}(F)$.

It is easy to see that $s_n = \{x_1, ..., x_n\}$ is shattered if and only if

$$\left\{ \big(f(x_1), ..., f(x_n)\big) \mid f \in F \right\} = \{0, 1\}^n.$$

For every sample σ denote by $P_\sigma F$ the coordinate projection of F, that is,

$$P_\sigma F = \left\{ \big(f(x_i)\big)_{x_i \in \sigma} \mid f \in F \right\},$$

and thus, the VC dimension is the largest cardinality of $\sigma \subset \Omega$ such that $P_\sigma F$ is the combinatorial cube of dimension $|\sigma|$.

The following lemma is known as the Sauer–Shelah Lemma ([S, Sh, VaC], see also [KaM] for a generalization of that result).

Lemma 3.2. *Let F be a class of binary valued functions and set $d = \mathrm{vc}(F)$. Then, for every finite subset $\sigma \subset \Omega$ of cardinality n,*

$$|P_\sigma F| \leq \sum_{i=0}^n \binom{n}{i} \leq \left(\frac{en}{d}\right)^d,$$

where the last inequality holds for $n \geq d$. In particular, for every $\varepsilon > 0$, $N\big(\varepsilon, F, L_\infty(\sigma)\big) \leq |P_\sigma F| \leq (en/d)^d$.

Using the Sauer–Shelah Lemma one can characterize the uniform Glivenko–Cantelli property of classes of binary valued functions in terms of the VC dimension.

Theorem 3.3. *A class of binary valued functions is a uniform Glivenko–Cantelli class if and only if it has a finite VC dimension.*

Proof. Assume that $\mathrm{vc}(F) = \infty$ and fix an integer $d \geq 2$. Then, there is a set $\sigma \subset \Omega$, $|\sigma| = d$ such that $P_\sigma F = \{0,1\}^d$, and let μ^σ be the uniform measure on σ (assigns a weight of $1/d$ to every point). For any $A \subset \sigma$ of cardinality $n \leq d/2$, let μ_n^A be the empirical measure supported on A. Since there is some $f_A \in F$ which is 1 on A and vanishes on $\sigma \backslash A$ then $|\mathbb{E}_{\mu^\sigma} f_A - \mathbb{E}_{\mu_n^A} f_A| = |1 - n/d| \geq 1/2$, and in particular, for any empirical measure μ_n supported on a subset of σ of cardinality smaller than $d/2$,

$$\sup_{f \in F} |\mathbb{E}_{\mu^\sigma} f - \mathbb{E}_{\mu_n} f| \geq \frac{1}{2}.$$

Hence, for any $n \leq d/2$,

$$\sup_\mu Pr\Big\{\sup_{f \in F} |\mathbb{E}_{\mu_n} f - \mathbb{E}_\mu f| \geq 1/2\Big\} = 1,$$

and since d can be made arbitrarily large, F is not a uGC class.

To prove the converse, recall that for every $0 < \varepsilon < 1$ and every empirical measure μ_n supported on a set σ of cardinality $n \geq d$, $N(\varepsilon, F, L_\infty(\sigma)) \leq |P_\sigma F| \leq (en/d)^d$. Since the empirical L_1 entropy is bounded by the empirical L_∞ one, $\log N_1(\varepsilon, F, n) \leq d \log(en/d)$. Thus, for every $\varepsilon > 0$, $\log N_1(\varepsilon, F, n) = o(n)$, showing that F is a uGC class. $\qquad\square$

Although the L_∞ entropy estimate depends on n and thus on the dimension of the coordinate projection, it is possible to derive dimension free L_p entropy bounds for $1 \leq p < \infty$. The first such bound was proved by Dudley [Du1] and is based on a combination of an extraction principle and the Sauer–Shelah Lemma. The extraction argument shows that if $K \subset F$ is "well separated" in $L_1(\mu_n)$ in the sense that every two points are different on a number of coordinated which is proportional to n, one can find a much smaller set of coordinates (whose size depends of the cardinality of K) onto which the coordinate projection of K is one-to-one.

Recall that a set is ε-separated with respect to a metric d if the distance between every two distinct points in the set is larger than ε. It is easy to see that the cardinality of a maximal ε-separated subset of F (denoted by $D(\varepsilon, F, d)$) is equivalent to the covering numbers of F, namely, for every $\varepsilon > 0$, $N(\varepsilon, F, d) \leq D(\varepsilon, F, d) \leq N(\varepsilon/2, F, d)$.

Theorem 3.4. *Let F be a class of binary valued functions and assume that* $\mathrm{vc}(F) = d$. *Then, for every $1 \leq p < \infty$ and every $0 < \varepsilon < 1$,*

$$N_p(\varepsilon, F) \leq \left(c_p \log \frac{2}{\varepsilon}\right)^d \left(\frac{1}{\varepsilon}\right)^{pd}.$$

Proof. Since the functions in F are $\{0,1\}$-valued it is enough to prove the claim for $p = 1$. The general case follows since for any $f, g \in F$ and any probability measure μ, $\|f - g\|_{L_p(\mu)}^p = \|f - g\|_{L_1(\mu)}$.

Fix $x_1, ..., x_n$ and $0 < \varepsilon < 1$, and let $\mu_n = n^{-1} \sum_{i=1}^{n} \delta_{x_i}$. Set K_ε to be any ε-separated subset of F with respect to the $L_1(\mu_n)$ norm and denote its cardinality by D.

Let $V = \{f_i - f_j | f_i \neq f_j \in K_\varepsilon\}$; thus, $|V| \leq D^2$ and since K_ε is ε-separated then every $v \in V$ has at least $n\varepsilon$ coordinates which belong to $\{-1, 1\}$.

Set $(X_i)_{i=1}^{t}$ to be independent $\{x_1, ..., x_n\}$-valued random variables, such that for every $1 \leq i \leq t$ and $1 \leq j \leq n$, $Pr(X_i = x_j) = 1/n$. For any $v \in V$, $Pr(\{\forall i, \ v(X_i) = 0\}) = \prod_{i=1}^{t} Pr(\{v(X_i) = 0\}) \leq (1 - \varepsilon)^t$, implying that

$$Pr\left(\{\exists v \in V, \ \forall i, \ v(X_i) = 0\}\right) \leq |V| \, (1 - \varepsilon)^t \leq D^2 (1 - \varepsilon)^t.$$

If $\sigma = (X_1, ..., X_t)$ then

$$Pr(\{P_\sigma K_\varepsilon \text{ is one} - \text{to} - \text{one}\}) \geq 1 - D^2 (1 - \varepsilon)^t,$$

and if the latter is greater than 0, there is a set $\sigma \subset \{1, ..., n\}$ such that $|\sigma| \leq t$ and

$$|P_\sigma K_\varepsilon| = \left|\{(f(x_i))_{i \in \sigma} \mid f \in K_\varepsilon\}\right| = D.$$

Selecting $t = \frac{2 \log D}{\varepsilon}$ suffices to ensure the existence of such a set σ. By the Sauer–Shelah Lemma,

$$D = |P_\sigma K_\varepsilon| \leq |P_\sigma F| \leq \left(\frac{e|\sigma|}{d}\right)^d \leq \left(\frac{2e \log D}{d\varepsilon}\right)^d. \tag{3.1}$$

To complete the proof, note that if $\alpha \geq 1$ and $\alpha \log^{-1} \alpha \leq \beta$ then $\alpha \leq \beta \log(e\beta \log \beta)$. By (3.1), $\log D \leq d \log(\frac{2e^2}{d\varepsilon} \log(\frac{2e}{d\varepsilon}))$, as claimed. $\square$

This result was strengthened by Haussler in [Hau] (see also [VW] for a different exposition).

Theorem 3.5. *There are constants C_p which satisfy that for every class F of binary valued functions with $\mathrm{vc}(F) = d$, any $1 \leq p < \infty$ and every $0 < \varepsilon < 1$,*

$$N_p(\varepsilon, F) \leq C_p d(4e)^d \varepsilon^{-pd}.$$

Real Valued Classes

Combinatorial parameters associated with real-valued classes can be obtained by extending the notion of a shattered set. Unlike the $\{0,1\}$ case, where every cube found in a coordinate projection of the class is the combinatorial cube of the appropriate dimension, in the real-valued case the size of the cube found in the projection is very important. The combinatorial dimension measures the tradeoff between the size of such a cube and its dimension.

Definition 3.6. *For every $\varepsilon > 0$, a set $\sigma = \{x_1, ..., x_n\} \subset \Omega$ is said to be ε-shattered by F if there is some function $s : \sigma \to \mathbb{R}$, such that for every $I \subset \{1, ..., n\}$ there is some $f_I \in F$ for which $f_I(x_i) \geq s(x_i) + \varepsilon$ if $i \in I$, and $f_I(x_i) \leq s(x_i) - \varepsilon$ if $i \notin I$. Let*

$$vc(F, \Omega, \varepsilon) = \sup \{|\sigma| \,|\, \sigma \subset \Omega, \ \sigma \text{ is } \varepsilon\text{-shattered by } F\}.$$

f_I is called the shattering function of the set I and the set $\{s(x_i)|x_i \in \sigma\}$ is called a witness to the ε-shattering. In cases where the underlying space is clear we denote the combinatorial dimension by $vc(F, \varepsilon)$.

Observe that if F is convex and symmetric and if $\sigma = \{x_1, ..., x_n\}$ is ε-shattered by F, then

$$\varepsilon B_\infty^n \subset \{(f(x_1), ..., f(x_n)) \,|\, f \in F\}.$$

Indeed, if σ is ε-shattered with a witness $(s(x_i))_{i=1}^n$ then for every $I \subset \{1, ..., n\}$ there is some f_I such that $f_I(x_i) \geq s(x_i) + \varepsilon$ if $i \in I$ and $f_I(x_i) \leq s(x_i) - \varepsilon$ for $i \in I^c$. For every such I, define $g_I = (f_I - f_{I^c})/2$ and since F is convex and symmetric, $g_I \in F$. It follows that $\{(f(x_1), ..., f(x_n)) \,|\, f \in F\} \supset \varepsilon B_\infty^n$, as claimed.

The first estimate on the empirical L_∞ covering numbers in terms of the combinatorial dimension was established in [ABCH] and was used to show that a class of uniformly bounded functions is a uGC class if and only if it has a finite combinatorial dimension for every ε. The proof that if F is a uGC class it has a finite combinatorial dimension for every ε follows from a similar argument to the one used in the binary valued case. For the converse, one requires empirical L_∞ entropy estimates combined with Theorem 2.2.

Other results on the L_∞ and L_p entropy estimates in terms of the combinatorial dimension were derived in [P, BL, Me2, Ta1, Ta3].

The following are the best known bounds on the uniform L_2 and L_∞ entropy in terms of the combinatorial dimension. The L_p estimates are from [MeV1] and the L_∞ ones can be found in [RuV].

Theorem 3.7. *There are constants K_p and c_p depending only on p such that the following holds. For every $F \subset B\big(L_\infty(\Omega)\big)$, every probability measure μ, every $1 \leq p < \infty$ and any $0 < \varepsilon < 1$,*

$$N\big(\varepsilon, F, L_p(\mu)\big) \leq \left(\frac{2}{\varepsilon}\right)^{K_p vc(F, c_p \varepsilon)}.$$

Moreover, for any $0 < \varepsilon, \delta < 1$ and every set $\sigma \subset \Omega$, $|\sigma| = n$,

$$\log N\big(\varepsilon, F, L_\infty(\sigma)\big) \leq K \cdot vc(F, c\varepsilon\delta) \log^{1+\delta} \left(\frac{n}{\delta\varepsilon}\right),$$

where K and c are absolute constants.

It is easy to show that the L_p bounds are optimal (up to the absolute constants) and that the optimal L_∞ bound (as a function of ε and n) is proportional to $vc(F, \varepsilon) \cdot \log(2n/\varepsilon)$.

By selecting empirical measures supported on the shattered sets, the uniform entropy can be bounded from below, demonstrating the sharpness of the upper bounds (up to a single power of a logarithm).

Theorem 3.8. *There are absolute constants K, K', c, c' such that for any $F \subset B\big(L_\infty(\Omega)\big)$ and every $0 < \varepsilon < 1$,*

$$K' \cdot vc(F, c'\varepsilon) \leq \log N_2(\varepsilon, F) \leq K \cdot vc(F, c\varepsilon) \log \left(\frac{2}{\varepsilon}\right).$$

3.2 Random Averages and the Combinatorial Dimension

In the past, the only known way to obtain complexity estimates was by bounding the uniform entropy via the combinatorial dimensions. Thus, much effort was put into the investigation of the combinatorial dimension of many interesting classes (see e.g. [AnB]).

It was noted in [Me2] that there are nontrivial connections between the combinatorial dimension and the Rademacher/gaussian averages associated with a class, and that the latter could be used to obtain sharper complexity bounds. To estimate the random averages using the combinatorial dimension, we introduce a uniform version of the gaussian averages. Let

$$\ell_n(F) \equiv \sup_{s_n} \ell(F/s_n) = \sup_{\{x_1, \ldots, x_n\} \subset \Omega} \frac{1}{\sqrt{n}} \mathbb{E} \sup_{f \in F} \left| \sum_{i=1}^{n} g_i f(x_i) \right|,$$

that is, the largest gaussian average associated with a coordinate projection on n points. Since $F \subset B\big(L_\infty(\Omega)\big)$ then $F/s_n \subset n^{-1/2} B_\infty^n$, and the largest projection one might encounter is when $F/s_n = n^{-1/2} B_\infty^n$, in which case $\ell(F/s_n) \sim \sqrt{n}$. We define a scale-sensitive parameter which measures for every $\varepsilon > 0$ the largest cardinality of a projection which has a "large" ℓ-norm. Let

$$t(F, \varepsilon) = \sup\{n \in \mathbb{N} | \ell_n(F) \geq \varepsilon\sqrt{n}\}.$$

$t(F, \varepsilon)$ was introduced in [Me2] where its connection to the uniform Glivenko–Cantelli condition and complexity estimates were investigated.

One can show that if $\ell(F/s_n) \sim \varepsilon\sqrt{n}$, there exists a set $\sigma \subset s_n$ of cardinality proportional to n, which is $c\varepsilon$ shattered by F, and thus the two "worst case" scale-sensitive parameters - the combinatorial dimension and $t(F, \varepsilon)$ are comparable.

Theorem 3.9. [MeV2] *There are absolute constants K and c such that for any $F \subset B(L_\infty(\Omega))$ and every $\varepsilon > 0$,*

$$\mathrm{vc}(F, c'\varepsilon) \le t(F, \varepsilon) \le K \frac{\mathrm{vc}(F, c\varepsilon)}{\varepsilon^2}.$$

This result can be used to estimate the combinatorial dimension of a convex hull of a class.

Corollary 3.10. [MeV2] *There are absolute constants K and c such that for any $F \subset B(L_\infty(\Omega))$ and every $\varepsilon > 0$,*

$$\mathrm{vc}(\mathrm{conv}(F), \varepsilon) \le \frac{K \cdot \mathrm{vc}(F, c\varepsilon)}{\varepsilon^2}.$$

Proof. Since the ℓ-norm of a set and of its convex hull are the same, then for any $\varepsilon > 0$, $t(F, \varepsilon) = t(\mathrm{conv}(F), \varepsilon)$. By Theorem 3.9,

$$\mathrm{vc}(\mathrm{conv}(F), \varepsilon) \le t(\mathrm{conv}(F), \varepsilon) = t(F, \varepsilon) \le \frac{K \mathrm{vc}(F, c\varepsilon)}{\varepsilon^2}. \qquad \square$$

It is possible to prove sharper estimates on the combinatorial dimension of convex hulls of F if one knows more on the structure of the class. For example, if F is a VC class of functions, then one has "global" information - the uniform entropy of the class at every scale ε. In the general case presented above, the information is "local" - the combinatorial dimension at a single scale, in which case it is possible to construct an example proving that the estimate in Corollary 3.10 is sharp. Indeed, in [MePR] it was shown that if $K_{n,N} \subset B_\infty^n$ is the symmetric convex hull of a random $\{0, 1\}$ polytope $F_{n,N}$ with N vertices, and if $N \ge cn^2$ then

$$\mathrm{vc}(K_{n,N}, \varepsilon) \sim \min \left\{ \frac{\log N\varepsilon^2}{\varepsilon^2}, n \right\}.$$

One can also show that $\mathrm{vc}(F_{n,N}) \sim \log N$ from which the sharpness of Corollary 3.10 follows.

3.3 Phase Transitions in GC Classes

The final remark in the previous section reveals a very significant (though not surprising) fact. If one is interested in scale-sensitive parameters (e.g. the uniform entropy) one can obtain much sharper bounds if one has some global information on the class. In this section we relate the combinatorial dimension to the random averages, given the combinatorial dimension at every scale. We focus on classes which have a power type p, that is, there exists a constant C such that for every $0 < \varepsilon < 1$, $\mathrm{vc}(F, \varepsilon) \le C\varepsilon^{-p}$.

Although it is natural to think that parameters which describe the complexity of uniform Glivenko–Cantelli classes behave smoothly in terms of the power type, there is a clear phase transition which occurs at $p = 2$. In some sense, if the class has power type p for $0 < p < 2$ it is "small", whereas if $p > 2$ the class is considerably larger. We will demonstrate this phenomenon in several examples.

The Uniform Central Limit Theorem

The first example of a phase transition can be seen in a uniform version of the Central Limit Theorem.

Definition 3.11. [Du3] *Let $F \subset B\big(L_\infty(\Omega)\big)$, set μ to be a probability measure on Ω and assume G_μ to be a gaussian process indexed by F, which has mean 0 and covariance*

$$\mathbb{E} G_\mu(f) G_\mu(g) = \int f g \, d\mu - \int f \, d\mu \int g \, d\mu.$$

F is called a universal Donsker class if for any probability measure μ, the law G_μ is tight in $\ell_\infty(F)$ and $\nu_n^\mu = n^{1/2}(\mu_n - \mu) \in \ell_\infty(F)$ converges in law to G_μ in $\ell_\infty(F)$, where μ_n a random empirical measure selected according to μ.

Stronger than the universal Donsker property is the uniform Donsker property. For such classes, ν_n^μ converges to G_μ uniformly in μ in some sense (see [Du3, VW] for more details). The following result of Giné and Zinn [GiZ2] is a relatively simple characterization of uniform Donsker classes.

For every probability measure μ on Ω, let $\rho_\mu^2(f,g) = \mathbb{E}_\mu(f - g)^2 - \big(\mathbb{E}_\mu(f - g)\big)^2$, and set for every $\delta > 0$, $F_\delta = \{f - g \,|\, f, g \in F, \ \rho_\mu(f,g) \le \delta\}$.

Theorem 3.12. [GiZ2] *A class F is a uniform Donsker class if and only if the following holds. For every probability measure μ on Ω, G_μ has a version with bounded, ρ_μ-uniformly continuous sample paths, and for these versions,*

$$\sup_\mu \mathbb{E} \sup_{f \in F} |G_\mu(f)| < \infty, \qquad \lim_{\delta \to 0} \sup_\mu \mathbb{E} \sup_{h \in F_\delta} |G_\mu(h)| = 0.$$

The main tool in the analysis of uniform Donsker classes is the Koltchinskii-Pollard entropy integral.

Theorem 3.13. [Du3] *If $F \subset B\big(L_\infty(\Omega)\big)$ satisfies that*

$$\int_0^\infty \sup_n \sup_{\mu_n} \sqrt{\log N\big(\varepsilon, F, L_2(\mu_n)\big)} \, d\varepsilon < \infty,$$

then it is a uniform Donsker class, where μ_n are empirical measures supported on at most n points.

On the other hand one can show [Du3] that if F is a uniform Donsker class then there is some constant C such that for every $\varepsilon > 0$,

$$\sup_{n} \sup_{\mu_n} \log N\left(\varepsilon, F, L_2(\mu_n)\right) \leq \frac{C}{\varepsilon^2}. \tag{3.2}$$

Therefore, the gap between the two estimates is at most logarithmic.

Corollary 3.14. [Me2] *Let* $F \subset B\left(L_\infty(\Omega)\right)$. *If there is some constant* C *such that* $\mathrm{vc}(F, \varepsilon) \leq C\varepsilon^{-p}$ *for* $0 < p < 2$ *then* F *is a uniform Donsker class. On the other hand, if* $\mathrm{vc}(F, \varepsilon) \geq c\varepsilon^{-p}$ *for* $p > 2$ *then* F *is not a uniform Donsker class.*

Proof. The first part of our claim follows from Lemma 3.7, since F satisfies the Koltchinskii-Pollard entropy condition. For the second part, recall that by Theorem 3.8, $N_2(F, \varepsilon) \geq C/\varepsilon^p$ for some $p > 2$. On the other hand, if F is a uniform Donsker class then $N_2(F, \varepsilon) = O(\varepsilon^{-2})$ which is a contradiction. □

Uniform ℓ-Norm Estimates

Another clear phase transition can be seen in the asymptotic behavior of $\ell_n(F)$.

Theorem 3.15. *Let* $F \subset B\left(L_\infty(\Omega)\right)$ *and assume that there is some* $\gamma > 0$ *such that for any* $\varepsilon > 0$, $\mathrm{vc}(F, \varepsilon) \leq \gamma\varepsilon^{-p}$. *Then, there are constants* C_p *which depend only on* p, *such that*

$$\ell_n(F) \leq \gamma^{1/2} C_p \begin{cases} 1 & if \quad 0 < p < 2, \\ \log^{3/2} n & if \quad p = 2, \\ n^{1/2-1/p} \log^{1/p} n & if \quad p > 2. \end{cases}$$

The proof is based on Dudley's entropy integral bound for $0 < p < 2$ and on a discrete version of that theorem for $p \geq 2$, combined with the uniform entropy estimate (see [Me2, Me6]). Note that in the case $p < 2$ and $p > 2$ the bound is tight up to a logarithmic factor. This is obvious for $p < 2$. As for $p > 2$, if $\mathrm{vc}(F, \varepsilon) = \gamma\varepsilon^{-p}$ then $t(F, \varepsilon) \geq K\gamma\varepsilon^{-p}$ for some absolute constant K. Hence, there is an integer $n \geq K\gamma\varepsilon^{-p}$ and a set s_n of cardinality n such that $\ell(F/s_n) \geq \varepsilon\sqrt{n} \geq (K\gamma)^{\frac{1}{p}} n^{\frac{1}{2}-\frac{1}{p}}$. The exact growth rate in the case $p = 2$ is not known.

Combining Theorem 3.15 with Theorem 2.10, we obtain the following estimate on the uGC sample complexity, which, as lower bounds show (see, e.g. [AnB]) is optimal rate wise, up to logarithmic factors.

Corollary 3.16. *Let* $F \subset B\left(L_\infty(\Omega)\right)$ *and assume that* $\mathrm{vc}(F, \varepsilon) \leq \gamma\varepsilon^{-p}$. *Then, there is a constant* $C_{p,\gamma}$ *such that*

$$S_F(\varepsilon, \delta) \leq C_{p,\gamma} \max\left\{\frac{1}{\varepsilon^p}, \frac{1}{\varepsilon^2} \log\frac{1}{\delta}\right\}$$

if $p \neq 2$. If $p = 2$ there is an additional logarithmic factor in $\frac{1}{\varepsilon}$.

In particular, if each $g \in G$ maps Ω into $[0,1]$ and $\mathrm{vc}(G, \varepsilon) \leq \gamma \varepsilon^{-p}$ for some $p \neq 2$, then

$$\sup_{T:\Omega \to [0,1]} S_{\mathcal{L}_q(G,T)}(\varepsilon, \delta) \leq C_{p,q,\gamma} \max \left\{ \frac{1}{\varepsilon^p}, \frac{1}{\varepsilon^2} \log \frac{1}{\delta} \right\}.$$

The proof of the latter part of the corollary follows from the fact that for every probability measure μ, $N(\varepsilon, \mathcal{L}_q(G, T), L_2(\mu)) \leq N(\varepsilon/q, G, L_2(\mu))$. Thus, the same asymptotic estimates on $\ell_n(G)$ hold for $\ell_n(\mathcal{L}_q(G, T))$, regardless of the specific selection of T.

Recall that the sample complexity bound obtained by the "classical" approach (Corollary 2.4) was $O(\varepsilon^{-(2+p)})$, which is considerably worse than the above result.

On the Size of Convex Hulls

Investigating the structure of convex hulls of classes has a natural origin in Machine Learning. In each learning procedure one has to solve, in one way or another, an optimization problem (locating an "almost" empirical minimizer). This is a reasonable task when the class is convex. Apriori, taking the convex hull of a nonconvex class might make it too complex for an efficient solution from the probabilistic point of view, in the sense that the sample complexity becomes huge. From the definition of the Glivenko–Cantelli condition it is clear that the uGC sample complexity of a class and of its convex hull are the same, but this is no longer true for the learning sample complexity, and thus it is important to at least obtain upper bounds on the way the combinatorial dimension, the metric entropy and the localized random averages grow when one takes the convex hull of F.

There are sharp estimates on the growth rate of the L_2 entropy of the convex hull of a set, given information on the growth rate of the entropy of the given set. These results are summarized in the next theorem, which is divided into two parts. The first is when the covering numbers of the class are polynomial in $1/\varepsilon$ while in the second they are exponential.

Theorem 3.17. *Let $F \subset B(L_\infty(\Omega))$ and set H to be the convex hull of F.*

1. *If there are $\gamma, p > 0$ such that $N(\varepsilon, F, L_2(\mu)) \leq \gamma \varepsilon^{-p}$ for every $\varepsilon > 0$, then there is an absolute constant C such that for every $\varepsilon > 0$,*

$$\log N(\varepsilon, H, L_2(\mu)) \leq C \gamma^{\frac{2}{p}} p \left(\frac{1}{\varepsilon}\right)^{\frac{2p}{2+p}}.$$

2. *If there are $\gamma > 0$ and p such that $\log N(\varepsilon, F, L_2(\mu)) \leq \gamma \varepsilon^{-p}$ for every $\varepsilon > 0$, then there is some constant $C_{p,\gamma}$ such that*

$$\log N\big(\varepsilon, H, L_2(\mu)\big) \leq C_{p,\gamma} \begin{cases} \varepsilon^{-2}\log^{1-2/p}(1/\varepsilon) & \text{if} \quad 0 < p < 2 \\ \varepsilon^{-2}\log^2(1/\varepsilon) & \text{if} \quad p = 2 \\ \varepsilon^{-p} & \text{if} \quad p > 2, \end{cases}$$

and all the estimates are sharp (up to the constants).

The fact that the rate of the entropy in part (1) is $\varepsilon^{-2p/(p+2)}$ was established by Dudley [Du2], and the constants are taken from [Me1]. Part (2) for $p \neq 2$ was proved in [CKP] (see [Me1] for a different proof), and for $p = 2$ in [G]. Another proof for the complete range $0 < p < \infty$ may be found in [BouKP].

As in the previous examples, the phase transition is apparent. If $0 < p < 2$ the entropy of the convex hull grows at a rate which is slightly better than $1/\varepsilon^2$. For "large" classes the power of the entropy does not change. Observe that there is a slight looseness in the analysis of Donsker classes using the entropy. If F is a Donsker class then its convex hull is also Donsker [Du3]. On the other hand, for any $2/3 < p < 2$ there is a probability measure μ and class $F \subset L_2(\mu)$ for which the $L_2(\mu)$ entropy is $O(\varepsilon^{-p})$, but the entropy of the convex hull could be of the order of $\varepsilon^{-2}\log^{1-2/p}(1/\varepsilon)$. In particular, the entropy integral diverges.

As a corollary to Theorem 3.17, we improve the estimates on the combinatorial dimension of convex hulls. For example, let F be a binary valued class with $\mathrm{vc}(F) = d$. Then, for any probability measure μ

$$\log N\big(\varepsilon, \mathrm{conv}(F), L_2(\mu)\big) \leq Cd\Big(\frac{1}{\varepsilon}\Big)^{\frac{2d}{1+d}},$$

where C is an absolute constant. In particular, by Theorem 3.8

$$\mathrm{vc}\big(\mathrm{conv}(F), \varepsilon\big) \leq K \cdot \log N_2\big(c\varepsilon, \mathrm{conv}(F)\big) \leq K \cdot d\Big(\frac{1}{\varepsilon}\Big)^{\frac{2d}{1+d}} = K \cdot \frac{\mathrm{vc}(F)}{\varepsilon^{2d/d+1}},$$

in which the exponent of ε is strictly smaller than 2.

Similar results can be established for classes with $\mathrm{vc}(F, \varepsilon) = O(\varepsilon^{-p})$, namely, that $\mathrm{vc}\big(\mathrm{conv}(F), \varepsilon\big) = O\big(\varepsilon^{-\max\{2,p\}}\big)$ for $p \neq 2$, while for $p = 2$ there is an additional logarithmic factor in $1/\varepsilon$. This estimate is better than the one in Corollary 3.10, because one has "global" information on the given class.

3.4 Concentration of the Combinatorial Dimension

This section is devoted to the concentration of measure phenomenon for the combinatorial dimension. The fact that the combinatorial dimension is highly concentrated (is a sense which will be explained below) is rather surprising and is due to Boucheron, Lugosi and Massart [BoLM2]. It is based on the following exponential inequality for general functions.

Theorem 3.18. [BoLM2] *Let $X_1, ..., X_n$ be independent random variables defined on some set Ω and set $Z = f(X_1, ..., X_n)$ to be a random variable. Assume that there is a function $g : \Omega^{n-1} \to \mathbb{R}$ such that for all $\{x_1, ..., x_n\} \subset \Omega$,*

1. $0 \leq f(x_1, ..., x_n) - g(x_1, ..., x_{i-1}, x_{i+1}, ..., x_n) \leq 1$ for all $1 \leq i \leq n$,
2. $\sum_{i=1}^{n} (f(x_1, ..., x_n) - g(x_1, ..., x_{i-1}, x_{i+1}, ..., x_n)) \leq f(x_1, ..., x_n)$,

then, for every $t > 0$,

$$Pr\{Z \geq \mathbb{E}Z + t\} \leq \exp\left[-\frac{t^2}{2\mathbb{E}Z + 2t/2}\right],$$

$$Pr\{Z \leq \mathbb{E}Z - t\} \leq \exp\left[-\frac{t^2}{2\mathbb{E}Z}\right].$$

For any fixed $\varepsilon > 0$ let $Z = \text{vc}(F, \{X_1, ..., X_n\}, \varepsilon)$ be the combinatorial dimension of the class F at scale ε on the random set $\{X_1, ..., X_n\}$, where $(X_i)_{i=1}^n$ are independent, distributed according to a measure μ. It is easy to see that by setting $g(x_1, ..., x_{n-1}) = \text{vc}(F, \{x_1, ..., x_{n-1}\}, \varepsilon)$, both conditions of Theorem 3.18 are satisfied. Indeed, removing one point can change the combinatorial dimension by at most one. By the same argument, the number of removals of points in $\{x_1, .., x_n\}$ which change the combinatorial dimension is exactly $\text{vc}(F, \{x_1, ..., x_n\}, \varepsilon)$. Thus, both conditions are verified and $Z = \text{vc}(F, \{X_1, ..., X_n\}, \varepsilon)$ satisfies the required concentration inequality.

Some other applications of Theorem 3.18 can be found in [BoLM1, BoLM2].

It is natural to ask whether the two other measures of complexity, the Rademacher averages and the L_p empirical entropy, display similar concentration phenomena. The first can be verified, simply by applying Talagrand's inequality (a fact which will be used in section 7). Whether the random variable $Z = \log N(\varepsilon, F, L_p(\mu_n))$ is concentrated is not known in general. As the following example [MeR] shows, the L_∞ log-covering numbers when the centers are taken from the set are not concentrated, in the sense that the best estimate on the concentration of the covering numbers is given by Markov's inequality. More precisely, Let $\tilde{N}(K, \varepsilon B_\infty^n)$ be the covering numbers of K by translates of εB_∞^n, centered at point in K. Then, we have the following

Theorem 3.19. *There exists an absolute constant C for which the following holds. For any $t \geq 2$ and $\varepsilon > 0$ there exist $N, n \in \mathbb{N}$ and a set $K \subset B_\infty^N$ such that the random variable*

$$Z(X_1, ..., X_n) = \log \tilde{N}(P_\sigma K, \varepsilon B_\infty^n)$$

satisfies $\mathbb{E}Z \leq 2 \log t$ and

$$Pr(Z \geq t) \geq \frac{C}{t},$$

where $(X_i)_{i=1}^n$ are independent, distributed according to the uniform measure on $\{1, ..., N\}$.

Proof. Without loss of generality assume that $d = t/\log 2 \in \mathbb{N}$. Set $N = 20d^2$ and let

$$K_1 = \left\{ \left\{ -\frac{2\varepsilon}{3}, \frac{2\varepsilon}{3} \right\}^d \times (0, ..., 0) \right\} \subset B_\infty^N$$

and $K_2 = \{e_1, ..., e_d\}$ where $(e_i)_{i=1}^N$ is the standard unit vector basis in $\mathbb{R}^N$. Define $K = K_1 \cup K_2$, observe that if $\{1, ..., d\} \subset \sigma$ then all the points in $P_\sigma K$ are ε-separated, and thus $N(P_\sigma K, \varepsilon B_\infty^\sigma) \geq 2^d$. On the other hand, if there is some $i \in \{1, ..., d\} \backslash \sigma$ then $P_\sigma e_i = 0$, and $\varepsilon B_\infty^\sigma$ is centered at $P_\sigma e_i$ and contains $P_\sigma K$, implying that $N(P_\sigma K, \varepsilon B_\infty^\sigma) \leq d$. Let μ be the uniform probability measure on $\{1, ..., N\}$ and denote

$$\alpha = \alpha_n(d, N) = Pr\left(\{\{1, ..., d\} \subset \{X_1, ..., X_n\}\}\right).$$

There exists an integer n such that

$$\frac{1}{20d} \leq \alpha \leq \frac{1}{10d}. \tag{3.3}$$

If the inequality (3.3) holds, then $Z \geq \log 2 \cdot d = t$ with probability $\alpha \geq \frac{1}{20d}$, while

$$\mathbb{E}Z \leq \alpha \cdot \log(2^d + d) + (1 - \alpha) \cdot \log d \leq 2 \log t,$$

which implies the Theorem.

To verify (3.3), observe that for fixed integers d and N, $\lim_{n \to \infty} \alpha_n(d, N) = 1$. Let $\sigma_n = \{X_1, ..., X_n\}$, set B_n be the event $\{\{1, ..., d\} \subset \sigma_n\}$ and put $A_i = \{1, ..., d\} \backslash \{i\}$. Clearly,

$$\alpha_{n+1}(d, N) = Pr\left(B_{n+1} \cap B_n\right) + Pr\left(B_{n+1} \cap B_n^c\right)$$

$$= \alpha_n(d, N) + \sum_{i=1}^d Pr\left(B_{n+1} \cap \{\{X_1, ..., X_n\} \cap \{1, ..., d\} = A_i\}\right)$$

$$= \alpha_n(d, N) + \frac{d}{N}\alpha_n(d - 1, N).$$

so,

$$0 \leq \alpha_{n+1}(d, N) - \alpha_n(d, N) \leq d/N = \frac{1}{20d}. \qquad \square$$

4 Learning Sample Complexity and Error Bounds

Bounding the learning sample complexity using the Glivenko–Cantelli condition is not the optimal strategy. In fact, the best that one can hope for using this approach is a sample complexity estimate of $\Omega(1/\varepsilon^2)$ (up to logarithmic factors), since this is the minimal number of examples needed to ensure that $|\mathbb{E}_\mu f - n^{-1} \sum_{i=1}^n f(X_i)| < \varepsilon$ with sufficiently high probability for a single function with a nontrivial variance.

The reason for the sub-optimality of this approach is that the random variable $\sup_{f \in F} |\mathbb{E}_\mu f - n^{-1} \sum_{i=1}^n f(X_i)|$ measures the worst deviation of empirical means from actual ones, going over the entire class; for the learning sample complexity problem, it suffices to control the deviation with respect to functions which are potential empirical minimizers (that is, functions which satisfy that $\mathbb{E}_{\mu_n} f$ is "small"), and this is a relatively small subclass of F. An added complication arises because this is a *random* subset of F.

The ability to obtain improved complexity estimates is based on two essential ingredients. Firstly, the class has to be "relatively small", otherwise, it is impossible to improve the uGC based bounds. Secondly, the class should have the property that the variance of each class member can be controlled by a power of its expectation.

Definition 4.1. F *is called a Bernstein class of type* $0 < \alpha \leq 1$ *with respect to the measure* μ, *if there is some constant* B *such that for any* $f \in F$, $\mathbb{E}_\mu f^2 \leq B(\mathbb{E}_\mu f)^\alpha$.

Note that this definition is measure dependent. To obtain uniform results one requires the class to have a Bernstein type α *for every measure*, and with the same constant.

Obviously, if F consists of functions bounded by $\sqrt{B}$ it has "Bernstein type 0" with a constant B. It is also evident that if F consists of nonnegative functions bounded by B then F has Bernstein type 1 with constant B with respect to any probability measure μ. Therefore, if the target T belongs to G then for any $1 \leq p < \infty$, $\mathcal{L}_p(G, T)$ has Bernstein type 1 with constant 1. In section 6 we show that even if $T \notin G$ and G is convex, then for any probability measure μ, $\mathcal{L}_p(G, T)$ has a Bernstein type 1 for any $1 < p \leq 2$ and Bernstein type $2/p$ if $2 < p < \infty$, and the constants do not depend on the measure μ.

The improved complexity bounds are based on the fact that if F has Bernstein type 1 with a constant B, the learning sample complexity at scale ε is controlled by the GC sample complexity of the intersection of the *star-shaped* hull of F with 0 (i.e., $\{\lambda f | 0 \leq \lambda \leq 1, \ f \in F\}$) with the ball $\{f | \mathbb{E}_\mu f^2 \leq B\varepsilon\}$. Similar results hold for any class with any nontrivial Bernstein type [Me4].

Lemma 4.2. [Me4] *Let* μ *be a probability measure on* Ω *and assume that* $F \subset B\big(L_\infty(\Omega)\big)$ *has Bernstein type 1 with respect to* μ, *with a constant* $B \geq 1$. *Then, for any* $\varepsilon > 0$,

$$Pr\left\{ \exists f \in F, \ \frac{1}{n} \sum_{i=1}^n f(X_i) \leq \frac{\varepsilon}{2}, \ \mathbb{E}_\mu f \geq \varepsilon \right\} \leq$$

$$2Pr\left\{ \sup_{h \in \text{star}(F,0), \mathbb{E}_\mu h^2 \leq B\varepsilon} |\mathbb{E}_\mu h - \mathbb{E}_{\mu_n} h| \geq \frac{\varepsilon}{2} \right\},$$

where $\text{star}(F, 0) = \{\lambda f | 0 \leq \lambda \leq 1, f \in F\}$.

Proof. Set $A = \{\exists f \in F, \frac{1}{n}\sum_{i=1}^{n} f(X_i) \leq \varepsilon/2, \mathbb{E}_\mu f \geq \varepsilon\}$ and observe that

$$
\begin{aligned}
Pr(A) &\leq Pr\{\exists f \in F, \mathbb{E}_\mu f \geq \varepsilon, \mathbb{E}_\mu f^2 < \varepsilon, \mathbb{E}_{\mu_n} f \leq \varepsilon/2 \} \\
&\quad + Pr\{\exists f \in F, \mathbb{E}_\mu f \geq \varepsilon, \mathbb{E}_\mu f^2 \geq \varepsilon, \mathbb{E}_{\mu_n} f \leq \varepsilon/2 \} \\
&\leq Pr\{\exists f \in F, \mathbb{E}_\mu f^2 < \varepsilon, |\mathbb{E}_\mu f - \mathbb{E}_{\mu_n} f| \geq \varepsilon/2\} \\
&\quad + Pr\Big\{\exists f \in F, \mathbb{E}_\mu f \geq \varepsilon, \mathbb{E}_\mu f^2 \geq \varepsilon, |\mathbb{E}_\mu f - \mathbb{E}_{\mu_n} f| \geq \frac{1}{2}\mathbb{E}_\mu f\Big\},
\end{aligned}
$$

because $\mathbb{E}_\mu f \geq \varepsilon$ and $\mathbb{E}_{\mu_n} f \leq \varepsilon/2$ imply that $|\mathbb{E}_\mu f - \mathbb{E}_{\mu_n} f| \geq \frac{1}{2}\mathbb{E}_\mu f \geq \varepsilon/2$. Let $H = \{\frac{\varepsilon f}{\mathbb{E}_\mu f} \mid f \in F, \mathbb{E}_\mu f \geq \varepsilon, \mathbb{E}_\mu f^2 \geq \varepsilon\}$ which is a subset of star$(F, 0)$ since $0 < \varepsilon/\mathbb{E}_\mu f \leq 1$. Every $f \in F$ satisfies that $\mathbb{E}_\mu f^2 \leq B\mathbb{E}_\mu f$, and thus for every $h \in H$, $\mathbb{E}_\mu h^2 \leq B\varepsilon$. In particular,

$$
\begin{aligned}
Pr(A) &\leq Pr\Big\{\exists f \in F, \mathbb{E}_\mu f^2 < \varepsilon, |\mathbb{E}_\mu f - \mathbb{E}_{\mu_n} f| \geq \frac{\varepsilon}{2}\Big\} \\
&\quad + Pr\Big\{\exists h \in H, \mathbb{E}_\mu h^2 \leq B\varepsilon, |\mathbb{E}_\mu h - \mathbb{E}_{\mu_n} h| \geq \frac{\varepsilon}{2}\Big\},
\end{aligned}
$$

which proves our claim. $\square$

The fact that the learning sample complexity can be interpreted as a GC estimate of a class consisting of functions with variance of the same order of magnitude as the required deviation, yields improved tail bounds by a straightforward application of Talagrand's inequality.

Theorem 4.3. [Me6] *There is an absolute constant C such that the following holds. Let $F \subset B(L_\infty(\Omega))$ be a class of Bernstein type 1 with a constant $B \geq 1$. Set $\mathcal{H} = \mathrm{star}(F, 0)$ and for every $\varepsilon > 0$ let $\mathcal{H}_\varepsilon = \mathcal{H} \cap \{h | \mathbb{E}_\mu h^2 \leq \varepsilon\}$. Then for every $0 < \varepsilon, \delta < 1$,*

$$
Pr\Big\{\exists f \in F, \frac{1}{n}\sum_{i=1}^{n} f(X_i) \leq \varepsilon/2, \mathbb{E}_\mu f \geq \varepsilon\Big\} \leq \delta
$$

Provided that

$$
n \geq C \max\Big\{\frac{R_n^2(\mathcal{H}_\varepsilon, \mu)}{\varepsilon^2}, \frac{B\log\frac{2}{\delta}}{\varepsilon}\Big\},
$$

where $(X_i)_{i=1}^{n}$ are selected according to μ and recall that

$$
R_n(\mathcal{H}_\varepsilon, \mu) = \frac{1}{\sqrt{n}}\mathbb{E} \sup_{h \in \mathcal{H}_\varepsilon} \Big|\sum_{i=1}^{n} \varepsilon_i h(X_i)\Big|.
$$

4.1 Error Bounds

An alternative presentation to the sample complexity problem is to use *error bounds*. The idea is to ensure that with high probability, every function in the class satisfies that $\mathbb{E}_\mu f \leq \frac{C}{n} \sum_{i=1}^n f(X_i) +$ "penalty term", which measures the richness of the given class. If $C = 1$, this is simply a one-sided GC condition and the penalty term can not decay faster than $\sigma_F/\sqrt{n}$, for $\sigma_F^2 = \sup_{f \in F} \text{var}(f)$. On the other hand, if $C > 1$ one can obtain faster rates. The rate by which the penalty term tends to 0 as a function of n is called the *error rate* and is analogous to the learning sample complexity. Indeed, if the error bound holds for any $f \in F = \mathcal{L}_p(G, T)$, it holds for any function which satisfies that $\mathbb{E}_{\mu_n} \ell_p(g) \leq \varepsilon/2$. For such functions $\mathbb{E}_\mu \ell_p(g) \leq C\varepsilon$ provided that n is large enough to ensure that the penalty term is sufficiently small. It can be shown (see, for example, [EHKV, AnB, DGL]) that the best error rate possible is $O(1/n)$ (which corresponds to a sample complexity estimate of $O(1/\varepsilon)$), and this holds for "very small" classes. For example, this is the rate of decay (up to a logarithmic factor) for losses associated with binary valued classes with a finite VC dimension.

There are various known error bounds in literature, and modern estimates for the penalty terms are based on the random averages associated with the class. It turns out that even if the class is small, in order to establish error rates faster than $1/\sqrt{n}$ (which can be obtained via a GC-style argument if the class is sufficiently small) one has to use the localized random averages instead of the global ones as penalty terms, as will be explained below. Recently, in [BM], a new method of obtaining error bounds was developed, based on the idea of comparing the empirical and the actual structures on the class.

4.2 Comparing Structures

One can consider the uniform law of large numbers as a way of comparing the empirical (random) structure and the structure endowed by μ uniformly over the class. Indeed, a typical statement would be that with probability larger than $1 - \delta$, for every $f \in F$, $|\mathbb{E}_\mu f - n^{-1} \sum_{i=1}^n f(X_i)| \leq \lambda_n(\delta)$. Hence, this condition is an additive notion of similarity between the two structures. It is possible to use a multiplicative notion of similarity - that for "most" functions in the class, a random coordinate projection is a good isomorphism. By this we mean that for a subset $G \subset F$ containing functions with "large" expectations and $0 < \rho < 1$, then with high probability, every $g \in G$ satisfies that

$$(1 - \rho)\mathbb{E}_\mu g \leq \frac{1}{n} \sum_{i=1}^n g(X_i) \leq (1 + \rho)\mathbb{E}_\mu g.$$

In this case we say that the coordinate projection P_σ onto $(X_1, ..., X_n)$ is a ρ-isomorphism of G. The connection to our problem is simple: suppose that $G_n = \{f \in F : \mathbb{E}f \geq \lambda_n\}$ and assume that P_σ is a ρ-isomorphism on G_n. Then, for every $f \in F$

$$\mathbb{E}f \leq \max \left\{ \frac{1}{1 - \rho} \mathbb{E}_n f, \lambda_n \right\},$$

(since if $\mathbb{E}f \geq \lambda_n$ then f is projected isomorphically onto σ), and this yields the desired error bound. The name ρ-isomorphism comes from the fact that if F consists of nonnegative functions then a projection is a ρ-isomorphism if and only if $P_\sigma : (G, L_1(\mu)) \to (G, L_1(\mu_n))$ is a bi-Lipschitz function.

In what follows we shall denote

$$\mathbb{E}\left\|\mu - \mu_n\right\|_F = \mathbb{E}\sup_{f \in F}\left|\mathbb{E}_\mu f - n^{-1}\sum_{i=1}^{n} f(X_i)\right|.$$

Theorem 4.4. [BM] *There is an absolute constant c for which the following holds. Let F be a class of functions, such that for every $f \in F$, $\mathbb{E}f = \lambda$ and $\|f\|_\infty \leq b$. Assume that F is a Bernstein class of type $0 < \beta < 1$ with a constant B, and suppose that $0 < \rho < 1$ and $0 < \alpha < 1$ satisfy*

$$\lambda \geq c\max\left\{\frac{bx}{n\alpha^2\rho}, \left(\frac{Bx}{n\alpha^2\rho^2}\right)^{1/(2-\beta)}\right\}.$$

1. If $\mathbb{E}\left\|\mu - \mu_n\right\|_F \geq (1+\alpha)\lambda\rho$, then

$$\Pr\left\{P_\sigma \text{ is not an } \rho\text{-isomorphism of } F\right\} \geq 1 - e^{-x}.$$

2. If $\mathbb{E}\left\|\mu - \mu_n\right\|_F \leq (1-\alpha)\lambda\rho$, then

$$\Pr\left\{P_\sigma \text{ is an } \rho\text{-isomorphism of } F\right\} \geq 1 - e^{-x}.$$

Observe that if the class is star-shaped around 0, and if the set $F_\lambda = \{f \in F : \mathbb{E}_\mu f = \lambda\}$ is ρ-isomorphically projected, then the same holds for the set $\{f \in F : \mathbb{E}_\mu f \geq \lambda\}$. This leads to the following error bound:

Corollary 4.5. *There is an absolute constant c for which the following holds. Let F be a class of functions bounded by b, which is star-shaped around 0 and is a Bernstein class of type β with a constant B. For $0 < \rho, \lambda, \alpha < 1$ and $x > 0$, if*

$$\lambda \geq c\max\left\{\frac{bx}{n\alpha^2\rho}, \left(\frac{Bx}{n\alpha^2\rho^2}\right)^{1/(2-\beta)}\right\},$$

and if $\mathbb{E}\left\|\mu - \mu_n\right\|_{F_\lambda} \leq (1-\alpha)\lambda\rho$, then with probability at least $1 - e^{-x}$, every $f \in F$ satisfies

$$\mathbb{E}f \leq \max\left\{\frac{\mathbb{E}_n f}{1-\rho}, \lambda\right\}.$$

In particular, if f^* is an ε-empirical minimizer (that is, if $n^{-1}\sum_{i=1}^{n} f^*(X_i) < \varepsilon$) then $\mathbb{E}f^* \leq \max\{\frac{\varepsilon}{1-\rho}, \lambda\}$.

Note that this approach improves on the results of previous section, mainly because the localized averages here are indexed by the constraint $F_r = \{f \in F : \mathbb{E}_\mu f = r\}$. By the Bernstein assumption, $F_r \subset \{f \in F : \mathbb{E}_\mu f^2 \leq Br\}$. In extreme cases, the resulting gap between the averages indexed by the F_r and $F \cap \{f : \mathbb{E}_\mu f^2 \leq r\}$ could be considerable.

Although the star-shape assumption is very mild, it means that the class is very regular in some sense. Indeed, every function encountered at level r will have a scaled down version for any $0 < t < r$. Thus, as the r decreases the sets F_r become richer in some sense - up to a critical value of r, beyond which it becomes impossible to compare the empirical and actual structures. Let us mention that the fact that a set can not be isomorphically projected should have a geometric interpretation, a point we shall return to later.

Although this approach yields the best known error bounds, and as such, the best estimate on the error of the empirical minimizer based on structural results, it is possible to obtain even better bounds on $\mathbb{E}f^*$. In [BM] it was shown that the error rate obtained using the structural argument presented above is a λ_n defined by $\mathbb{E}\|\mu_n - \mu\|_{F_r} \sim r$, while sharp estimates on $\mathbb{E}f^*$ are of the order of r' which minimizes the function $\mathbb{E}\|\mu_n - \mu\|_{F_r} - r$.

5 Estimating the Localized Averages

It is clear from the sample complexity estimates that the behavior of $\mathbb{E}\|\mu_n - \mu\|_F$ or alternatively, that of the Rademacher/gaussian averages is the determining factor in the tail estimates which lead to sharp bounds for both the GC and the learning sample complexities. When global averages are used, that is, when the set indexing the process is the entire class, one can estimate the averages via Dudley's entropy integral. This is particularly useful because there are natural connections between the entropy of the base class G and the entropy of the loss class $\mathcal{L}_p(G, T)$, implying a similar upper bound on the global averages. Another possibility is to use comparison theorems for the averages (e.g. Slepian's Lemma), since p-loss classes are Lipschitz images of the base class.

There are many interesting examples of base classes used in practice that can be written as compositions of simpler base classes [AnB]. In most cases, there are structural results which enable one to bound the global random averages associated with G using those of the original classes $\mathcal{H}_i$ without resorting to entropy estimates. The most extreme case, in which an entropic argument would clearly be the wrong approach, is when G is the convex hull of elements of $\mathcal{H}$. The random averages associated with G are the same as those of $\mathcal{H}$ - whereas the metric entropy of G is much larger than the metric entropy of $\mathcal{H}$.

The situation becomes more difficult when one has to estimate the localized averages, that is, the expectation of the empirical or Rademacher process indexed by $F_r = \{f \in F : \mathbb{E}_\mu f = r\}$. As a first step in the analysis, one can use the assumption that F is a Bernstein class and replace the "L_1" constraint with an L_2 one, which is easier to handle in the context of subgaussian processes. Indeed, as mentioned before, if F is a Bernstein class of type 1 with a constant B then

$$F_r \subset \{f \in F : \mathbb{E}_\mu f^2 \leq Br\},$$

but the latter set could be considerably larger than the former. The next section is devoted to the question of estimating the expectation of the process indexed by $\{f \in F : \mathbb{E}_\mu f^2 \leq r\}$.

5.1 L_2 Localized Averages

The complexity of the learning problem is determined by the function which measures the localized averages and not the global one. Unfortunately, there are no clear structural properties which enable one to compare the localized averages of a "complicated class" to those of simpler classes used to compose it. Let us give two examples to demonstrate this point; firstly, suppose that G is the convex hull of $\mathcal{H}$, which is assumed to be star-shaped around 0 and symmetric. Given an L_2 structure on the indexing sets $\mathcal{H}$ and G, the aim is to compare

$$\mathbb{E} \sup_{\{h \in \mathcal{H}: \|h\| \leq r\}} X_h, \quad \text{and} \quad \mathbb{E} \sup_{\{g \in G: \|g\| \leq r\}} X_g,$$

where $\{X_t : t \in T\}$ is the gaussian process indexed by T.

The two quantities are clearly different; the former corresponds to the averages associated with an index set generated by the convex hull of the intersection of $\mathcal{H}$ and a ball of radius r, while the latter is generated by the intersection of the convex hull of $\mathcal{H}$ with a ball of radius r, and may be considerably larger.

Comparing the two averages is equivalent to the problem of estimating the modulus of continuity of a gaussian process indexed by a set, and the one indexed by its convex hull. There are some results which connect the two (see, for example, [BouKP]), all of which are based on additional parameters of the original class. For example, in the approach presented in [BouKP], additional information on the Kolmogorov numbers or the metric entropy was required. It is not clear whether such an additional information is really needed to compare the two averages.

The other example is when $G = \phi(\mathcal{H})$, where ϕ is a Lipschitz function. Although it is possible to apply Slepian's inequality and thus compare the two expectations, one has to identify the pre image $\phi^{-1}\{g \in G | \|g\| \leq r\}$ which may be difficult in practical examples.

When it is possible to estimate the uniform entropy of the class, the L_2 localized averages can be bounded by combining Dudley's entropy bound and an estimate on the average diameter of a random coordinate projection. Since the upper limit in Dudley's entropy integral can be taken to be the radius of a ball centered at the origin which contains the set, then

$$\frac{1}{n}\mathbb{E}_{\mu \times \varepsilon} \sup_{\{f \in F : \mathbb{E}_\mu f^2 \leq r\}} \left| \sum_{i=1}^{n} \varepsilon_i f(X_i) \right| \leq \frac{C}{\sqrt{n}} \mathbb{E}_\mu \int_0^{\sqrt{Y}} \sqrt{\log N\left(\varepsilon, F, L_2(\mu_n)\right)} d\varepsilon,$$

where $Y = \frac{1}{n} \sup_{\{f \in F : \mathbb{E}_\mu f^2 \leq r\}} \sum_{i=1}^{n} f^2(X_i)$. The expected value of the radius can be estimated by the following inequality from [Ta2], which states that

$$EY \leq r + \frac{8}{\sqrt{n}} \mathbb{E} \sup_{\{f \in F : \mathbb{E}_\mu f^2 \leq r\}} \left| \sum_{i=1}^{n} \varepsilon_i f(X_i) \right|,$$

and the localized averages may be bounded by bootstrapping. This line of argumentation together with Theorem 4.3 yields the following upper bound.

Corollary 5.1. [BBM, Me6] *Let G be a convex class of functions which map Ω into $[0,1]$, and assume that $\log N_2(\varepsilon, G) \leq \gamma \varepsilon^{-p}$ for some $0 < p < \infty$. Then, the error rate of the class $\mathcal{L}_2(G, T)$ is $O\big(n^{-2/(2+p)}\big)$ [resp. a learning sample complexity estimate of $O\big(\varepsilon^{-(1+p/2)}\big)$] if $0 < p < 2$ and of $O\big(n^{-1/p}\big)$ [resp. $O\big(\varepsilon^{-p}\big)$] if $p > 2$. Both results are uniform in the target T.*

This result is independent of the underlying measure μ and as such, is sharp. If $0 < p < 2$, the error rate is faster than the $1/\sqrt{n}$ (which one can prove via a uGC argument), while if $p > 2$ the "localized" approach yields the same rates as the uGC one.

5.2 Data Dependent Bounds

Next, we turn our attention to data dependent error bounds. Recall that the critical value of r in the isomorphic coordinate projections approach is given by the solution r' to the equation $\xi(r) \sim r$ for $\xi(r) = \mathbb{E}\|\mu - \mu_n\|_{F_r}$. If F happens to be a Bernstein class of type 1 with a constant B, and if we set

$$\psi(r) = \mathbb{E} \sup_{\{f \in F \; : \; \mathbb{E}_\mu f^2 \leq Br^2\}} \left| \sum_{i=1}^{n} \varepsilon_i f(X_i) \right|,$$

then the equation $\psi(r) = r$ has a unique positive solution, which we denote by r^*, and satisfies that $r^* \geq r'$. Thus, r^* is an upper bound on the error rate established in Corollary 4.5. From the practical point of view, it is advantageous to bound r' (or r^*) using only empirical data (that is, the values of class members on the given sample). This is a useful strategy because one can, in principle, discover the error rate without knowing the "size" of the class. The fact that the "L_1" constraint in the definition of the indexing set F_r was replaced with an L_2 constraint (which is used in the definition of ψ) enables one to estimate r^* efficiently [BBM]. To that end, we introduce random functions $\hat{\psi}$ which have similar properties to ψ, but can also be computed from the given data. Moreover, these functions have the following property: if $\hat{r}^*$ is the random variable defined as the unique positive solution of the equations $\hat{\psi}(r) = r$, then $\hat{r}^*$ satisfies that $r^* \leq \hat{r}^* + 1/n$ with high probability.

One can show [BBM] that the random functions

$$\hat{\psi}_{X_1,\ldots,X_n}(r) = \frac{C_1}{n} \mathbb{E}_\varepsilon \sup_{\{f \in \mathrm{star}(F,0) \; : \; n^{-1} \sum_{i=1}^{n} f^2(X_i) \leq r\}} \left| \sum_{i=1}^{n} \varepsilon_i f(X_i) \right| + \frac{C_2 x}{n}$$

have the desired properties, where C_1, C_2 are appropriately selected absolute constants. These functions are computable since they depend on the empirical

Rademacher averages associated with the intersection of the star-shaped hull of F and 0 with an *empirical ball* of radius $\sqrt{r}$ (unlike ψ where the ball was an $L_2(\mu)$ ball). Hence, $\hat{\psi}$ is determined by the Euclidean structure endowed on F by the coordinate projection, and thus is potentially computable.

From the statistical point of view, there is a practical method of estimating $\hat{r}^*$, called the *iterative procedure*, an idea which was introduced in [KoP] and refined in [BBM].

Set $\hat{r}_{k+1} = \min\{1, \hat{\psi}(\hat{r}_k)\}$ where $\hat{r}_0 = 1$. It is possible to show that this (computable) sequence converges quickly to $\hat{r}^*$; it takes $N = O(\log\log n)$ steps to ensure that $\hat{r}^* \leq \hat{r}_N \leq \hat{r}^* + 1/n$. Thus, with high probability, the iterative procedure gives a quick estimate of r^*.

Passing to the L_2 constraint implies that $\hat{\psi}(r)/\sqrt{r}$ is non-increasing, a fact which plays a significant role in the proof of the fast convergence of $\hat{r}_k$. The price is that r^* is only an upper bound on r'. Although it should be possible to obtain an iterative procedure scheme using the "L_1" localized averages (and not their L_2 version), such results are not yet known.

Though the iterative procedure yields very positive results in many cases, the problem of determining which parameters govern the localized averages associated with a class is far from being fully understood, and its solution will have far reaching implications.

5.3 Geometric Interpretation

A question arising from the discussion is the geometric interpretation of $\hat{r}^*$ or its "L_1" analog $\hat{r}'$. First, we consider F to be class of nonnegative functions, and note that for any set $\sigma = \{x_1, ..., x_n\}$

$$\mathbb{E}_\varepsilon \sup_{\{f \in F \,:\, n^{-1}\sum_{i=1}^n f(x_i) \leq r\}} \left| \sum_{i=1}^n \varepsilon_i f(x_i) \right| \equiv R(V \cap rnB_1^n),$$

where

$$V = \{(f(x_1), ... f(x_n)) : f \in F\} \subset [0,1]^n.$$

Since $R(rnB_1^n) = rn$, the critical radius $\hat{r}'$ defines an intersection body which is extremal - a subset of F (and thus of B_∞^n), whose Rademacher average (or, in a similar fashion, its ℓ-norm) is of the same order of magnitude as the "regular body" containing it. This assumption seems to apply that at that scale, the original body is "large", e.g., in the sense that a random coordinate projection will not be a good isomorphism. One possible explanation of that fact could be that this intersection body contains a large cube in a high dimensional coordinate projection, and leads to the following

Question 5.2. Let $K \subset B_\infty^n$ be such that $R(K \cap rn^{1/p}B_p^n) \sim nr$. Does it mean that K has a coordinate projection σ of proportional dimension which contains $crB_\infty^{|\sigma|}$?

This question is a generalization of Elton's Theorem [MeV1], which states that if $K \subset B_\infty^n$ such that $R(K) \geq \delta n$, then there is $\sigma \subset \{1, ..., n\}$, which satisfies that $|\sigma| \geq c_1\delta^2 n$, and $c_2\delta B_\infty^{|\sigma|} \subset P_\sigma K$. In our scenario, one has the additional information that $K_{r,p} = K \cap rn^{1/p}B_p^n$ is an intersection body, but in return one hopes for a cube in a projection of K_r whose dimension does not depend on r. Hence, one should think of it as "Elton's Theorem" for small scales, since r can be as small as $1/n$, and at a scale smaller than $1/\sqrt{n}$, Elton's Theorem is vacuous. Note that by taking $K' = K_r/r$, Elton's Theorem implies the case $p = \infty$. Moreover, the case $p \geq 2$ can be verified (up to logarithmic factors in n) in a similar fashion to the proof of Elton's Theorem in [MeV1]. Of course, for a general class of functions, the constraint which defines the localized averages is not an ℓ_p^n constraint, and thus a different argument might be required altogether to exhibit that the intersection body is "large".

A question of a similar flavor (again, in the simplified context of classes of nonnegative functions) is how to estimate the functions $H(r) = M^*(V \cap rB_p^n)$ for a given set $V \subset B_\infty^n$, where $M^*(K) = \int_{S^{n-1}} \|x\|_{K^*}$. Even when V is very regular, e.g. $V = B_q^n$ this task is nontrivial, though possible. In fact, in that case the dual to the norm endowed by the intersection body (which is the J-functional of the two spaces) can be estimated up to an absolute constant, which leads to sharp bounds on the integral. The general case seems to be considerably more difficult.

6 Bernstein Type of L_p Loss Classes

The assumption which makes the proof of Theorem 4.3 (and analogous results) possible, is that the class in question has a nontrivial Bernstein type. Thus, one needs to check whether this condition holds for various classes of functions. As stated before, if the target function T belongs to the base class G then $\mathcal{L}_p(G, T)$ is a Bernstein class of type 1 with constant 1. If $T \notin G$ the situation is less obvious. Recall that we are interested in base classes which consist of functions whose ranges are contained in $[0, 1]$, that the range of T is also contained in the same interval and that the p loss functional associated with the target T is $\ell_p(g) = |T - g|^p - |T - P_GT|^p$, where P_GT is the best approximation of T in G with respect to the $L_p(\mu)$ norm. In general, the geometry of G and the location of T determine whether the loss class has a nontrivial Bernstein type and influences the constants in a manner which will be specified below.

If the base class is convex the situation is simpler to handle, since the best approximation map onto a convex set in a strictly convex and smooth space is well defined and has a geometric interpretation. In particular if $X = L_p(\mu)$ for $1 < p < \infty$, then every target T will have a unique best approximation in G and the loss class $\mathcal{L}_p(G, T)$ has a nontrivial Bernstein type with a constant which depends only on p and not on the underlying measure.

Theorem 6.1. [Me4] *Let* $1 < p < \infty$ *and assume that* G *is a compact convex subset of* $L_p(\mu)$. *If* $2 \leq p < \infty$ *then for every* $f \in \mathcal{L}_p(G, T)$,

$$\mathbb{E}_\mu f^2 \leq 4p^2 (\mathbb{E}_\mu f)^{\frac{2}{p}}.$$

If $1 < p < 2$ *then*

$$\mathbb{E}_\mu f^2 \leq \frac{4p \cdot 2^p}{p-1} \mathbb{E}_\mu f.$$

We shall present a proof in the case $1 < p < 2$. The other case follows a similar path.

We begin by bounding $\mathbb{E}_\mu f^2$ from above. For any $1 < p < \infty$ and any $f \in \mathcal{L}_p(G, T)$, $\mathbb{E}_\mu f^2 \leq p^2 \mathbb{E}_\mu |g - P_G T|^2$, which follows from the fact that $|x|^p$ is a Lipschitz function on $[0, 1]$ with a constant p.

To bound $\mathbb{E}_\mu |g - P_G T|^2$ from above using $\mathbb{E}_\mu f$, we use the uniform convexity of $L_p(\mu)$. Recall that the modulus of convexity of $L_p(\mu)$ is given by $\delta_p(\varepsilon) = 1 - (1 - (\varepsilon/2)^p)^{1/p}$, while for $1 < p < 2$, $\delta_p(\varepsilon) \geq (p-1)\varepsilon^2/2 \cdot 2^p$ [HHZ, Ha] (in fact δ_{L_p} is equivalent to $c_p \varepsilon^2$ as $\varepsilon \to 0$).

The main observation required for the proof is based on a standard separation argument.

Lemma 6.2. *Let* X *be a uniformly convex, smooth Banach space with a modulus of convexity* δ_X *and let* $G \subset X$ *be compact and convex. Set* $T \notin G$ *and put* $d = \|T - P_G T\|$. *Then, for every* $g \in G$,

$$\delta_X \left(\frac{\|g - P_G T\|}{d_g} \right) \leq 1 - \frac{d}{d_g},$$

where $d_g = \|T - g\|$.

Proof (of Theorem 6.1). We estimate $\|g - P_G T\|^2$ from above by $\mathbb{E}_\mu f = d_g^p - d^p$. Applying Lemma 6.2 to $X = L_p$,

$$\|g - P_G T\|^2 \leq \frac{2 \cdot 2^p}{p-1} d_g (d_g - d).$$

Since G consists of functions into $[0, 1]$ and T takes valued in $[0, 1]$ then $0 \leq d, d_g \leq 1$. Observe that for any $0 \leq y \leq x \leq 1$, $x(x - y) \leq \frac{2}{p}(x^p - y^p)$, and by selecting $x = d_g$ and $y = d$,

$$\mathbb{E}_\mu f^2 \leq p^2 \|g - P_G T\|^2 \leq \frac{2p^2 \cdot 2^p}{p-1} d_g (d_g - d) \leq \frac{4p \cdot 2^p}{p-1} \mathbb{E}_\mu f. \qquad \square$$

Let us mention that our analysis does not pass smoothly to the nonconvex case. To start with, one has a problem with the definition of the loss class, because it is no longer true that every target has a unique best approximation in G. Even if $P_G T$ a is well defined function, the geometric characterization of the nearest point map is no longer valid. However, it is still possible to obtain partial estimates on the Bernstein type of $\mathcal{L}_2(G, T)$. To that end, we require

Lemma 6.3. [MeW] *Assume that there is some $0 \leq \beta < 1$ such that for every $g \in G$*

$$\langle P_G T - T, P_G T - g \rangle \leq \frac{\beta}{2} \|P_G T - g\|^2_{L_2(\mu)}. \tag{6.1}$$

Then $\mathcal{L}_2(G, T)$ is a Bernstein class of type 1 with a constant $16/(1 - \beta)$.

In cases where T has a unique best approximation in G, one can find the smallest value of β for which (6.1) holds, which leads to a bound on the constant in the Bernstein type of the 2-loss class.

Without loss of generality, assume that G is a compact subset of $L_2(\mu)$. Denote by $\text{nup}(G, \mu)$ the set of points with more than a unique best approximation in G with respect to the $L_2(\mu)$ norm. Suppose $T \in L_2(\mu) \backslash (G \cup \text{nup}(G, \mu))$, and let

$$r_{G,\mu}(T) := \inf \left\{ \|g - P_G T\| : g \in \{\lambda(T - P_G T) : \lambda > 0\} \cap \text{nup}(G, \mu) \right\}.$$

Intuitively, $r_{G,\mu}(T) = \|f_{\text{nup}} - P_G T\|$ where f_{nup} is the point in $\text{nup}(G, \mu)$ found by extending a ray from $P_G T$ through T until reaching $\text{nup}(G, \mu)$. One can show that the smallest possible value of β is

$$\beta_{G,\mu}(T) := \frac{\|T - P_G T\|}{r_{G,\mu}(T)} = \frac{\|T - P_G T\|}{\|f_{\text{nup}} - P_G T\|}.$$

Clearly, if G is convex then $\text{nup}(G, \mu)$ is the empty set; hence for all $T \in L_2(\mu)$, $r_{G,\mu}(T) = \infty$ and $\beta_{G,\mu}(T) = 0$, which is exactly the result obtained in Theorem 6.1 - that $\mathcal{L}_2(G, T)$ has Bernstein type 1 with a constant 16.

The analysis of the Bernstein type of loss classes when the target has more than a unique best approximation is incomplete. Moreover, the Bernstein type of p-loss classes in the nonconvex case is far from being understood.

Let us point out that learning is inherently more difficult if the base class G is nonconvex [LeBW], in the sense that the best uniform upper bound on the sample complexity that one can obtain (at least in the agnostic case) is $\Omega(1/\varepsilon^2)$ regardless of the "size" of the base class G.

7 Classes of Linear Functionals

Here we investigate classes which seemingly have an easy structure - classes which consist of linear functionals on a given domain. The simplest example we consider is when the class G is the dual unit ball of some infinite dimensional Banach space X, and the domain of the functionals is the unit ball B_X. Thus,

$$G = \{x^* : B_X \to \mathbb{R} \mid \|x^*\| \leq 1\}.$$

It was shown in [DuGZ] that G is a uniform Glivenko–Cantelli class of functions if and only if X has a nontrivial type. Actually, if X is infinite dimensional, the exact asymptotic behavior of the combinatorial dimension of G is determined by the Rademacher type of X. This follows from the next observation:

Lemma 7.1. $\{x_1, ..., x_n\} \subset B_X$ *is ε-shattered by B_{X^*} if and only if $(x_i)_{i=1}^n$ are linearly independent and ε-dominate the ℓ_1^n unit-vector basis, that is, for every $(a_i)_{i=1}^n \subset \mathbb{R}$, $\varepsilon \sum_{i=1}^n |a_i| \leq \| \sum_{i=1}^n a_i x_i \|$.*

Using this lemma, it is possible to upper bound the combinatorial dimension of B_{X^*}.

Theorem 7.2. [MeS] *Let X be a Banach space. Then, for every $\varepsilon > 0$,*

$$\mathrm{vc}(B_{X^*}, B_X, \varepsilon) \leq \left(\frac{T_p(X)}{\varepsilon} \right)^{\frac{p}{p-1}} + 1,$$

where $T_p(X)$ is the Rademacher type p constant of X.

If X happens to be infinite dimensional, then this upper bound can be complemented. This is due to the fact that if $p^* = \sup\{p | X \text{ has type } p\}$ then $\ell_{p^*}^n$ is finitely represented in X.

Theorem 7.3. [MeS] *Let X be an infinite dimensional Banach space. Then $\mathrm{vc}(B_{X^*}, B_X, \varepsilon)$ is finite for every $\varepsilon > 0$ if and only if X has a nontrivial type. In addition,*

$$\left(\frac{1}{\varepsilon} \right)^{\frac{p^*}{p^*-1}} - 1 \leq \mathrm{vc}(B_{X^*}, B_X, \varepsilon).$$

A more general problem is when the domain of the class of functionals is not B_X but some other convex symmetric set. Let K and L be two convex symmetric bodies in $\mathbb{R}^m$ and consider elements of the polar body L° as functions on K using the fixed inner product in $\mathbb{R}^m$. It turns out that computing the combinatorial dimension $\mathrm{vc}(L^\circ, K, \varepsilon)$ in equivalent to a factorization problem. For every two convex, symmetric sets K and L in $\mathbb{R}^m$ and every integer $n \leq m$, let $\Gamma_n(K, L) = \inf \|A\| \|B\|$, where the infimum is taken with respect to all subspaces of $E \subset \mathbb{R}^m$ of dimension n and all operators $B : (E, \| \ \|_{L \cap E}) \to \ell_1^n$, $A : \ell_1^n \to (E, \| \ \|_{K \cap E})$, such that $AB = id : E \to E$.

Lemma 7.4. [MeS] *For every integer n and any convex and symmetric sets K and L,*

$$\frac{1}{\Gamma_n(K, L)} = \sup \left\{ \varepsilon | \exists \{x_1, ..., x_n\} \subset K, \ \varepsilon(L \cap E) \subset \mathrm{absconv}(x_1, ..., x_n) \right\} \quad (7.1)$$

where $E = \mathrm{span}\{x_1, ..., x_n\}$ and $\mathrm{absconv}(x_1, ..., x_n)$ is the symmetric convex hull of $\{x_1, ..., x_n\}$.

Thus, $1/\Gamma_n(K, L)$ is the discrete inverse of $\mathrm{vc}(L^\circ, K, \varepsilon)$.

In [MeS] the factorization constants $\Gamma_n(B_p^m, B_q^m)$ were investigated. However, from the Machine Learning perspective, the significant point is that one can estimate the combinatorial dimension of classes of functions which can be viewed as linear functionals on some domain. One such case is when the base class G is the unit ball of some Sobolev space X which is embedded in $C(\Omega)$.

Hence, for every $g \in G$ and every $x \in \Omega$, $g(x) = \delta_x(g)$. If the point evaluation functionals are uniformly bounded in X^* (say by 1) then

$$
\mathrm{vc}(G, \Omega, \varepsilon) \leq \mathrm{vc}(B_{X^{**}}, B_{X^*}, \varepsilon) \leq \left(\frac{T_p(X)}{\varepsilon} \right)^{\frac{p}{p-1}},
$$

which is often tighter than results obtained by more direct approaches [Me3].

The next example, which plays a central role in modern Machine Learning, has a similar flavor. Let Ω be a compact set and set $K : \Omega \times \Omega \to \mathbb{R}$ to be a positive definite, continuous function. Given a probability measure ν on Ω, set $T_K : L_2(\nu) \to L_2(\nu)$ by $(T_K f)(x) = \int K(x,y) f(y) d\nu(y)$. According to the spectral Theorem, T_K has a diagonal representation, and denote by $(\phi_n(x))_{n=1}^{\infty}$ the orthonormal basis and by $(\lambda_i)_{i=1}^{\infty}$ the non-increasing sequence of eigenvalues corresponding to this diagonal representation. Mercer's Theorem implies that $\nu \times \nu$ almost surely,

$$
K(x,y) = \sum_{n=1}^{\infty} \lambda_n \phi_n(x) \phi_n(y),
$$

and since K is continuous, then under mild assumptions on ν this representation of K holds for every $x, y \in \Omega$.

Let F_K be the class consisting of all the functions of form $\sum_{i=1}^{\infty} a_i K(x_i, \cdot)$ where $x_i \in \Omega$ and $a_i \in \mathbb{R}$ satisfy that $\sum_{i,j=1}^{\infty} a_i a_j K(x_i, x_j) \leq 1$.

There is extensive literature on learning algorithms based on the class F_K and on similar classes (in which $(a_i)_{i=1}^{\infty}$ satisfy slightly different constraints), all called kernel classes [ScS, CuS]. These are probably the most frequently used classes in real-life applications such as optical character recognition, DNA sequence analysis, Time Series Prediction and many more.

It is easy to see that F_K is the unit ball of the Hilbert space associated with the kernel, called the *reproducing kernel Hilbert space* and denoted by $\mathcal{H}$. An alternative way to define the reproducing kernel Hilbert space which makes the interpretation of F_K as a class of linear functionals clearer, is via the *feature map*. Define $\Phi : \Omega \to \ell_2$ by $\Phi(x) = \left(\sqrt{\lambda_i} \phi_i(x) \right)_{i=1}^{\infty}$. Then,

$$
F_K = \left\{ f(\cdot) = \langle \beta, \Phi(\cdot) \rangle_{\ell_2} \mid \|\beta\|_2 \leq 1 \right\}.
$$

In other words, the feature map is a way of embedding the space Ω in ℓ_2 and F_K can be represented as the unit ball in the space. Each $\beta \in \ell_2$ acts as a functional on the image of the Ω via the feature map, denoted by $\Phi(\Omega)$, and in particular, if B_2 denotes the unit ball in ℓ_2 then $\mathrm{vc}(F, \Omega, \varepsilon) = \mathrm{vc}(B_2, \Phi(\Omega), \varepsilon)$ for any $\varepsilon > 0$. Using the volumetric methods of [MeS] it is possible to estimate the combinatorial dimension of F_K and of other families of kernel classes. For F_K one can show that if, for some measure with a strictly positive density, the eigenfunctions of the integral operator T_K are uniformly bounded, then the combinatorial dimension of the class can be bounded above in terms of the eigenvalues of T_K. In general, it is possible to connect the fact that $\{x_1, ..., x_n\}$ is ε-shattered to the eigenvalues of the Gram matrix $\left(K(x_i, x_j)\right)_{i,j=1}^n$. However, spectral information is not enough to determine the largest ε for which $\{x_1, ..., x_n\}$ is ε-shattered by F_K, and for that one needs to analyze the geometry of TB_1^n, where T is the linear operator which maps the unit vectors e_i to $\Phi(x_i)$.

Because of the relatively simple geometry of F_K, the L_2-localized averages associated with F_K can be computed exactly. Observe that the index set $H_r = F_K \cap \{f | \mathbb{E}_\mu f^2 \le r\}$ with respect to which the supremum is taken, is an intersection of a ball and an ellipsoid, which makes the computation of the L_2 norm of $\sup_{f \in H_r} |\sum_{i=1}^n \varepsilon_i f(X_i)|$ possible. Indeed, if $\mathcal{E}$ and $\mathcal{E}'$ are ellipsoids with the same principal directions and axes $(a_i)_{i=1}^\infty$ and $(b_i)_{i=1}^\infty$ respectively, then the ellipsoid $\mathcal{B}$ whose axes are $(\min\{a_i, b_i\})_{i=1}^\infty$ satisfies that $\mathcal{B} \subset \mathcal{E} \cap \mathcal{E}' \subset \sqrt{2}\mathcal{B}$. Therefore, one can replace the set $\mathcal{E} \cap \mathcal{E}'$ indexing the Rademacher process by $\mathcal{B}$, losing only a multiplicative factor, and it is a straightforward exercise to compute the L_2 norm of the supremum of the process indexed by $\mathcal{B}$. Next, one shows that the L_2 norm of $\sup_{f \in H_r} |\sum_{i=1}^n \varepsilon_i f(X_i)|$ is equivalent to its L_1 norm, which follows from concentration.

Theorem 7.5. [Me5] *For every $1 < p < \infty$ there is a constant c_p for which the following holds. Let $F \subset B\left(L_\infty(\Omega)\right)$, set μ to be a probability measure on Ω and put $\sigma_F^2 = \sup_{f \in F} \mathbb{E}_\mu f^2$. If n satisfies that $\sigma_F^2 \ge 1/n$ then*

$$c_p \left(\mathbb{E} \sup_{f \in F} \left| \sum_{i=1}^n \varepsilon_i f(X_i) \right|^p \right)^{\frac{1}{p}} \le \mathbb{E} \sup_{f \in F} \left| \sum_{i=1}^n \varepsilon_i f(X_i) \right| \le \left(\mathbb{E} \sup_{f \in F} \left| \sum_{i=1}^n \varepsilon_i f(X_i) \right|^p \right)^{\frac{1}{p}},$$

where $(X_i)_{i=1}^n$ are independent random variables distributed according to μ.

The upper bound is just Hölder's inequality. The condition on the σ_F is needed only for the lower bound, which is more delicate than the Kahane-Khintchine inequality because of the additional randomness due to $(X_i)_{i=1}^n$.

Combining these facts one can prove the following:

Corollary 7.6. [Me5] *There are absolute constants C and c such that if $(\lambda_i)_{i=1}^\infty$ is the nondecreasing sequence of eigenvalues of the integral operator $(T_K f)(x) = \int K(x, y) f(y) d\mu(y)$ and if $\lambda_1 \ge 1/n$, then for every $r \ge 1/n$,*

$$\frac{1}{\sqrt{n}} \mathbb{E} \sup_{\{f \in F: \, \mathbb{E}_\mu f^2 \le r\}} \left| \sum_{i=1}^n \varepsilon_i f(X_i) \right| \sim \left(\sum_{j=1}^\infty \min\{\lambda_j, r\} \right)^{\frac{1}{2}},$$

where $(X_i)_{i=1}^\infty$ are independent, distributed according to μ.

It is evident that the localized averages depend on the underlying measure μ, since the spectrum of the integral operator may change dramatically when the measure changes. Moreover, since the underlying measure is *unknown* it is impossible to estimate the eigenvalues of the integral operator. This leads to another question: is it possible to replace the eigenvalues of the integral opera-tor with those of the normalized Gram matrix $(n^{-1}K(X_i, X_j))_{i,j=1}^n$? In other words, is there a concentration of measure phenomenon for the eigenvalues of $n^{-1}(K(X_i, X_j))_{i,j=1}^n$ around those of T_K? A positive answer would enable one to estimate the L_2-localized averages from empirical (and thus computable) data.

There are some partial asymptotic results in that direction due to Koltchinskii and Giné [Ko, KoG]. In our case, T_K is trace class (because K is continuous), and for such operators they proved that the spectrum of $(n^{-1}K(X_i, X_j))_{i,j=1}^n$ converges almost surely (as sets) to the spectrum of the integral operator T_K. There are few known results on the rate of convergence, and those are less than satisfactory.

In a similar way one can bound the empirical local Rademacher averages, $\frac{1}{\sqrt{n}}\mathbb{E}_\varepsilon \sup_{\{f\in F:\ \mathbb{E}_{\mu_n} f^2 \leq r\}} |\sum_{i=1}^n \varepsilon_i f(X_i)|$ where μ_n is the (random) empirical measure supported on $(X_1, ..., X_n)$, in which case the eigenvalues of the inte-gral operator in Corollary 7.6 are replaced by the normalized eigenvalues of the Gram matrix - making the iterative procedure presented in the previous section feasible.

8 Concluding Remarks

Although this article focused in its entirety on the sample complexity problem, we would like to end it by briefly mentioning several other questions which are significant in Learning Theory and have a mathematical flavor.

Firstly, there is the question of the learning algorithm. Empirical mini-mization is not the only possibility to attack the learning problem. Even if one chooses empirical minimization, there is an algorithmic problem of finding an almost minimizer, and the *computational complexity* of the algorithm has to be estimated. Secondly, there is an issue of the degree of approximation; a solution of the learning problem is a function which is almost the optimal in the class. However, even the optimal can be "very far" from the target. This is a classical tradeoff between large classes for which the approximation error is small for many targets, but the sample complexity is huge, and small classes for which a small sample is needed for the probabilistic construction but the approximation error is very large. Balancing the statistical error and the ap-proximation error simultaneously can be achieved via *model selection*, in which a very large class G is divided into an increasing family of classes G_i such that $\bigcup_{i=1}^\infty G_i = G$, and one wants to find the "best class" from which the target T can be approximated. A formulation similar in nature involves regularization. For example, if G is the entire reproducing kernel Hilbert space with norm $\| \ \|_K$ then one can consider $\ell(g) = (g - T)^2 + \gamma\|g\|_K$. This loss functional balances the way g approximates T on the sample and the "smoothness" of g, captured by $\|g\|_K$.

All of the issues mentioned above and others that were not mentioned are intriguing mathematically. Our hope is that this article would encourage more mathematicians to investigate questions arising in this rapidly developing field, boosting the theoretical side of Machine Learning.

References

[ABCH]   Alon, N., Ben-David, S., Cesa-Bianchi, N., Haussler, D.: Scale sensitive dimensions, uniform convergence and learnability. J. of ACM, **44** (4) 615–631 (1997)

[AnB]   Anthony, M., Bartlett, P.L.: Neural Network Learning. Cambridge University Press (1999)

[BBM]   Bartlett, P.L., Bousquet, O., Mendelson, S.: Localized Rademacher complexities. Ann. Stat., to appear

[BL]   Bartlett, P.L., Long, P.M.: Prediction, learning, uniform convergence, and scale-sensitive dimensions. J. Comput. System Sci., **56**, 174–190 (1998)

[BM]   Bartlett, P.L., Mendelson, S.: Empirical risk minimization. Preprint

[BaBM]   Barron, A., Birgé, L., Massart, P.: Risk bounds for model selection via penalization. Probab. Theory Related Fields, **113** (3), 301–413 (1999)

[BiM]   Birgé, L., Massart, P.: From model selection to adaptive estimation. In: Festschrift for Lucien Le Cam, 55–87. Springer, New York (1997)

[BoLM1]   Boucheron, S., Lugosi, G., Massart, P.: A sharp concentration inequality with applications in random combinatorics and learning. Random Struct. Algor., **16**, 277–292 (2000)

[BoLM2]   Boucheron, S., Lugosi, G., Massart, P.: Concentration inequalities using the entropy method. Ann. Probab., **31** (3), 1583–1614 (2003)

[Bou]   Bousquet, O.: A Bennett concentration inequality and its application to suprema of empirical processes. C.R. Acad. Sci. Paris, Ser. I, **334**, 495–500 (2002)

[BouKP]   Bousquet, O., Koltchinkii, V., Panchenko, D.: Some local measures of complexity of convex hulls and generalization bounds. In: Kivinen, J., Sloan, R.H. (eds) Proceedings of the 15th Annual Conference on Computational Learning Theory COLT02. Lecture Notes in Computer Sciences, **2375**, Springer, 59–73 (2002)

[CKP]   Carl, B., Kyrezi, I., Pajor, A.: Metric entropy of convex hulls in Banach spaces. J. London Math. Soc., **60** (2), 871–896 (1999)

[CuS]   Cucker, P., Smale, S.: On the mathematical foundations of learning. B. Am. Math. Soc., **39** (1) 1–49 (2002)

[DGL]   Devroye, L., Györfi, L., Lugosi, G.: A Probabilistic Theory of Pattern Recognition. Springer (1996)

[DL]   Devroye, L., Lugosi, G.: Combinatorial Methods in Density Estimation. Springer (2000)

[Du1]   Dudley, R.M.: Central limit theorems for empirical measures. Ann. Probab., **6** (6), 899–929 (1978)

[Du2]   Dudley, R.M.: Universal Donsker classes and metric entropy. Ann. Probab., **15**, 1306–1326 (1987)

[Du3]   Dudley, R.M.: Uniform central limit theorems. Cambridge Studies in Advanced Mathematics 63. Cambridge University Press (1999)

[DuGZ]   Dudley, R.M., Giné, E., Zinn, J.: Uniform and universal Glivenko–Cantelli classes. J. Theoret. Prob., **4**, 485–510 (1991)

[EHKV]   Ehrenfeucht, A., Haussler, D., Kearns, M.J., Valiant, L.G.: A general lower bound on the number of examples needed for learning. Inform. Comput., **82**, 247–261 (1989)

[HHZ]   Habala, P., Hajek, P., Zizler, V.: Introduction to Banach Spaces, vols. I and II. Matfyzpress, Univ. Karlovy, Prague (1996)

[Ha]   Hanner, O.: On the uniform convexity of L^p and l^p. Ark. Math., **3**, 239–244 (1956)

[Hau]   Haussler, D.: Sphere packing numbers for subsets of Boolean n-cube with bounded Vapnik–Chervonenkis dimension. J. Comb. Theory A, **69**, 217–232 (1995)

[Ho1]   Hoffman–Jørgensen, J.: Sums of independent Banach space valued random variables. Aarhus University, Preprint series 1972/1973, **15** (1973)

[Ho2]   Hoffman–Jørgensen, J.: Probability in Banach spaces. Lecture Notes in Mathematics, **598**, 164–229, Springer-Verlag (1976)

[G]   Gao, F.: Metric entropy of convex hulls. Isr. J. Math., **123**, 359–364 (2001)

[GiZ1]   Giné, E., Zinn, J.: Some limit theorems for empirical processes. Ann. Probab., **12** (4), 929–989 (1984)

[GiZ2]   Giné, E., Zinn, J.: Gaussian characterization of uniform Donsker classes of functions. Ann. Probab., **19** (2), 758–782 (1991)

[K]   Kahane, J.P.: Some random series of functions. Heath-Lexington, 1968, 2nd edition, Cambridge University Press (1985)

[KaM]   Karpovsky, M.G., Milman, V.D.: Coordinate density of sets of vectors. Discrete Math., **24**, 177–184 (1978)

[Ko]   Koltchinskii, V.: Asymptotics of spectral projections of some random matrices approximating integral operators. In: Progress in Probability, **43**, Birkhauser (1998)

[KoG]   Koltchinskii, V., Giné, E.: Random matrix approximation of spectra of integral operators. Bernoulli **6** (1), 113–167 (2000)

[KoP]   Koltchinskii, V., Panchenko, D.: Rademacher processes and bounding the risk of function learning. High Dimensional Probability, II (Seattle, WA, 1999), 443–457. In: Progress in Probability, **47**, Birkhauser.

[L]   Ledoux, M.: The concentration of measure phenomenon. Mathematical Surveys and Monographs, **89**, AMS (2001)

[LeBW]   Lee, W.S., Bartlett, P.L., Williamson, R.C.: The importance of convexity in learning with squared loss. IEEE Trans. Info. Theory, **44** (5), 1974–1980 (1998)

[Lu]   Lugosi, G.: Pattern classification and learning theory. In: Györfi, L. (ed) Principles of Nonparametric Learning. Springer, 1–56 (2002)

[M]   Massart, P.: About the constants in Talagrand's concentration inequality for empirical processes. Ann. Probab., **28** (2) 863–884 (2000)

[Me1]   Mendelson, S.: On the size of convex hulls of small sets. J. Mach. Learn. Res., **2**, 1–18 (2001)

[Me2]   Mendelson, S.: Rademacher averages and phase transitions in Glivenko–Cantelli class. IEEE Trans. Info. Theory, **48** (1), 251–263 (2002)

[Me3]   Mendelson, S.: Learnability in Hilbert spaces with reproducing kernels. J. Complexity, **18** (1), 152–170 (2002)

[Me4]   Mendelson, S.: Improving the sample complexity using global data. IEEE Trans. Info. Theory, **48** (7), 1977–1991 (2002)

[Me5]   Mendelson, S.: Geometric parameters of kernel machines. In: Kivinen, J., Sloan, R.H. (eds), Proceedings of the 15th annual conference on Computational Learning Theory COLT02. Lecture Notes in Computer Sciences, **2375**, Springer, 29–43 (2002)

[Me6]   Mendelson, S.: A few notes on statistical learning theory. In: Mendelson, S., Smola, A.J. (eds), Proceedings of the Machine Learning Summer School, Canberra 2002. Lecture Notes in Computer Sciences, **2600**, Springer (2003)

[MePR]   Mendelson, S., Pajor, A., Rudelson, M.: Work in progress

[MeR]   Mendelson, S., Rudelson, M.: Concentration of covering numbers. Unpublished notes

[MeS]   Mendelson, S., Schechtman, G.: The shattering dimension of sets of linear functionals. Ann. Probab., to appear

[MeV1]   Mendelson, S., Vershynin, R.: Entropy and the combinatorial dimension. Invent. Math., **152** (1), 37–55 (2003)

[MeV2]   Mendelson, S., Vershynin, R.: Remarks on the geometry of coordinate projections in $\mathbb{R}^n$. Isr. J. Math., to appear

[MeW]   Mendelson, S., Williamson, R.C.: Agnostic learning of non-convex classes of functions. In: Kivinen, J., Sloan, R.H. (eds), Proceedings of the 15th Annual Conference on Computational Learning Theory COLT02. Lecture Notes in Computer Sciences, **2375**, Springer, 1–13 (2002)

[MiT-J] Milman, V.D., Tomczak-Jaegermann, N.: Sudakov type inequalities for convex bodies in $\mathbb{R}^n$. In: Geometric Aspects in Functional Analysis 85-85, Lecture Notes in Mathematics, **1267**, 113–121, Springer (1987)

[P]   Pajor, A.: Sous espaces ℓ_1^n des espaces de Banach. Hermann, Paris (1985)

[R]   Rio, E.: Une ingalit de Bennett pour les maxima de processus empiriques. (French) [A Bennet-type inequality for maxima of empirical processes]. Ann. Inst. H. Poincar Probab. Statist., **38** (6) 1053–1057 (2002)

[RuV]   Rudelson, M., Vershynin, R.: Preprint

[S]   Sauer, N.: On the density of families of sets. J. Comb. Theory A, **13**, 145–147 (1972)

[ScS]   Schölkopf, B., Smola, A.J.: Learning with Kernels. MIT Press (2001)

[Sh]   Shelah, S.: A combinatorial problem: stability and orders for models and theories in infinitary languages. Pac. J. Math., **41**, 247–261 (1972)

[Ta1]   Talagrand, M.: Type, infratype, and the Elton-Pajor theorem. Invent. Math., **107**, 41–59 (1992)

[Ta2]   Talagrand, M.: Sharper bounds for Gaussian and empirical processes. Ann. Probab., **22** (1), 28–76 (1994)

[Ta3]   Talagrand, M.: Vapnik–Chervonenkis type conditions and uniform Donsker classes of functions. Ann. Probab., to appear

[T-J]   Tomczak–Jaegermann, N.: Banach–Mazur distance and finite-dimensional operator ideals. Pitman Monographs and Surveys in Pure and Applied Mathematics, **38** (1989)

[VW]   Van der Vaart, A.W., Wellner, J.A.: Weak Convergence and Empirical Processes. Springer-Verlag (1996)

[Va]   Vapnik, V.: Statistical Learning Theory. Wiley (1998)

[VaC]   Vapnik, V.N., Chervonenkis, A.Ya.: Necessary and sufficient conditions for uniform convergence of means to mathematical expectations. Theory Prob. Applic., **26** (3), 532–553 (1971)

[Vi]   Vidyasagar, A.: The Theory of Learning and Generalization. Springer-Verlag (1996)

[Z]   Zhou, D.X.: The covering number in learning theory. J. Complexity, **18** (3), 739–767 (2002)

[ZS]   Zhou, D.X., Smale, S.: Estimating the approximation error in learning theory. Anal. Appl. (Singap.), **1** (1), 17–41 (2003)

Essential Uniqueness of an M-Ellipsoid of a Given Convex Body

V.D. Milman[1*] and A. Pajor[2]

[1] School of Mathematical Sciences, Tel Aviv University, Tel Aviv 69778, Israel
milman@post.tau.ac.il
[2] Equipe d'Analyse et Mathématiques Appliquées, Université de Marne-la-Vallée,
5 boulevard Descartes, Champs sur Marne, 77454 Marne-la-Vallee cedex 2,
France pajor@math.univ-mlv.fr

Summary. We show that different M-ellipsoids associated with a convex body, that are introduced in Asymptotic Convexity, are "essentially" unique, in the sense that they admit equivalent large dimensional sections.

1 Introduction

This is a short note, just an observation, but curious and important enough to be broadly known. So we decided to write it down.

We use the following standard notation. The space $\mathbb{R}^n$ is equipped with the canonical Euclidean scalar product $(.\,,.)$. Its unit ball is denoted by D.

Let K be a convex body in $\mathbb{R}^n$ such that 0 is in its interior. Its polar K° is defined as usual by

$$K^\circ = \left\{ x \in \mathbb{R}^n \,:\, (x,y) \le 1 \text{ for every } y \in K \right\}.$$

Let A and B be two subsets of $\mathbb{R}^n$. The covering number $N(A,B)$ is defined as usual as

$$N(A,B) = \min \left\{ {}^\sharp \Lambda : \Lambda \subset \mathbb{R}^n, A \subset \Lambda + B \right\}$$

where $\Lambda + B = \{ x + y \,:\, x \in \Lambda, y \in B \}$.

The volume of a measurable subset A of $\mathbb{R}^n$ is denoted by $|A|$. Let $\sigma > 0$ and let K be a convex compact subset of $\mathbb{R}^n$ with 0 in its interior. We say that an ellipsoid $\mathcal{E}$ of $\mathbb{R}^n$ is an M-ellipsoid of K with constant σ, or shortly an M-ellipsoid of K, if, setting $\lambda = (|K|/|\mathcal{E}|)^{1/n}$ in order that $|K| = |\lambda \mathcal{E}|$, we have

$$N(K, \lambda \mathcal{E}) \le e^{\sigma n}.$$

* Partially supported by the Israel Science Foundation founded by the Academy of Sciences and Humanities.

We assume below in the paper that ellipsoids are centered at 0.

It is proved in [M1] (see also [M2] and [Pi] for simplified proofs) that there exists a universal constant such that for every n, every n-dimensional symmetric convex body has an M-ellipsoid with respect to this constant. Many interesting properties of centrally symmetric convex bodies and corresponding normed spaces were revealed using M-ellipsoids. The existence of an M-ellipsoid was established even for non centrally symmetric bodies in [M-P1]. We refer to the survey in [G-M].

It is proved in [M-P1] that if $\mathcal{E}$ is an M-ellipsoid of $K \subset \mathbb{R}^n$ with constant σ, and if 0 is the barycenter of K, then there exists $\sigma_1 \geq \sigma$ depending only on σ such that

$$N(\lambda \mathcal{E}, K) \leq e^{\sigma_1 n} \ , N\big(K^\circ, (\lambda \mathcal{E})^\circ\big) \leq e^{\sigma_1 n} \text{ and } N\big((\lambda \mathcal{E})^\circ, K^\circ\big) \leq e^{\sigma_1 n}.$$

Of course, we cannot expect the uniqueness of such an ellipsoid and since the definition is of isomorphic type, it should at least depend on the parameter σ. At the same time, one can construct different M-ellipsoids, having deep and non-trivial additional properties. We say that an ellipsoid $\mathcal{E}$ is a regular M-ellipsoid of order α with constant σ for a convex body $K \subset \mathbb{R}^n$, if $|\mathcal{E}| = |K|$ and for any $t > 0$,

$$N(K, t\mathcal{E}) \leq e^{\sigma n / t^\alpha} \text{ and } N(\mathcal{E}, tK) \leq e^{\sigma n / t^\alpha}. \tag{1}$$

It is proved in [Pi] that for any $0 < \alpha < 2$, there exists $\sigma(\alpha) = O\big((2 - \alpha)^{1/2}\big)$ as $\alpha \to 2$, such that every centrally symmetric convex body K has an ellipsoid $\mathcal{E}_\alpha$ which is a regular M-ellipsoid of order α with constant $\sigma(\alpha)$ for K. Note that the covering numbers in (1) do not depend on the choice of the center. The previous assumption on the barycenter of K was needed because this statement used polarity.

These ellipsoids depend on α and the purpose of this note is to show that surprisingly, in some "essential" way, these M-ellipsoids are unique. This question arises from consideration of different M-ellipsoids in [M-P2].

2 Essential Uniqueness of M-Ellipsoids

We begin by a definition of essential uniqueness.

Definition. *Let c be a positive function on $(0, 1)$. Two ellipsoids $\mathcal{E}_1$ and $\mathcal{E}_2$ in $\mathbb{R}^n$ are called essentially c-equivalent if for every $0 < \varepsilon < 1$, there is a subspace $E \subset \mathbb{R}^n$ of dimension larger than $(1 - \varepsilon)n$, such that*

$$\frac{1}{c(\varepsilon)} \, \mathcal{E}_2 \cap E \subset \mathcal{E}_1 \cap E \subset c(\varepsilon) \, \mathcal{E}_2 \cap E. \tag{2}$$

As we will actually compare two ellipsoids in a given n-dimensional space, we put "$c(\varepsilon)$" in the definition. However, in Asymptotic Convexity, one compares two families of ellipsoids, say $\mathcal{E}_1(n)$ and $\mathcal{E}_2(n)$, $n \in \mathbb{N}$, and call these families *essentially* equivalent if for every $0 < \varepsilon < 1$, there exists $c(\varepsilon) > 0$ such that for every integer n, there is a subspace $E(\varepsilon) \subset \mathbb{R}^n$ of dimension larger than $(1 - \varepsilon)n$ such that

$$\frac{1}{c(\varepsilon)} \, \mathcal{E}_2(n) \cap E(\varepsilon) \subset \mathcal{E}_1(n) \cap E(\varepsilon) \subset c(\varepsilon) \, \mathcal{E}_2(n) \cap E(\varepsilon).$$

We can now state our result on the uniqueness of an M-ellipsoid.

Theorem. *Let $K \subset \mathbb{R}^n$ be any convex compact body. Let $\mathcal{E}_1$ and $\mathcal{E}_2$ be two M-ellipsoids of K, with the same volume $|K| = |\mathcal{E}_1| = |\mathcal{E}_2|$ and the same constant σ. Then there exists a function $c > 0$ on $(0, 1)$, depending only on σ, such that $\mathcal{E}_1$ and $\mathcal{E}_2$ are essentially c-equivalent.*

Proof. In the proof, we use notation $\sigma_1, \sigma_2, ..$ to denote positive numbers depending only on the parameter σ associated with the M-ellipsoids $\mathcal{E}_1$ and $\mathcal{E}_2$ given above. As noticed previously, there exists $\sigma_1 > 0$ such that

$$N(K, \mathcal{E}_i) \leq e^{\sigma_1 n} \text{ and } N(\mathcal{E}_i, K) \leq e^{\sigma_1 n}, \; i = 1, 2.$$

Therefore,

$$N(\mathcal{E}_1, \mathcal{E}_2) \leq N(\mathcal{E}_1, K) . N(K, \mathcal{E}_2) \leq e^{\sigma_2 n}$$

for some new constant σ_2. This means that $\mathcal{E}_2$ is an M-ellipsoid of $\mathcal{E}_1$, with constant σ_2.

It is a well-known property of an M-ellipsoid $\mathcal{E}$ of a symmetric convex body C with constant σ and same volume as C, that

$$|C \cap \mathcal{E}| \geq \sigma_3^n \, |C|$$

where $\sigma_3 > 0$ depends only on σ (see [M2]). Applying this property to the pair $\mathcal{E}_1$ and $\mathcal{E}_2$, we have

$$|\mathcal{E}_1 \cap \mathcal{E}_2| \geq \sigma_3^n \, |K|.$$

At the same time, the intersection of two ellipsoids is $\sqrt{2}$-equivalent to an ellipsoid, which means that there exists an ellipsoid $\mathcal{E}$ such that

$$\mathcal{E} \subset \mathcal{E}_1 \cap \mathcal{E}_2 \subset \sqrt{2} \, \mathcal{E}.$$

Thus $N(\mathcal{E}_i, \mathcal{E}) \leq e^{\sigma_4 n}$, $i = 1, 2$, for some new constant σ_4. We conclude that

$$\mathcal{E} \subset \mathcal{E}_i \text{ and } N(\mathcal{E}_i, \mathcal{E}) \leq e^{\sigma_4 n}, \quad i = 1, 2.$$

It is now standard and trivial to show that for any $\varepsilon > 0$, there are subspaces $E_i(\varepsilon)$, $i = 1, 2$, of dimension larger than $(1 - (\varepsilon/2))n$ such that

$$\mathcal{E} \cap E_i(\varepsilon) \subset \mathcal{E}_i \cap E_i(\varepsilon) \subset e^{\sigma_5/\varepsilon} \, \mathcal{E} \cap E_i(\varepsilon)$$

for some new constant σ_5. Indeed, let $\mathbb{R}^n$ be equipped with the Euclidean structure associated with $\mathcal{E}$. Then for each $i = 1, 2$, $E_i(\varepsilon)$ is spanned by the $(1 - (\varepsilon/2))n$ axes of $\mathcal{E}_i$ corresponding to the $(1 - (\varepsilon/2))n$ smallest eigenvalues. Now let $E(\varepsilon) = E_1(\varepsilon) \cap E_2(\varepsilon)$. It has dimension larger than $(1 - \varepsilon)n$ and (2) is satisfied with $c(\varepsilon) = e^{\sigma_5/\varepsilon}$, which shows that $\mathcal{E}_1$ and $\mathcal{E}_2$ are essentially c-equivalent. □

Corollary 1. *Let $K \subset \mathbb{R}^n$ be a convex body containing 0. Let $\mathcal{E}$ be an M-ellipsoid of K with constant σ and let $0 < \alpha < 2$. Then for every $\varepsilon > 0$, there exists a subspace $E \subset \mathbb{R}^n$ with dimension larger than $(1 - \varepsilon)n$, such that for every $t > 0$, one has*

$$N\big(K \cap E, t\,c(\varepsilon)\,\mathcal{E} \cap E\big) \leq e^{\sigma_1 n/t^\alpha}$$

where σ_1 and $c(\varepsilon)$ depend only on α and σ.

Proof. We first assume that K is centrally symmetric. Then as we recalled above, there exists a regular M-ellipsoid $\mathcal{E}_\alpha$ of K of order α and with same volume as $\mathcal{E}$. Since it is also an M-ellipsoid for K, it follows from the previous Theorem that $\mathcal{E}$ and $\mathcal{E}_\alpha$ are essentially equivalent. Therefore, using (1) and (2) there exists a subspace E with dimension larger than $(1 - \varepsilon)n$, such that for every $t > 0$, we have

$$N\big(K \cap E, t\,c(\varepsilon)\,\mathcal{E} \cap E\big) \leq N(K \cap E, t\,\mathcal{E}_\alpha \cap E) \leq e^{\sigma_1 n/t^\alpha}$$

where σ_1 and $c(\varepsilon)$ depend only on α and σ. To get the right-hand side inequality above, observe that because $\mathcal{E}_\alpha$ is centrally symmetric, one has $N(K \cap E, t\,\mathcal{E}_\alpha \cap E) \leq N(K, (t/2)\,\mathcal{E}_\alpha)$.

When K is not centrally symmetric the existence of a regular M-ellipsoid of K of order α is still an open problem. Let $K' = K - K$ be the difference body. Since we assume that $0 \in K$, it follows that $K \subset K'$. It is proved in [M-P1], that an M-ellipsoid for K is also an M-ellipsoid for K' and vice-versa, only the constants associated with them differ by a universal factor. Let $\mathcal{E}$ be an M-ellipsoid for K with same volume as K'. Recall that from [R-S], we have $|K| \leq |K'| \leq 4^n |K|$. Thus without loss of generality we can assume that $1 \leq |K|/|\mathcal{E}| \leq 4^n$. We now use the first part of the proof with K', which is centrally symmetric. Let $\mathcal{E}'_\alpha$ be a regular M-ellipsoid for K' of order α and with same volume as $\mathcal{E}$. These are M-ellipsoids for K'. From the above Theorem, they are essentially equivalent. Therefore, there exists a subspace E with dimension larger than $(1 - \varepsilon)n$, such that for every $t > 0$, we have

$$N\big(K' \cap E, t\,c(\varepsilon)\,\mathcal{E} \cap E\big) \leq N(K' \cap E, t\,\mathcal{E}'_\alpha \cap E) \leq e^{\sigma'_1 n/t^\alpha}$$

where σ'_1 and $c'(\varepsilon)$ depend on α and σ. The result follows from $K \subset K'$. □

Corollary 2. *Let $K \subset \mathbb{R}^n$ be a convex body with 0 as barycenter. Let $\mathcal{E}_1$ be an M-ellipsoid of $K \cap (-K)$ and $\mathcal{E}_2$ be an M-ellipsoid of $K - K$, with the same volume $|K| = |\mathcal{E}_1| = |\mathcal{E}_2|$ and the same constant σ. Then there exists a function $c > 0$ on $(0,1)$, depending only on σ, such that $\mathcal{E}_1$ and $\mathcal{E}_2$ are essentially c-equivalent.*

Proof. It is proved in [M-P1] that when 0 is the barycenter of K, both ellipsoids $\mathcal{E}_1$ and $\mathcal{E}_2$ are M-ellipsoids of K. The result follows from the Theorem.

$\square$

References

[G-M]   Giannopoulos, A., Milman, V.: Asymptotic convex geometry: A short overview. In: Donaldson, S.K., Eliashberg, Ya., Gromov, M. (eds) Different Faces of Geometry. International Mathematical Series, Kluwer/Plenum Publishers, vol. 3 (2004), to appear

[M1]   Milman, V.: An inverse form of Brunn-Minkowski inequality with applications to local theory of normed spaces. C.R. Acad Sci. série I, **302**, 25–28 (1986)

[M2]   Milman, V.: Isomorphic symmetrizations and geometric inequalities. GAFA Seminar Notes, Springer Lectures Notes, **1317**, 107–131 (1988)

[M-P1] Milman, V., Pajor, A.: Entropy and asymptotic geometry of non-symmetric convex bodies. Advances in Math., **152**, 314–335 (2000)

[M-P2] Milman, V., Pajor, A.: Regularization of star bodies by random hyperplane cut off. Studia Math, to appear

[Pi]   Pisier, G.: The Volume of Convex Bodies and Banach Space Geometry. Cambridge University Press (1989)

[R-S]   Rogers, C.A., Shephard, G.C.: The difference body of a convex body. Arch. Math., **8**, 220–233 (1957)

On the Thermodynamic Limit
for Disordered Spin Systems

L. Pastur

Department of Mathematics, University Paris 7, Paris, France
pastur@math.jussieu.fr

Summary. We present basic facts on the existence of the thermodynamic limit for the Ising model, whose interaction and the external field possess the ergodic structure. Most of the facts are well known, although we discuss also the recent progress in establishing the thermodynamic limit in the mean field type models with random interaction. We consider also simplest models of statistical physics, displaying phase transitions.

1 Frameworks and Models

1.1 Generalities

The goal of statistical mechanics is to derive properties of macroscopic bodies from detailed data on their microscopic constituents (particles, spins, etc.). The data are encoded in the choice of parameters, characterizing the constituents and usually called coordinates, and in the form of the energy of the body, expressed via the coordinates and called the Hamiltonian. To avoid technicalities we will confine ourselves to the lattice spin systems, where the coordinates assume values ± 1 and are indexed by the sites of the d-dimensional lattice $\mathbb{Z}^d$. We denote the coordinates s_x, $x \in \mathbb{Z}$, $s_x = \pm 1$. Given a set $\Lambda \subset \mathbb{Z}^d$, we denote

$$s_\Lambda = \{s_x\}_{x \in \Lambda} \qquad (1.1)$$

the configuration of the system, confined to the "box" Λ. In this case a typical example of the Hamiltonian is

$$H_\Lambda(s_\Lambda) = -\frac{1}{2} \sum_{x,y \in \Lambda} J_{xy} s_x s_y - \sum_{x \in \Lambda} h_x s_x , \qquad (1.2)$$

where $J : \mathbb{Z}^d \times \mathbb{Z}^d \to \mathbb{R}$ is symmetric ($J_{xy} = J_{yx}$), and is called the pair interaction energy, and $h : \mathbb{Z}^d \to \mathbb{R}$ is called the external field.

The basic hypothesis of statistical mechanics is that the macroscopic properties are described by expectations of certain functions of configurations, called the observables, with respect to the special probability measure, called the Gibbs measure, and defined as

$$\mu_\Lambda(s_\Lambda) = e^{-\beta H_\Lambda(s_\Lambda)}/Z_\Lambda \tag{1.3}$$

where

$$Z_\Lambda = \sum_{\{s_x = \pm 1,\ x \in \Lambda\}} e^{-\beta H_\Lambda(s_\Lambda)} \tag{1.4}$$

is called the partition function and $\beta \geq 0$ is called the inverse temperature. We will denote the expectation with respect to the Gibbs measure $\langle \ldots \rangle_{H_\Lambda}$.

A simple example of observable is the magnetization per size, defined as

$$m_\Lambda(s_\Lambda) = |\Lambda|^{-1} M_\Lambda(s_\Lambda), \quad M_\Lambda(s_\Lambda) = \sum_{x \in \Lambda} s_x . \tag{1.5}$$

A quantity of primary interest in statistical mechanics is the free energy

$$f_\Lambda = -(\beta|\Lambda|)^{-1} \log Z_\Lambda \tag{1.6}$$

as a function of β and parameters of the Hamiltonian ($|\Lambda|$ in (1.5)–(1.6) and everywhere below denotes cardΛ). In particular, replacing in (1.2) h_x by $h_x + \varepsilon$, we obtain for the expectation of magnetization per site

$$\langle m_\Lambda \rangle_{H_\Lambda} = -\left.\frac{\partial f_\Lambda}{\partial \varepsilon}\right|_{\varepsilon=0} \tag{1.7}$$

and

$$\beta|\Lambda|\langle (m_\Lambda - \langle m_\Lambda \rangle_{H_\Lambda})^2 \rangle_{H_\Lambda} = -\left.\frac{\partial^2 f_\Lambda}{\partial \varepsilon^2}\right|_{\varepsilon=0} \tag{1.8}$$

for its variance. Likewise, we have

$$|\Lambda|^{-1}\langle H_\Lambda \rangle_{H_\Lambda} = \frac{\partial(\beta f_\Lambda)}{\partial \beta}, \tag{1.9}$$

for the mean energy per site, and

$$|\Lambda|^{-1}\langle (H_\Lambda - \langle H_\Lambda \rangle_{H_\Lambda})^2 \rangle_{H_\Lambda} = -\left.\frac{\partial^2(\beta f_\Lambda)}{\partial \beta^2}\right|_{\varepsilon=0} \tag{1.10}$$

for the variance of the energy.

The notion of a macroscopic body (system) implies that the typical length scale ℓ, determining Λ, is much bigger than the distance between microscopic constituents (atoms, spins, etc.). To formalize this, we consider an infinite collection of rectangular boxes, defined as follows. Let $a_1, \ldots, a_d$ and $b_1, \ldots, b_d$ be integers such that $-\infty < a_i < b_i < \infty$. Denote $L_i = [a_i, b_i) \subset \mathbb{Z}$ and

$$\Lambda = L_1 \times L_2 \times \cdots \times L_d \subset \mathbb{Z}^d \ . \tag{1.11}$$

We write that

$$\Lambda \to \infty \ , \tag{1.12}$$

if

$$\ell_i := b_i - a_i \to \infty, \quad i = 1, \ldots, d \ . \tag{1.13}$$

We say that the system, described by the Hamiltonian (1.2), admits the thermodynamic limit if there exists

$$\lim_{\Lambda \to \infty} f_\Lambda := f \ . \tag{1.14}$$

For a definition of the limit $\Lambda \to \infty$ for more general class of sets of $\mathbb{Z}^d$ see [Ru1].

The initial step of any statistical mechanics study is to prove the existence of the thermodynamic limit. In fact, since certain physical quantities are expressed via derivatives of the free energy with respect to β and parameters of the Hamiltonian (see e.g. (1.7)–(1.10)), we need the convergence in (1.14) in the class C^k in these parameters for some $k > 0$. Denote $n - 1$ the maximum value of k for which it is the case with respect to a certain parameter. Then we say that the system undergoes a phase transition of the order n with respect to the parameter.

Study of phase transitions is one of the main goals of statistical mechanics. We are not going to discuss here phase transition (except simple models of Section 4). We refer the reader to review works [Am, Bax, Bo, Bo-Pi, Fr, LL, MPV, Mi, Ne, Pat, Ru1, Ru2, Sim, Sin, Ta2, Th, TKS, Wi, ZJ] and references therein.

1.2 Interactions

Returning to the thermodynamic limit, we note first that there are two basic classes of statistical mechanics models, depending on the range of the interaction in (1.2). The first class, often called the short range models, can be defined by the interaction $\{J_{xy}\}_{x,y \in \mathbb{Z}^d}$ that decays sufficiently fast as $|x - y| \to \infty$. A simple, but already fairly non-trivial example is the nearest neighbor interaction, where

$$J_{xy} = 0, \quad |x - y| > 1 \ . \tag{1.15}$$

An important subclass of the short range interaction comprises translation invariant interactions for which

$$J_{xy} = J_{x-y}, \quad h_x = h \ , \tag{1.16}$$

where $J : \mathbb{Z}^d \to \mathbb{R}$ is such that

$$\sum_{x \in \mathbb{Z}^d} |J_x| < \infty \ . \tag{1.17}$$

Respective statistical mechanics models are rather realistic but their detailed analysis, especially concerning the existence and properties of phase transitions is possible only in certain cases (see [Bax, Fr, LL, Mi, Ne, Ru1, Pat, Sim, Sin, TKS]).

The second class of interactions and respective statistical mechanics models is called the mean field type (or the long range) models. The simplest case is the Kac model, where

$$J_{xy} = \frac{J}{|\Lambda|}, \quad x, y \in \Lambda, \quad J > 0 . \tag{1.18}$$

The model is less realistic, because it prescribes the interaction of the same strength between all the spins, independently of the distance between them (cf. (1.17)). An advantage of the model is that it can be analyzed in detail and gives a satisfactory qualitative description of the ferromagnetic and certain other phase transitions (see Section 4). Besides, the model can be obtained as a limiting case of short range models, whose interaction is of the form

$$J_{x-y} = JR^{-d}\varphi\left(\frac{x-y}{R}\right) , \tag{1.19}$$

where $R > 0$ and $\varphi : \mathbb{R} \to \mathbb{R}^+$ is an integrable function, normalized as

$$\int_{\mathbb{R}^d} \varphi(x)dx = 1 . \tag{1.20}$$

It is easy to see that (1.19)–(1.20) satisfy (1.17) for any $R > 0$.

Assume that for this interaction there exists the thermodynamic limit (1.14) (see Theorem 2.1 below), which we denote f_R^{sr}, making explicit its dependence on the "interaction radius" R. Then, according to [He-Le, KKPS],

$$\lim_{R\to\infty} f_R^{sr} = f^{Kac},$$

where f^{Kac} is the free energy of the Kac model (see formula (4.2) of Section 4).

Consider now the disordered case, where the external field is an ergodic random field on $\mathbb{Z}^d$ and the interaction is an ergodic random matrix. This means that there exist a probability space $(\Omega, \mathcal{F}, \mathbf{P})$, where Ω is a set, $\mathcal{F}$ is a σ-algebra of events, and $\mathbf{P}$ is a probability measure, and commuting automorphisms $T_1, \ldots, T_d$ of Ω such that

$$\mathbf{P}(T_i^{-1}A) = \mathbf{P}(A), \quad \forall A \in \mathcal{F} \quad i = 1, \ldots, d ; \tag{1.21}$$
$$T_i^{-1}A = A, \quad A \in \mathcal{F}, \quad i = 1, \ldots, d \Rightarrow \mathbf{P}(A) = 0, 1 .$$

In this case any measurable function $\widetilde{h} : \Omega \Rightarrow \mathbb{R}$ determines the ergodic field on $\mathbb{Z}^d$ by the formula

$$h(x, \omega) = \widetilde{h}(T^x\omega), \quad T^x = T_1^{x_1} \cdots T_d^{x_d} , \tag{1.22}$$

where $x = (x_1, \ldots, x_d) \in \mathbb{Z}^d$. Likewise, any function $\tilde{J} : \mathbb{Z}^d \times \Omega \to \mathbb{R}$ measurable for any $x \in \mathbb{Z}^d$ determines the ergodic matrix

$$J_{x,y}(\omega) = \tilde{J}_{x-y}(T^x \omega) . \tag{1.23}$$

The above relations imply

$$h_x(T^a \omega) = h_{x+a}(\omega), \quad J_{x,y}(T^a \omega) = J_{x+a,y+a}(\omega) . \tag{1.24}$$

In the non-random case, where Ω consists of a single point, (1.24) reduces to (1.16). One can say that h and J possess the translation symmetry in the mean, because, if $\mathbf{E}\{\ldots\}$ denotes the mathematical expectation with respect to the measure $\mathbf{P}$ and h and J possess all moments, then

$$\mathbf{E}\{h_{x_1+a} \cdot \ldots \cdot h_{x_m+a}\} \quad \text{and} \quad \mathbf{E}\{J_{x_1+a,y_1+a} \cdots J_{x_m+a,y_m+a}\} ,$$

do not depend on $a \in \mathbb{Z}^d$ for all $m \in \mathbb{N}$ and all collections $x_1, \ldots, y_m \in \mathbb{Z}^{2md}$.

A simple but important class of the ergodic random matrices, providing short range interactions for spin models, consists of matrices

$$J_{xy}(\omega) = \varphi_{x-y} I_{xy}, (\omega) \tag{1.25}$$

where $\varphi \in \ell^1(\mathbb{Z}^d)$ and $\varphi_x = \varphi_{-x}$, and I is a bounded ergodic matrix. Here are two typical examples of I:

1.
$$I_{xy} = \sum_{\mu=1}^{p} \xi_x^\mu \xi_y^\mu , \tag{1.26}$$

where ξ^μ, $\mu = 1, \ldots, p$, are bounded ergodic fields, in particular $\{\xi_x^\mu\}_{\mu=1, x \in \mathbb{Z}^d}^p$ can be i.i.d. random variables.

2. $\{I_{xy}\}_{x,y \in \mathbb{Z}^d}$ is a collection of identically distributed bounded random variables, independent modulo the symmetry condition

$$I_{xy} = I_{yx} . \tag{1.27}$$

The condition of boundedness of I in (1.25) is often too restrictive. We will replace the condition by the following

$$\sup_{x,y \in \mathbb{Z}^d} \mathbf{E}\{|I_{xy}|\} < \infty . \tag{1.28}$$

Likewise, we will assume that in the general case of the ergodic interaction matrix (1.23)

$$\sum_{x \in \mathbb{Z}^d} \mathbf{E}\{|J_{0x}|\} < \infty . \tag{1.29}$$

Noting that in view of (1.24) $\mathbf{E}\{|J_{xy}|\}$ depends only on $x - y$, and we can view this conditions as an analog of (1.17).

Passing to the mean field models we note first that unlike the translation invariant case, where the Kac model (1.18) is practically unique, in the disordered case there are several models. The first one is known as the Hopfield model. Its interaction has the form (cf. (1.25)–(1.26))

$$J_{xy} = N^{-1} \sum_{\mu=1}^{p(N)} \xi_x^\mu \xi_y^\mu, \quad x, y \in \Lambda, \ N = |\Lambda|, \tag{1.30}$$

where $\{\xi_x^\mu\}_{\mu=1, x \in \mathbb{Z}^d}^p$ are as in (1.26). The model was proposed and solved in [Lu, Pa-Fi] in the case of an N-independent p. Of particular interest is the case, where $p = [\alpha N]$, $\alpha > 0$ and $\{\xi_x^\mu\}$ are the Bernoulli random variables, i.e., i.i.d. random variables, assuming values ± 1 with probability $1/2$. This case can be used as a model of associative memory. In this interpretation the coordinates $\{s_x\}_{x \in \Lambda}$ model neurons and $\xi^\mu = \{\xi_x^\mu\}_{x \in \Lambda}$ model the patterns to be memorized. Physics arguments predict, in particular, a sharp "failure of memory" transition at $\alpha \simeq 0,14$, $\alpha = \lim_{N \to \infty} p/N$ [Am, MPV].

The second model is known as the Sherrington–Kirkpatrick model. Here (cf. (1.25)–(1.27))

$$J_{xy} = \frac{I_0}{N} + \frac{I_{xy}}{N^{1/2}} \tag{1.31}$$

where $\{I_{xy}\}_{x,y \in \mathbb{Z}^d}$ is a collection of identically and symmetrically distributed random variables, independent modulo the symmetry condition (1.27) and such that

$$\mathbf{E}\{I_{xy}\} = 0, \quad \mathbf{E}\{I_{xy}^2\} = I^2 > 0 . \tag{1.32}$$

Of particular interest is the case of Gaussian I_{xy}'s, that are uniquely determined by (1.32).

The Hopfield model and the Sherrington-Kirkpatrick models have a number of other applications outside of statistical physics (integer programming, assignment problem, combinatorial optimization, etc., see [Am, Bo, Ho, MPV, Ta2].

1.3 Thermodynamic Limit of the Ground State Energy

The paper is devoted to a discussion of the thermodynamic limit of the free energy of the above models. Of certain interest is also the thermodynamic limit of the ground state energy of the models. The ground state energy is defined as the minimum value of the Hamiltonian of the model, i.e.

$$E_\Lambda = \min_{s_\Lambda \in \{-1,1\}^{|\Lambda|}} H_\Lambda(s_\Lambda) . \tag{1.33}$$

The thermodynamic limit is defined as (cf. (1.12)–(1.14))

$$e = \lim_{\Lambda \to \infty} |\Lambda|^{-1} E_\Lambda . \tag{1.34}$$

Writing f_Λ as

$$f_\Lambda = |\Lambda|^{-1} E_\Lambda - (\beta|\Lambda|)^{-1} \log \left(\sum_{s_\Lambda \in \{-1,1\}^\Lambda} e^{-\beta(H_\Lambda(s_\Lambda) - E_\Lambda)} \right), \qquad (1.35)$$

we find that f_Λ is a non-decreasing function of β. Hence

$$|\Lambda|^{-1} E_\Lambda = \lim_{\beta \to \infty} f_\Lambda = \sup_\beta f_\Lambda. \qquad (1.36)$$

The next theorem shows that (1.34) follows from (1.14) [Gu-To] and that the limiting ground state energy per site e is related to the limiting free energy f via the analog of (1.36).

Theorem 1.1. *Consider a spin model, defined by (1.2) and such that*

$$\sup_\Lambda |\Lambda|^{-1} \left(\sum_{x,y \in \Lambda} |J_{xy}| + \sum_{x \in \Lambda} |h_x| \right) < \infty.$$

Assume that the thermodynamic limit (1.14) of the free energy exists. Then the thermodynamic limit (1.34) of the ground state energy also exists and

$$e = \lim_{\beta \to \infty} f = \sup_{\beta > 0} f .$$

Proof. We have from (1.4), (1.6), and (1.33):

$$f_\Lambda \le \frac{E_\Lambda}{\Lambda} \le f_\Lambda + \frac{\ln 2}{\beta}.$$

Besides, according to (1.35), f_Λ is a non-decreasing function of β. Hence, passing first to the limit $\Lambda \to \infty$ and then to the limit $\beta \to \infty$, we obtain the assertion of the theorem. $\qquad \square$

2 Thermodynamic Limit in the Case of Short Range Interaction

We will begin from the translation invariant models, defined by (1.2)–(1.16).

Theorem 2.1. *Consider a spin model, defined by (1.2),(1.16), and (1.17). Then the model admits the thermodynamic limits (1.11)–(1.14) and the limiting free energy satisfies the bound*

$$|f| \le \frac{1}{2} \sum_{x \in \mathbb{Z}^d} |J_x| + |h| + \beta^{-1} \log 2 . \qquad (2.1)$$

Proof. Consider first the case of the finite range interaction, where there exists $R < \infty$ such that (cf. (1.15))

$$J_x = 0, \quad |x| > R . \tag{2.2}$$

Denote

$$F_\Lambda = -\beta^{-1} \log Z_\Lambda, \tag{2.3}$$

where the partition function Z_Λ is defined in (1.4). It is easy to see that because of (1.16) F_Λ depends only on the lengths $\ell_1, \ldots, \ell_d$ of the "sides" $L_1, \ldots, L_d$ of Λ of (1.11): $F_\Lambda = F(\ell_1, \ldots, \ell_d)$. Besides, we have the bound for any $\Lambda \subset \mathbb{Z}^d$

$$|F_\Lambda| \leq |\Lambda| \big(\overline{J}/2 + |h| + \beta^{-1} \log 2 \big), \tag{2.4}$$

where

$$\overline{J} = \sum_{x \in \mathbb{Z}^d} |J_x| . \tag{2.5}$$

Then (2.2) implies in the one-dimensional case $d = 1$:

$$-C_1 + F(\ell') + F(\ell'') \leq F(\ell' + \ell'') \leq F(\ell') + F(\ell'') + C_1, \tag{2.6}$$

where

$$C_1 = (2R + 1)\overline{J}$$

is independent of ℓ, ℓ' and ℓ''. Fix a positive integer λ and write any positive integer ℓ as

$$\ell = p\lambda + \mu, \quad 0 \leq \mu < \lambda - 1, \quad p \in \mathbb{N} . \tag{2.7}$$

We obtain from (2.6)

$$-\frac{C_1 p}{\ell} \frac{F(\mu)}{\ell} + \frac{\lambda p}{\ell} \frac{F(\lambda)}{\lambda} \leq \frac{F(\ell)}{\ell} \leq \frac{\lambda p}{\ell} \cdot \frac{F(\lambda)}{\lambda} + \frac{F(\mu)}{\ell} + \frac{C_1 p}{\ell} . \tag{2.8}$$

Since λ is arbitrary, we conclude in view of (1.6),(2.3), and (2.4) that the thermodynamic limit exists in the one-dimensional case of (2.2). In the multidimensional case we can use a similar argument, based on the inequality

$$F(\ell_1, \ldots, \ell_{i-1}, \ell_i' + \ell_i'', \ell_{i+1}, \ldots, \ell_d) \leq F(\ell_1, \ldots, \ell_{i-1}, \ell_i', \ell_{i+1}, \ldots, \ell_d)$$
$$+ F(\ell_1, \ldots, \ell_{i-1}, \ell_i'', \ell_{i+1}, \ldots, \ell_d) + C_{i,d}, \quad i = 1, \ldots, d, \tag{2.9}$$

and the opposite inequality with $C_{i,d}$ replaced by $-C_{i,d}$, where $C_{i,d} = (2R + 1)\ell_1 \cdots \ell_{i-1}\ell_{i+1} \cdots \ell_d \overline{J}$, $i = 1, \ldots, d$. This and (2.4) allows us to prove the existence of the thermodynamic limit and the bound (2.4) in the case of the finite range interaction (2.2) (see e.g. [Mi, Mi-Po, Ru1]).

In the general case of interaction, satisfying condition (1.17), we introduce

$$J_x(t) = J_x \chi_R(x) + t J_x \big(1 - \chi_R(x) \big), \quad t \in [0, 1] , \tag{2.10}$$

where χ_R is the indicator of the ball of radius R in $\mathbb{Z}^d$. Denote $H_\Lambda(t)$ the respective Hamiltonian (1.2) and $Z_\Lambda(t)$ the respective partition function (1.4). Then $Z_\Lambda(1)$ is the partition function, corresponding to the interaction J, and $Z_\Lambda(0) := Z_\Lambda^{(R)}$ corresponds to the finite range interaction $J\chi_R$. We have then

$$\log Z_\Lambda(1) - \log Z_\Lambda(0) = \frac{\beta}{2} \sum_{x,y\in\Lambda} J_{x-y}\big(1 - \chi_R(x-y)\big) \int_0^1 \langle s_x s_y \rangle_{H_\Lambda(t)} dt.$$

Since $|s_x| = 1$, $|\langle s_x s_y \rangle_{H_\Lambda(t)}| \le 1$, we obtain the bound

$$\left|(\beta|\Lambda|)^{-1}\log Z_\Lambda - (\beta|\Lambda|)^{-1}\log Z_\Lambda^{(R)}\right| \le \frac{1}{2}\sum_{|x|>R} |J_x| . \tag{2.11}$$

Passing subsequently to the limits $\Lambda \to \infty$ and then $R \to \infty$, we obtain the theorem in full generality. $\qquad\square$

Remark. According to (1.8) and (1.10) f_Λ is a concave function of h and βf_Λ is a concave function of β. This and bounds (2.1) and (2.4) imply that the limiting free energy f is continuous in h and β.

Consider now the disordered case.

Theorem 2.2. *Consider a spin system with an ergodic interaction and external field, satisfying the conditions (1.29). Then there exists a non-random quantity f such that we have with probability 1 for the free energy f_Λ of the system*

$$\lim_{\Lambda\to\infty} f_\Lambda = f$$

and

$$f \le \frac{1}{2}\sum_{x\in\mathbb{Z}} \mathbf{E}\{|J_{0x}|\} + \mathbf{E}\{|h_0|\} + \beta^{-1}\log 2 .$$

Proof. Assume first that the interaction is bounded and of a finite range, i.e. that there exist $A < \infty$, $R < \infty$, such that

$$\sup_{x,y\in\mathbb{Z}} |J_{xy}| \le A, \quad J_{x,y} = 0, \quad |x-y| > R . \tag{2.12}$$

Note that in the disordered case $F_\Lambda := -\beta\log Z_\Lambda$ depends on Λ via the random variables $\{J_{xy}\}_{x,y\in\Lambda}$, $\{h_x\}_{x\in\Lambda}$. Since they are ergodic (see (1.24)), we have

$$F_\Lambda(T^a\omega) = F_\Lambda(\omega), \quad \forall a \in \mathbb{Z}^d . \tag{2.13}$$

Consider first the case $d = 1$. According to (1.11) we write for $\Lambda = [a, a+\ell)$, $a \in \mathbb{Z}$, $\ell \in \mathbb{N}$

$$\Lambda = \bigcup_{k=0}^{p-1} \Lambda_k + \mathcal{R}, \quad \Lambda_k = \big[a+\lambda k, a+\lambda(k+1)\big), \quad \mathcal{R} = [a+\lambda p, a+\lambda p+\mu) .$$

Now, arguing as in deriving (2.8), we obtain

$$|\Lambda|^{-1}F_\Lambda \leq \frac{\lambda p}{\ell} \cdot \frac{1}{p} \sum_{k=0}^{p-1} \frac{F_{\Lambda_k}}{|\Lambda_k|} + \frac{JR^2}{\lambda} + \frac{J\lambda^2}{\ell} \qquad (2.14)$$

and the opposite inequality in which the sign of the second term of the r.h.s. is opposite.

In view of (2.13) we have $F_{\Lambda_{k+1}}(\omega) = F_{\Lambda_k}(T^\lambda\omega)$. Hence $\{F_{\Lambda_k}\}_{k\in\mathbb{N}}$ is a stationary sequence. This allows us to use the Birkhoff ergodic theorem while passing to the limit $\Lambda \to \infty$ ($\ell \to \infty$) in (2.14) and to obtain with probability 1:

$$-\frac{JR^2}{\lambda} + \mathbf{E}\left\{\frac{F_{\Lambda_1}}{|\Lambda_1|}\Big|\mathcal{F}_\lambda\right\} \leq \liminf_{\Lambda\to\infty} \frac{F_\Lambda}{|\Lambda|} \leq \limsup_{\Lambda\to\infty} \frac{F_\Lambda}{|\Lambda|} \leq \mathbf{E}\left\{\frac{F_{\Lambda_1}}{|\Lambda_1|}\Big|\mathcal{F}_\lambda\right\} + \frac{JR^2}{\lambda} ,$$

where $\mathcal{F}_\lambda$ is the σ-algebra of events, invariant with respect to $(T^\lambda)^q$, $q \in \mathbb{Z}$, and $\mathbf{E}\{\cdots|\mathcal{F}_\lambda\}$ is respective conditional mathematical expectation. Since λ is arbitrary, the last inequalities imply the existence of the limit $\lim_{\Lambda\to\infty} |\Lambda|^{-1}F_\Lambda = \lim_{\Lambda\to\infty} f_\Lambda$ with probability 1. It is easy to check that the limit is invariant with respect to T^q, $q \in \mathbb{Z}$, hence it is non-random because of the ergodicity (1.21). This proves the theorem in the case (2.12) of finite range interaction and $d = 1$. Multidimensional case $d \geq 1$ can be treated analogously [Kh-Si, Pa-Fi, Vu], by using the multiparameter ergodic theorem (see e.g. [Kr], Section 6.2) .

In a general case of the Hamiltonian satisfying (1.29) and $d \neq 1$ we introduce the "truncated" interaction

$$J_{xy}\chi_R(x - y)\chi_A(J_{xy}) , \qquad (2.15)$$

where χ_R is the indicator of the ball of radius R in $\mathbb{Z}^d$ and χ_A is the indicator of the interval $[-A, A]$ of $\mathbb{R}$. Since (2.15) satisfies (2.12), the respective free energy $f_\Lambda^{(A,R)}$ possesses the thermodynamic limit $f^{(A,R)}$ with probability 1. By introducing the "interpolating" interaction, analogous to that of (2.10) we obtain (cf. (2.11))

$$|f_\Lambda - f_\Lambda^{(A,R)}| \leq \frac{1}{2|\Lambda|} \sum_{x\in\Lambda} j_x^{(A)} + \frac{1}{2|\Lambda|} \sum_{x\in\Lambda} j_x^{(R)} , \qquad (2.16)$$

where

$$j_x^{(A)} = \sum_{y\in\mathbb{Z}} |J_{xy}|\chi_A(J_{xy}), \quad j_x^{(R)} = \sum_{y\in\mathbb{Z}} |J_{xy}|\chi_R(x - y) .$$

These formulas define two ergodic fields, finite with probability 1, because of the Fubini theorem and the inequalities

$$\mathbf{E}\{|j_x^{(A)}|\} \leq \sum_{y\in\mathbb{Z}} \mathbf{E}\{|J_{0y}|\} := \hat{J} < \infty, \quad \mathbf{E}\{|j_x^{(R)}|\} \leq \hat{J} < \infty . \qquad (2.17)$$

Passing to the limit $\Lambda \to \infty$ in (2.16), we obtain, again in view of the ergodic theorem,

$$\limsup_{\Lambda \to \infty} |f_\Lambda - f^{(A,R)}| \le \frac{1}{2} \sum_{x \in \mathbb{Z}} \mathbf{E}\{|J_{0x}|\chi_A(J_{0x})\} + \frac{1}{2} \sum_{x \in \mathbb{Z}} \mathbf{E}\{|J_{0x}|\chi_R(x)\} \ .$$

The terms of the series of the r.h.s. are bounded from above by the terms of the sequence $\{\mathbf{E}\{|J_{0x}|\}\}_{x \in \mathbb{Z}}$, belonging to $\ell^1(\mathbb{Z})$ in view of conditions (1.29) of the theorem. Besides, the terms vanish as $A \to \infty$ and $R \to \infty$. Hence the limit of the r.h.s. of the last inequality is zero as $A \to \infty$, $R \to \infty$. This proves the theorem in the full generality. $\qquad\square$

Remarks. 1) Note that we have proved not only the existence with probability 1 of the limit of the free energy, but also that the limit is non-random, despite that for any finite system the free energy is random due to the randomness of the parameters of the Hamiltonian. This property of macroscopic characteristics of disordered systems, satisfying (1.21), (1.24), the free energy first of all, is known as selfaveraging.

2) We mention a slightly different scheme of proof of the existence of the thermodynamic limit. Consider again the one-dimensional translation invariant case. Denoting

$$\Phi_\ell = F_\ell + C_1, \tag{2.18}$$

where F_ℓ and C_1 are the same as in (2.6), we obtain from the right-hand side of (2.6) the inequality

$$\Phi(\ell' + \ell'') \le \Phi(\ell') + \Phi(\ell'') \ , \tag{2.19}$$

showing that $\Phi : \mathbb{N} \to \mathbb{R}$ is a subadditive sequence. Recalling that Φ is bounded below (see (2.4)), we conclude in view of the well-known property of subadditive sequences (see [Po-Sz], Problem 99) that

$$\lim_{\ell \to \infty} \Phi(\ell)/\ell = \inf_{\ell \in \mathbb{N}} \Phi(\ell)/\ell > -\infty \ . \tag{2.20}$$

This and (2.18) imply that $\lim_{\ell \to \infty} F(\ell)/\ell = \inf_\ell \Phi(\ell)/\ell$. In the disordered case (1.24), (1.29) we obtain the inequality

$$\Phi_{\Lambda'+\Lambda''} \le \Phi_{\Lambda'} + \phi_{\Lambda''} \ , \tag{2.21}$$

where $\Lambda' = [a, a+b')$, $\Lambda'' = [a+b', a+b'+b'')$, $a \in \mathbb{Z}$, $b', b'' \in \mathbb{N}$. Hence, to prove the existence with probability 1 of the limit $\lim_{\Lambda \to \infty} \Phi_\Lambda/|\Lambda|$ we have to use the subadditive ergodic theorem (see [Kr], Theorem 2.9 of Chapter 6).

3 Thermodynamic Limit for Mean Field Type Models

3.1 Weak Selfaveraging Property

The mean field type models require different means to prove the existence of the thermodynamic limit. Indeed, the physical idea behind the proofs of the previous section is quite simple: if a large system is split into a number of still large parts, then the free energy of the system is the sum of the parts plus the correction (cross) terms due to the interaction of adjacent parts across their boundary. The cross terms are roughly proportional to the boundary area and vanish as the surface-volume tends to zero (see e.g. inequality (2.9)). This is seen already on the level of the Hamiltonian (1.2) itself. Indeed, if $\Lambda = \Lambda' \cup \Lambda''$, $\Lambda \cap \Lambda'' = \emptyset$, then we have

$$H_\Lambda = H_{\Lambda'} + H_{\Lambda''} + H_{\Lambda'\Lambda''} , \quad H_{\Lambda'\Lambda''} = \sum_{x \in \Lambda'} \sum_{y \in \Lambda''} J_{xy} s_x s_y ,$$

and since the interaction is of short range (see (1.17),(1.29)), the cross term $H_{\Lambda'\Lambda''}$ is of the order the area of the boundary between Λ' and Λ'', provided that the boundary is sufficiently regular (see e.g. (1.11)).

On the other hand, even in the simplest case of the Kac model (1.18), $H_{\Lambda'\Lambda''}$ is of the same order as $H_{\Lambda'}$ and $H_{\Lambda''}$. This is why we divide the proof of existence of the thermodynamic limit for disordered mean-field type models into two parts. First, we prove that fluctuations of the free energy vanish as $\Lambda \to \infty$. In the simplest case this will be the limiting relation [Pa-Sh]

$$\lim_{\Lambda \to \infty} \mathbf{E}\left\{ \left(f_\Lambda - \mathbf{E}\{f_\Lambda\} \right)^2 \right\} = 0 . \tag{3.1}$$

We will say in this case that the free energy possesses the weak selfaveraging property. This is a rather simple fact, that can be proven by several methods (see below). To have the selfaveraging property of the free energy we have to prove in addition that its expectation $\mathbf{E}\{f_\Lambda\}$ converges to a certain limit as $\Lambda \to \infty$. This proved to be a rather nontrivial problem, that was solved recently for the Sherrington–Kirkpatrick (see [Gu-To] and Theorem 3.5 below).

Theorem 3.1. *Consider the Sherrington–Kirkpatrick model, defined by interaction (1.31) in which $\{I_{xy}\}_{x,y \in \mathbb{Z}^d}$ are the Gaussian random variables, verifying (1.32). Then its free energy admits the bound*

$$\mathbf{E}\left\{ \left(f_\Lambda - \mathbf{E}\{f_\Lambda\} \right)^2 \right\} \le \frac{C\beta^2 I^4}{|\Lambda|} \tag{3.2}$$

where C is an absolute constant.

The proof is based on the following

Proposition 3.2. *Let $\{\xi_j\}_{j=1}^N$ be a collection of independent random variables, having probability laws $P_1, \ldots, P_N$, and let φ_N be a Borel function of $\{\xi_j\}_{j=1}^N$. Set*

$$\varphi_{N,j} = \int \varphi_N(\xi_1, \ldots, \xi_j, \xi_{j+1}, \ldots, \xi_N) P(d\xi_{j+1}) \cdots P(d\xi_N) ,$$

so that

$$\varphi_{N,N} = \varphi_N(\xi_1, \ldots, \xi_N), \quad \varphi_{N,0} = \mathbf{E}\{\varphi_N(\xi_1, \ldots, \xi_N)\} \ .$$

Then

$$\mathbf{E}\{|N^{-1}\varphi_N - \mathbf{E}\{N^{-1}\varphi_N\}|^2\} = N^{-2} \sum_{j=1}^{N} \mathbf{E}\{|\varphi_{N,j} - \varphi_{N,j-1}|^2\} \ , \qquad (3.3)$$

where $\mathbf{E}\{\ldots\}$ *denotes the expectation with respect to the product measure* $P_1 \cdots P_N$.

Proof. Write the difference $\varphi_N - \mathbf{E}\{\varphi_N\} = \varphi_{N,N} - -\varphi_{N,0}$ as the sum

$$\sum_{j=1}^{N} (\varphi_{N,j} - \varphi_{N,j-1}) \ .$$

Then it is easy to find that the expectation of the square of the difference is equal to the r.h.s. of (3.3), because the expectation of the product of different terms of the sum is always zero due to definition of $\varphi_{N,j}$. □

The proposition is a simple application of the martingale idea. Its use in our case is facilitated by

Corollary 3.3. *Assume that under the condition of the previous proposition we have* $\mathbf{E}\{|\varphi_N - \varphi_N|_{\xi_j=0}|^2\} \leq C, \ j = 1, \ldots, N,$ *where* C *is independent of* j *and* N. *Then*

$$\mathbf{E}\{|N^{-1}\varphi_N - \mathbf{E}\{N^{-1}\varphi_N\}|^2\} \leq C/N \ .$$

Proof of Theorem 3.1. We will use the corollary, taking $F_\Lambda = -\beta^{-1}\log Z_\Lambda$ as φ_N, $N = |\Lambda|$ and $\xi_z = \{J_{x,z}\}_{x \in \Lambda \setminus \{z\}}$ as ξ_j. Introduce the "interpolating" Hamiltonian

$$H_\Lambda(t) = H_{\Lambda_z} - h_z s_z - N^{-1/2}I_{zz}/2 + H_z t, \ t \in [0,1],$$

where

$$\Lambda_z = \Lambda \setminus \{z\}, \quad H_z = -N^{-\frac{1}{2}} \sum_{x \in \Lambda_z} I_{xz} s_x s_z \ .$$

We obtain

$$F_\Lambda - F_\Lambda|_{\xi_z=0} = \int_0^1 \langle H_z \rangle_t dt = -N^{-1/2} \sum_{x \in \Lambda_z} \int_0^1 I_{xz} \langle s_x s_z \rangle_t dt,$$

where $\langle \ldots \rangle_t$ denotes the expectation with respect to the Gibbs measure (1.3), corresponding to the Hamiltonian $H_\Lambda(t)$. By using the Schwarz inequality we have

$$\mathbf{E}\{|F_\Lambda - F_\Lambda|_{\xi_z=0}|^2\} \leq \int_0^1 \langle H_z \rangle_t^2 dt$$

$$= N^{-1} \sum_{x,y \in \Lambda_z} \int_0^1 \mathbf{E}\{I_{xz}I_{yz}\langle s_x s_z \rangle_t \langle s_y s_z \rangle_t\} dt \ . \ (3.4)$$

We will use now the formula, valid for any Gaussian variable ξ with zero mean and variance I^2 and a C^1 function $\psi : \mathbb{R} \to \mathbb{R}$ that is polynomially bounded:

$$\mathbf{E}\{\xi\psi(\xi)\} = I^2\mathbf{E}\{\psi'(\xi)\} . \tag{3.5}$$

The formula can be easily proved by the integration by parts.

Applying the formula twice to exclude the Gaussian random variables I_{xz}, I_{yz} from the integrand of (3.4), we obtain the product of $\beta^2 I^4 N^{-1}$ and a polynomial of degree 4 in variables $\langle s_x s_z \rangle_t$, $\langle s_y s_z \rangle_t$, and $\langle s_x s_y \rangle_t$ with numerical coefficients. This yields the bound $N^{-1}C\beta^2 I^4$ for the integrand of (3.4) with an absolute constant C, hence the bound $\mathbf{E}\{|F_\Lambda - F_\Lambda|_{\xi_x=0}|^2\} \leq C\beta^2 I^4$, that allows us to use Corollary 3.3. This proves the theorem. $\square$

Remarks. 1) We formulated and proved the theorem for the case of independent identically and symmetrically distributed Gaussian interactions $\{I_{xy}\}_{x,y\in\mathbb{Z}}$ in (1.31)–(1.32). In fact practically the same argument proves an N-independent bound for $\mathbf{E}\{|F_\Lambda - F_\Lambda|_{\xi_x=0}|^2\}$, hence the theorem, in a more general case of independent (except the symmetry condition (1.27)) random variables $\{I_{xy}\}_{x,y\in\mathbb{Z}}$, having zero mean, finite variance I^2 (see (1.32)), and such that

$$\mu_3 := \max_{x,y\in\mathbb{Z}} \mathbf{E}\{|I_{xy}|^3\} < \infty . \tag{3.6}$$

In this case we use the formula

$$\mathbf{E}\{\xi\psi(\xi)\} = \sum_{a=1}^{p} \frac{\kappa_{a+1}}{a!}\mathbf{E}\{\psi^{(a)}(\xi)\} + \varepsilon_p , \tag{3.7}$$

instead of (3.5), where κ_a, $a = 1, 2, \ldots$ are the semi-invariants (cumulants) of a random variable ξ,

$$|\varepsilon_p| \leq C_p \sup_t |\psi^{(p+1)}(t)| \cdot \mathbf{E}\{|\xi|^{p+2}\},$$

and C_p is a numerical constant. The formula for $p = 1$ yields for the r.h.s. of (3.4) the bound $O(\mu_3 N^{1/2})$, hence the bound $O(\mu_3 N^{-1/2})$ for the variance of the free energy, i.e. an analog of (3.2) with $O(|\Lambda|^{-1/2})$ on its r.h.s. To obtain the bound whose r.h.s is of the same order $O(|\Lambda|^{-1})$ as in (3.2) we have to assume that $\mathbf{E}\{I_{xy}^3\} = 0$, $\mu_4 := \max_{x,y\in\mathbb{Z}} \mathbf{E}\{|I_{x,y}|^4\} < \infty$.

2) Theorem 3.1 implies that $f_\Lambda - \mathbf{E}\{f_\Lambda\}$ converges to zero in probability as $|\Lambda| = N \to \infty$. By applying a slightly more sophisticated than in Proposition 3.2 analysis to the expectation of the fourth power of $\varphi_N - \mathbf{E}\{\varphi_N\}$, we obtain (cf.(3.3))

$$\mathbf{E}\{|N^{-1}\varphi_N - \mathbf{E}\{N^{-1}\varphi_N\}|^4\} \leq CN^{-4}\left(\sum_{j=1}^{N} \left(\mathbf{E}\{(\varphi_{N_{i,j}} - \varphi_{N,j-1})^4\}\right)^{1/2}\right)^2 .$$

This inequality, which is a particular case of the Burkholder inequalities (see e.g. [Shi], p. 499), allows us to prove the bound of the order $O(N^{-2})$ for

$\mathbf{E}\{|f_\Lambda - \mathbf{E}\{F_\Lambda\}|^4\}$, and, as a result, the convergence to zero of $f_\Lambda - \mathbf{E}\{f_\Lambda\}$ with probability 1 as $\Lambda \to \infty$.

3) An analogous result is valid for the Hopfield model with interaction (1.30). Namely, we have [Sh-Ti]

Theorem 3.4. *Consider the Hopfield model, defined by* (1.2), (1.30), *in which* $\{\xi_x^\mu\}_{\mu=1,x\in\mathbb{Z}}^p$ *are independent Bernoulli random variables, i.e., assume values* ± 1 *with probability* $1/2$ *and* $p = [\alpha N]$, $0 < \alpha < \infty$. *Then the free energy* f_Λ *of the model admits the bound*

$$\mathbf{E}\{|f_\Lambda - \mathbf{E}\{f_\Lambda\}|^2\} \leq C/|\Lambda| . \tag{3.8}$$

4) By using the Tchebyshev inequality we can obtain from (3.2) and (3.8) and from more strong bounds of Remark 2 polynomially decaying in $|\Lambda|$ bounds for the probability $\mathbf{P}\{|f_\Lambda - Ef_\Lambda| \geq \varepsilon\}$. More advanced techniques, in particular those, based on the concentration phenomenon, yield exponential bounds in Λ for this probability [Bo-Pi, Ta2].

3.2 Strong Selfaveraging Property of the Free Energy

According to the previous subsection, to prove the existence of the non-random limit of the free energy (in probability, with probability 1) we have to prove the existence of the limit of the expectation $\mathbf{E}\{f_\Lambda\}$ as $\Lambda \to \infty$:

$$\lim_{\Lambda \to \infty} \mathbf{E}\{f_\Lambda\} = f . \tag{3.9}$$

We will present an elegant recent proof [Gu-To] of this fact for the Sherrington-Kirkpatrick model. The ideas, used in the proof, provide also a crucial ingredient in justifying the explicit form of the limiting free energy [Gu, Ta1], proposed by Parisi in the early 1980's (see [MPV] for an extensive physics presentation and discussion).

Theorem 3.5. *Consider the Sherrington–Kirkpatrick model defined by* (1.31)–(1.32), *in which* $\{I_{xy}\}_{x,y\in\mathbb{Z}}$ *are independent (modulo*(1.27)*) Gaussian random variables. Then the relation* (3.9) *holds.*

Proof. We are going to prove that if $F_\Lambda = -\beta^{-1}\log Z_\Lambda$, where Z_Λ is the partition function of the Sherrington–Kirkpatrick model, then the quantity

$$\overline{F}_N := \mathbf{E}\{F_\Lambda\} \tag{3.10}$$

that depends only on $N = |\Lambda|$ is a subadditive sequence in N. To this end we introduce, following [Gu-To], the "interpolating" Hamiltonian

$$H_\Lambda(t) = \sqrt{t}H_\Lambda + \sqrt{1-t}H'_{\Lambda'} + \sqrt{1-t}H''_{\Lambda''} , \tag{3.11}$$

where $\Lambda = \Lambda' \cup \Lambda''$, $\Lambda' \cap \Lambda'' = \emptyset$, and $H_\Lambda, H'_{\Lambda'}, H''_{\Lambda''}$ are given by (1.2), (1.31), (1.32) in which $\{I_{xy}\}_{x,y\in\Lambda}$, $\{I'_{xy}\}_{x,y\in\Lambda'}$ and $\{I''_{xy}\}_{x,y\in\Lambda''}$ are three independent identically distributed collections of the Gaussian random variables, that are independent modulo condition (1.27). It is easy to see that if Z_Λ, $Z_{\Lambda'}$ and $Z_{\Lambda''}$ are the corresponding partition functions (1.4) and $Z_\Lambda(t)$ is the partition function, corresponding to Hamiltonian (3.11), then

$$Z_\Lambda(1) = Z_\Lambda, \quad Z_\Lambda(0) = Z_{\Lambda'} Z_{\Lambda''} . \tag{3.12}$$

We have now

$$\frac{d}{dt}\mathbf{E}\{ -\beta \log Z_\Lambda(t)\} = -\frac{1}{4\sqrt{Nt}} \sum_{x,y\in\Lambda} \mathbf{E}\{I_{xy}\langle s_x s_y\rangle_t\}$$
$$+\frac{1}{4\sqrt{N'(1-t)}} \sum_{x,y\in\Lambda'} \mathbf{E}\{I'_{xy}\langle s_x s_y\rangle_t\} + \frac{1}{4\sqrt{N''(1-t)}} \sum_{x,y\in\Lambda'} \mathbf{E}\{I''_{xy}\langle s_x s_y\rangle_t\} , \tag{3.13}$$

where $\langle \dots \rangle_t$ denotes the expectation with respect to the Gibbs measure (1.3), corresponding to the Hamiltonian (3.11), $N = |\Lambda|$, $N' = |\Lambda'|$, $N'' = |\Lambda''|$, $N = N' + N''$. Now, by using the integration by parts formula (3.5), we obtain for the r.h.s. of (3.13)

$$-\frac{\beta I^2}{8}\mathbf{E}\Big\{ \frac{1}{N} \sum_{x,y\in\Lambda} \left(1 - \langle s_x s_y\rangle_t^2\right) - \frac{1}{N'} \sum_{x,y\in\Lambda'} \left(1 - \langle s_x s_y\rangle_t^2\right)$$
$$-\frac{1}{N''} \sum_{x,y\in\Lambda''} \left(1 - \langle s_x s_y\rangle_t^2\right)\Big\} . \tag{3.14}$$

It is convenient to introduce the product of two copies of the Gibbs measure (1.3) for $H_\Lambda(t)$ of (3.11) that allows us to write $\langle s_x s_y\rangle_t^2 = \langle\langle s_x^{(1)} s_y^{(1)} s_x^{(2)} s_x^{(2)}\rangle\rangle_t$ where $\{s_x^{(1)}\}_{x\in\Lambda}$ and $\{s_x^{(2)}\}_{x\in\Lambda}$ are two copies (replicas) of the spin variables and $\langle\langle \dots \rangle\rangle_t$ denotes the expectation with respect to the product Gibbs measure. We obtain instead of (3.14)

$$-\frac{\beta I^2 N}{8}\mathbf{E}\Big\{\Big\langle\Big\langle S_\Lambda^2 - \frac{N'}{N}S_{\Lambda'}^2 - \frac{N''}{N}S_{\Lambda''}^2 \Big\rangle\Big\rangle_t\Big\} , \tag{3.15}$$

where

$$S_\Lambda := \frac{1}{N} \sum_{x\in\Lambda} s_x^{(1)} s_x^{(2)} = \frac{N'}{N}S_{\Lambda'} + \frac{N''}{N}S_{\Lambda''} .$$

This and the convexity of the function $x \mapsto x^2$ imply that expression (3.15) is non-positive, hence the function $\mathbf{E}\{-\beta \log Z_\Lambda(t)\}$ is non-increasing. Taking into account (3.12), we conclude finally that $\overline{F}_N$ of (3.10) is subadditive:

$$\overline{F}_{N'+N''} \le \overline{F}_{N'} + \overline{F}_{N''} \tag{3.16}$$

(cf. (2.19)). Now, to obtain the existence of the limit $\lim_{N\to\infty} N^{-1}\overline{F}_N$ we have only to check that $N^{-1}\overline{F}_N$ is bounded from below. We have

$$\frac{1}{N}\frac{d}{d\beta}\beta\overline{F}_N = -\frac{1}{2N^{3/2}}\sum_{x,y\in\Lambda}\mathbf{E}\{I_{xy}\langle s_x s_y\rangle_{H_\Lambda}\} .$$

By using again formula (3.5) we obtain for the r.h.s. of this equality

$$-\frac{\beta I^2}{4N^2}\sum_{x,y\in\Lambda}\left(1 - \langle s_x s_y\rangle^2_{H_\Lambda}\right) .$$

Since $\overline{F}_N|_{\beta=0} = -\beta^{-1}N\log 2$, we obtain finally the bound

$$N^{-1}\overline{F}_N \geq -\beta^{-1}\log 2 - \frac{\beta I^2}{4} .$$

We use now the standard subadditivity lemma (see [Po-Sz], Problem 99), that yields the assertion of the theorem:

$$\lim_{N\to\infty}\overline{F}_N/N = \inf_N \overline{F}_N/N . \qquad\qquad \square$$

Remark. We remind the reader, that the existence of the thermodynamic limit for the translation invariant short range case can also be proved, basing on the subadditivity of the free energy (see Remark 2 after Theorem 2.1).

4 Two Simple Models with Phase Transitions

4.1 Kac Model

The model is defined by the Hamiltonian (1.2)–(1.18)

$$H_\Lambda = -\frac{1}{2N}\sum_{x,y\in\Lambda} s_x s_y - h\sum_{x\in\Lambda} s_x = -\frac{1}{2N}M_\Lambda^2 - hM_\Lambda,$$

where M_Λ is given in (1.5). The respective partition function (1.4) is

$$Z_\Lambda = \sum_{\{s_x=\pm 1,\ x\in\Lambda\}} \exp\left\{\frac{\beta}{2N}M_\Lambda^2 + \beta hM_\Lambda\right\}. \qquad (4.1)$$

By using the identity

$$e^{\frac{\sigma x^2}{2}} = \frac{1}{\sqrt{2\pi\sigma}}\int_{-\infty}^{\infty} e^{-\frac{a^2}{2\sigma} - xa}da ,$$

valid for $\sigma > 0$, $x \in \mathbb{R}$, and choosing here $\sigma = (\beta N)^{-1}$, $x = \beta M_\Lambda$ to "linearize" the term $\beta M_\Lambda^2/2N$ in the exponent of (4.1), we rewrite the partition function as

$$Z_\Lambda = \sqrt{\frac{\beta N}{2\pi}} \int_{-\infty}^{\infty} e^{-\frac{\beta N t^2}{2}} \sum_{\{s_x = \pm 1,\, x \in \Lambda\}} \exp\left\{ \beta(t+h) \sum_{x \in \Lambda} s_x \right\} dt$$

$$= \sqrt{\frac{\beta N}{2\pi}} \int_{-\infty}^{\infty} \exp\left\{ -\beta N\left(\frac{t^2}{2} - \beta^{-1} \log 2 \cosh \beta(t+h) \right) \right\} dt.$$

Now a standard application of the Laplace method yields the existence of the thermodynamic limit of the free energy and its explicit form

$$f = \min_{t \in \mathbb{R}} \left\{ \frac{t^2}{2} - \frac{1}{\beta} \log 2 \cosh \beta(t+h) \right\}. \tag{4.2}$$

Note that we obtained here the existence of the thermodynamic limit for the free energy as a by-product of its explicit calculation.

It is easy to find that if $h \neq 0$, then the minimum in (4.2) is unique and solves the equation

$$t = \tanh \beta(t+h). \tag{4.3}$$

The solution is real analytic in $\beta > 0$ and in $h \neq 0$. We have also

$$-\frac{\partial f}{\partial h} = t.$$

On the other hand, we have from (4.2) (see also (1.7))

$$\langle N^{-1} M_\Lambda \rangle = -\frac{\partial f_\Lambda}{\partial h}.$$

Recall now a simple property of convex functions, known in statistical mechanics as the Griffith lemma (see e.g. [Ru1])

Proposition 4.1. *Let I be an interval of $\mathbb{R}$ and $\{f_n\}$ be a sequence of convex C^1 functions $f_n : I \to \mathbb{R}$ that converges point-wise to a (convex) function f. If f is differentiable at a point $x_0 \in I$, then*

$$\lim_{n \to \infty} f_n'(x_0) = f'(x_0).$$

By using the proposition, we can identify the solution $t(\beta, h)$ of (4.3) with the limiting expectation of the magnetic moment per site, an important physical quantity, known as the magnetization of the system. We will denote this quantity $m(\beta, h)$.

Consider now the behavior of $m(\beta, h)$ near $h = 0$ for $\beta > 0$. It is easy to see that if $\beta < 1 := \beta_c$, then $m(\beta, 0)$ is zero and $f(\beta, h)$ is real analytic in $h \in \mathbb{R}$. If, however, $\beta > \beta_c$, then (4.3) has two solutions $\pm t(\beta, 0)$ and

$$\lim_{h \to 0^\pm} m(\beta, h) = \pm t(\beta, 0).$$

In other words, the magnetization is real analytic in h for $\beta < \beta_c$ and has the jump $2t(\beta, 0)$ at $h = 0$, if $\beta > \beta_c$.

Since $m(\beta, h)$ is the first derivative of the free energy, we say that the model displays the phase transition of the fist order at $h = 0$, if the temperature is lower than the critical temperature $T_c = \beta_c^{-1} = 1$. The quantity $m(\beta, 0^+)$ is known as the spontaneous magnetization.

This is the simplest model of the ferromagnetic phase transition, proposed by P. Curie and Weiss.

The model has one more phase transition, this time in temperature. Namely, consider the behavior of $m(\beta, 0^+) := m(\beta)$ for $\beta_c \leq \beta \leq \infty$. It is easy to find from (4.3) that $m(\beta)$ is monotone increasing in β, $m(\infty) = 1$, $m(\beta_c) = 0$, and that

$$m(\beta) = \sqrt{3(\beta - \beta_c)}(1 + o(1)), \quad \beta \searrow \beta_c.$$

We see that $m(\beta)$ is non-analytic at β_c. Denote

$$\chi(\beta, h) = \frac{\partial m}{\partial h}. \tag{4.4}$$

$\chi(\beta, h)$ is known as the magnetic susceptibility. Then we have for $\beta < \beta_c$

$$\chi(\beta, h) = \frac{\beta}{\cosh^2 \beta m(\beta, h) - \beta},$$

in particular, $\chi^{-1}(\beta) = (\beta_c - \beta)(1 + o(1))$, $\beta \nearrow \beta_c$, i.e., the second derivative of the free energy with respect to the external magnetic field h has a singularity at $\beta = \beta_c$. We say that for these values of the parameters (the temperature and the external field) the model has the phase transition of the second order (or the critical point at $\beta = \beta_c, h = 0$).

Another manifestation of this phase transition is the singularity of the heat capacity

$$c(\beta, h) := -T \frac{\partial^2 f}{\partial T^2}, \quad T = \beta^{-1}.$$

It is easy to show that $c(\beta, 0)$ has a jump at β_c:

$$c(\beta, 0) = \begin{cases} 0, & \beta < \beta_c, \\ c(\beta_c^+, 0) - \text{const} \cdot (\beta - \beta_c), & \beta \searrow \beta_c, \end{cases}$$

where $c(\beta_c^+, 0) > 0$.

It is widely believed and proved in certain cases that more realistic models, in particular, models with the nearest neighbor interaction (1.15)) have a qualitatively the same behavior of basic quantities m, χ, and c and others. The only difference is numerical values of indices of the power low singularities at $T_c = \beta_c^{-1}$. Note that, according to the above, we can write for $\tau = (T - T_c)/T_c$:

$$m(\beta, 0) = C_{m1}|\tau|^\beta, \quad \tau \to 0^-, \tag{4.5}$$

$$m(\beta_c, h) = C_{m2}|\tau|^{1/\delta}, \quad h \to 0, \tag{4.6}$$

$$\chi(\beta, 0) = C_\chi |\tau|^{-\gamma}, \quad \tau \to 0^+, \tag{4.7}$$

$$c(\beta, 0) = C_\pm |\tau|^{\alpha_\pm}, \quad \tau \to 0^\pm, \tag{4.8}$$

where

$$\beta = 1/2, \ \gamma = 1, \ \delta = 3, \ \alpha_{\pm} = \text{not defined} \qquad (4.9)$$

On the other hand, for the nearest neighbor positive interaction in the two-dimensional case $d = 2$ we have (see e.g. [TKS])

$$\beta = 1/8, \ \gamma = 7/4, \ \delta = 15, \ \alpha_{\pm} = 0, \qquad (4.10)$$

more precisely, the heat capacity has the logarithmic singularity at $\tau = 0$.

The numbers α, β, γ, δ are known as critical indices of a model (for one more collection of critical indices see (4.21)). According to the universality property of phase transitions of the second order the critical indices of the short range models depend only on the dimension d of the lattice $\mathbb{Z}^d$ and do not depend on the interaction provided that it decays fast enough at infinity. Proofs of the conjecture as well as calculation of critical indices is a goal of the theory of critical phenomena (see [LL, Pat, Sin, TKS, Wi, ZJ] and references therein).

Similar argument allows us to find the free energy for a simplest random version of the Kac model [Lu, Pa-Fi], i.e., the Hopfield model (1.30) with $p = 1$ (in fact, any finite p [Pa-Fi]) and $\mathbf{E}\{(\xi_x^\mu)^2\} = 1$. We obtain (cf. (4.2))

$$f = \min_t \left\{ \frac{t^2}{2} - \beta^{-1}\mathbf{E}\{\log 2 \cosh \beta(t\xi_0^\mu + h)\} \right\} \qquad (4.11)$$

It is easy to see that if $\mathbf{E}\{\xi_x^\mu\} \neq 0$, the qualitative behavior of the model is similar to that of the Kac model. If, however, the probability law of ξ_0^μ is symmetric with respect to zero, then there exists a phase transition at $\beta_c^{-1} = T_c = \mathbf{E}\{(\xi_x^\mu)^2\} = 1$, but $m(\beta, h)$ is continuous in h for all β (recall, that in the Kac model $m(\beta, h)$ has a jump at $h = 0$ for $\beta > \beta_c$). On the other hand, the susceptibility (4.4) as function of β for $h = 0$ has a cusp at $\beta = \beta_c$, i.e. its derivative has a jump at $\beta = \beta_c$. Hence, we have here the phase transition of the third order. This is a toy model of the spin glass phase transition, where the system possesses a non-trivial magnetic structure below T_c, but the macroscopic magnetic moment is absent [MPV].

4.2 Spherical Model

The model that was also proposed by M. Kac, is defined by the same Hamiltonian (1.2) in which now $\{s_x\}_{x \in \Lambda}$ vary over the sphere S_N of the radius $\sqrt{N}$, $N = |\Lambda|$ in $\mathbb{R}^N$, but not over the vertices $\{s_x = \pm 1\}_{x \in \Lambda}$ of the N-dimensional cube, as in the Ising model, considered above. Hence, the partition function of the model is

$$\int_{\sum_{x \in \Lambda} s_x^2 = N} e^{-\beta H_\Lambda(s)} \lambda_N(ds),$$

where λ_N is the Lebesgue measure on S_N.

The model can be obtained as a limiting case of the so called n-vector (or classical Heisenberg) model, in which s_x, $x \in \mathbb{Z}^d$ are n-dimensional vectors, uniformly distributed over the unit sphere in $\mathbb{R}^n$ (the Ising model can be seen as the $n = 1$ case of the n-vector model). It can be shown (see [Ka-Th, KKPS]) that the free energy of the spherical model (see formula (4.18)) is the limit as $n \to \infty$ of the free energy of the n-vector model.

To make the subsequent argument simpler we replace the model by another, in fact asymptotically equivalent, model in which the integration is carried out over the N-dimensional ball of the radius $\sqrt{N}$, i.e., we consider the partition function

$$Z_\Lambda = \int_{\sum_{x \in \Lambda} s_x^2 \leq N} e^{-\beta H_\Lambda(s)} \prod_{x \in \Lambda} ds_x.$$

We outline now a derivation of form of the free energy of the model in the translation invariant case (1.16)–(1.17). Consider a more general quantity

$$Z_\Lambda(r) = \int_{\sum_{x \in \Lambda} s_x^2 \leq r^2 N} e^{-\beta H_\Lambda(s)} \prod_{x \in \Lambda} ds_x, \quad r > 0.$$

We have evidently

$$Z_\Lambda = Z_\Lambda(1). \tag{4.12}$$

In view of (1.17) the convolution operator J, defined in $l^2(\mathbb{Z}^d)$ by the matrix $\{J_{x-y}\}_{x,y \in \mathbb{Z}^d}$, is bounded. Denoting its norm $\widehat{J}$, we can consider the integral

$$\Xi_\Lambda(z) = \int_0^\infty e^{-z\beta N r^2/2} 2r Z_\Lambda(r) dr, \tag{4.13}$$

defined for $z > \widehat{J}$. An easy calculation shows that

$$\Xi_\Lambda(z) = N^{-3/2} \left(\frac{2\pi}{\beta} \right)^{1/2} \left(\det(z - J_\Lambda) \right)^{-1/2} e^{\frac{\beta}{2}((z-J_\Lambda)^{-1} h_\Lambda, h_\Lambda)}, \tag{4.14}$$

where J_Λ is the finite dimensional operator, defined in $l^2(\Lambda)$ by the matrix $\{J_{x-y}\}_{x,y \in \Lambda}$ and $h_\Lambda = \{h, ..., h\} \in l^2(\Lambda)$.

Consider now a sequence of infinitely expanding rectangular domains $\Lambda \to \infty$ of (1.11) and use the standard results on restrictions of the convolution operators to these domains [Gr-Sz]. This leads to the following formula for $z > \widehat{J}$:

$$\sigma(z) := \lim_{\Lambda \to \infty} (\beta N)^{-1} \log \Xi_\Lambda(z) \tag{4.15}$$

$$= \frac{1}{2\beta} \log \frac{2\pi}{\beta} - \frac{1}{2\beta} \int_{\mathbb{T}^d} \log(z - \widehat{J}(k)) dk + \frac{h^2}{2(z - \widehat{J}(0))},$$

where $\widehat{J}(k)$ is the symbol of the convolution operator J, i.e.,

$$\widehat{J}(k) = \sum_{x \in \mathbb{Z}^d} J_x e^{2\pi i x k}, \quad k \in \mathbb{T}^d, \tag{4.16}$$

and $\mathbb{T}^d$ is the d-dimensional torus.

On the other hand, we can rewrite (4.13) as

$$\Xi_\Lambda(z) = \int_0^\infty e^{-N\beta(f_\Lambda(\sqrt{t})+zt/2)} dt,$$

where

$$f_\Lambda(r) = -\frac{1}{\beta N} \log Z_\Lambda(r),$$

and in view of (4.12) we have

$$f_\Lambda = f_\Lambda(1), \tag{4.17}$$

where f_Λ is the free energy of the model. By using now the Tauberian theorem of the paper [Mi-Po] we find from (4.15) for any $r > 0$

$$\lim_{\Lambda \to \infty} f_\Lambda(r) = f(r),$$

where

$$f(r) = -\min_{z > \hat{\jmath}} \left\{ \sigma(z) + \frac{zr^2}{2} \right\}.$$

It can be shown that the derivative of $f_\Lambda(r)$ with respect to r is uniformly bounded in N in a certain neighborhood of $r = 1$. This implies that the convergence f_Λ to f is uniform in r and we obtain from (4.17) the following formula for the free energy of the spherical model

$$f = -\min_{z > \hat{\jmath}} \left\{ \sigma(z) + \frac{z}{2} \right\}, \tag{4.18}$$

where $\sigma(z)$ is defined in (4.15).

Assume now that the symbol (4.16) has a unique and non-degenerate maximum at $k = 0$. In particular, this will be the case for the nearest neighbor non-negative interaction (cf. (1.15))

$$J_x = \begin{cases} J, & |x| = 1, \\ 0, & |x| > 1, \end{cases} \tag{4.19}$$

modelling a ferromagnet. Indeed, in this case

$$\widehat{J}(k) = 2J \sum_{l=1}^d \cos k_l, \quad k = (k_1, ..., k_d).$$

This assumption implies that $\hat{\jmath} = \widehat{J}(0)$, and we have from (4.18) the equation for the value of z at which the r.h.s of (4.18) has the minimum:

$$\beta = \int_{\mathbb{T}^d} \frac{dk}{z - \hat{J}(k)} + \frac{\beta h^2}{2(z - \hat{J}(0))^2}, \ z > \hat{J}(0). \tag{4.20}$$

The integral here is finite at $z = \hat{J}(0)^+$ if $d \geq 3$, and is infinite if $d = 1, 2$. Basing on this, it can be shown that the solution $z(\beta, h)$ of (4.20), hence the free energy, is real analytic for $\beta > 0$, $h \neq 0$ and all $d \geq 1$, for $\beta > 0$, $h = 0$ and $d = 1, 2$, and for $\beta < \beta_c$, $h = 0$ and $d \geq 3$, where

$$\beta_c = \int_{\mathbb{T}^d} \frac{dk}{\hat{J}(0) - \hat{J}(k)}.$$

Hence, there is no phase transitions in the spherical model in those domain of parameters β, h and d.

If, however, $\beta > \beta_c$, and $d \geq 3$, then the limiting magnetization

$$m(\beta, h) = -\frac{\partial f}{\partial h} = \frac{h}{z - \hat{J}(0)}$$

has the jump

$$2m(\beta, 0^+), \ m(\beta, 0^+) = \sqrt{(\beta - \beta_c)/\beta_c}, \ \beta \geq \beta_c$$

at $h = 0$ as function of h.

Hence, the qualitative behavior of the spherical model with interaction (4.19) (in fact, in much more general case) and for $d \geq 3$ is analogous to that of the Kac model. A new important property of the spherical model is the absence of phase transition in low dimensions $d = 1, 2$. The dependence of the existence of phase transition on the dimension of the lattice is an important property, expected in realistic models.

Here are the critical indices of the spherical model (see e.g. [Pat, TKS]):

$$\alpha = 1/3, \ \beta = 1/2, \ \gamma = 2, \ \delta = 5, \quad d = 3 \tag{4.21}$$
$$\alpha = 0, \ \beta = 1/2, \ \gamma = 1, \ \delta = 3, \quad d > 4.$$

Viewing s_x, $x \in \Lambda$ as a family of random variables, whose probability law is given by the Gibbs measure, it is instructive to compute their covariance in the thermodynamic limit, i.e

$$C(x - y) = \lim_{\Lambda \to \infty} \left(\langle s_x s_y \rangle_{H_\Lambda} - \langle s_x \rangle_{H_\Lambda} \langle s_y \rangle_{H_\Lambda} \right), \tag{4.22}$$

It can be shown that

$$C(x - y) = \beta (z - J)^{-1}_{x-y}, \quad z > \hat{J}. \tag{4.23}$$

Since

$$(z - J)^{-1}_{x-y} = \int_{\mathbb{T}^d} \frac{e^{2\pi i(x-y)} dk}{z - \hat{J}(k)}, \tag{4.24}$$

it follows (see e.g. [Pat]) that in the case (4.19) and $h = 0$ the asymptotic form of the covariance as $|x - y| \to \infty$ is

$$C(x - y) = \text{Const} \cdot e^{-|x-y|/\xi(\beta)}/|x - y|^{(d-1)/2}. \qquad (4.25)$$

The quantity $\xi(\beta)$ is called the correlation length of the model. It is easy to see that for $d \geq 3$

$$\xi(\beta) = C_\xi(\beta - \beta_c)^{-\nu}, \ \nu = 1/(d - 2), \ d > 2, \ \beta \nearrow \beta_c. \qquad (4.26)$$

The number ν is one more critical index of the model. Its value for the Ising model with the nearest neighbor interaction is 1 for $d = 2$ and 0.638 for $d = 3$ (see e.g. [TKS, ZJ]).

The above shows that the "spins" are more and more correlated as the temperature approaches the critical temperature, at which we have only the power law decay of correlations: $C(x) = O(|x|^{-(d-2)})$, $|x| \to \infty$. We conclude that the phase transition is related to a strong increase of correlations between the coordinates of the system.

4.3 Concluding Remarks on Phase Transitions

Phase transitions arise only in the thermodynamic limit, for the free energy of a finite system is evidently analytic in temperature and parameters of the Hamiltonian. Hence, it is natural to ask what are the effects of restrictions of our calculations to finite systems and can one exploit these effects to extract the limiting singularities from analytic or numerical calculations on small systems or from the Monte-Carlo calculations on larger but still finite systems. It is generally believed that the answer to these questions is positive and there exists extensive literature on the subject, discussing precursors of the true phase transitions in infinite systems, their dependence on the boundary conditions, their modification by making some of the dimensions of the system finite (films, layers), and their manifestation in various numerical and modelling procedures. We refer the reader to early [Fi, Ba], and more recent [Bi-He, Ne-Ba] reviews of the subject.

References

[Am]   Amit, D.J.: Modeling Brain Function: The World of Attractor Neural Networks. Cambridge University Press (1989)

[Ba]   Barber, M.: Finite-size scaling. In: Domb, C., Lebowitz, J.L. (eds) Phase Transitions and Critical Phenomena, Vol. 8, 146–266. Academic Press (1983)

[Bax]   Baxter, R.J.: Exactly Solved Models in Statistical Mechanics. Academic Press (1989)

[Bi-He] Binder, K., Heermann, D.W.: Monte Carlo Simulation in Statistical Physics: An Introduction. Springer-Verlag (2002)

[Bo]   Bovier, A.: Statistical mechanics of disordered systems. Lecture Notes at Centre for Mathematical Physics and Stochastic University of Aarhus (2001). Available via http://www.maphysto.dk/oldpages/publications/publications2001-static.html

[Bo-Pi]   Bovier, A., Picco, P. (eds): Mathematical Aspects of Spin Glasses and Neural Networks. Birkhäuser Inc. (1998)

[Fi]   Fisher, M.E.: The theory of critical point singularity. In: Green, M.S. (ed) Proceedings of the International School of Physics "Enrico Fermi", Course LI: Critical Phenomena, 1–99. Academic Press (1971)

[Fr]   Fröhlich, J.: Mathematical aspects of the physics of disordered systems. In: Phénomènes Critiques, Systèmes Aléatoires, Théories de Jauge, Part I, II (Les Houches, 1984), 725–893. North-Holland (1986)

[Fr-Ze]   Fröhlich, J., Zegarlinski, B.: The high-temperature phase of long-range spin glasses. Comm. Math. Phys., 110, 121–155 (1987). Some comments on the Sherrington–Kirkpatrick model of spin glasses, 112, 553–566 (1987); Spin glasses and other lattice systems with long range interactions, 120, 665–688 (1989)

[Gr-Sz]   Grenander, U., Szhegö, G.: Töplitz Forms and Their Applications. University of California Press (1958)

[Gu]   Guerra, F.: Broken replica symmetry bounds in the mean field spin glass model. Comm. Math. Phys., 233, 1–12 (2003)

[Gu-To]   Guerra, F., Toninelli, F.L.: The thermodynamic limit in mean field spin glass models. Comm. Math. Phys., 230, 71–79 (2002)

[He-Le]   Hemmer, P.C., Lebowitz, J.L.: Systems with weak long-range potentials. In: Green, M.S., Domb, C. (eds) Phase Transitions and Critical Phenomena, Vol. 5b, 107–203. Academic Press, (1976)

[Ho]   Hopfield, J.: Neural networks and physical systems with emergent collective computational abilities. Proc. Nat. Ac. Sci., 79, 2554–2558 (1982)

[Ka-Th]   Kac, M., Thompson, C.J.: Spherical model and the infinite spin dimensionality limit. Phys. Norveg., 5, 163–168 (1971)

[Kh-Si]   Khanin, K.M., Sinai, Ya.G.: Existence of free energy for models with long-range random Hamiltonians. J. Stat. Phys., 20, 573–584 (1987)

[KKPS]   Khorunzhy, A., Khoruzhenko, B., Pastur, L., Shcherbina, M.: Large-n limit in the statistical mechanics and the spectral theory of disordered systems. In: Domb, C., Lebowitz, J.L. (eds) Phase Transitions and Critical Phenomena, Vol. 15, 74–239. Academic Press (1992)

[Kr]   Krengel, U.: Ergodic Theorems. Walter de Gruyter (1985)

[LL]   Landau, L.D., Lifshitz, E.M.: Statistical Physics, Part 1. Pergamon Press (1986)

[Lu]   Luttinger, J.M.: Exactly soluble model of spin glass. Phys. Rev. Lett., 37, 778–782 (1977)

[MPV]   Mézard, M., Parisi, G., Virazoro, M.: Spin Glasses Theory and Beyond. World Scientific (1988)

[Mi]   Minlos, R.A.: Introduction to Mathematical Statistical Physics. AMS (2000)

[Mi-Po]   Minlos, R.A., Povzner, A.Ya.: On the thermodynamic limit for entropy. Proceedings of the Moscow Math. Soc., 17, 269–300 (1967)

[Ne]   Newman, C.M.: Topics in Disordered Systems. Birkhäuser Verlag (1997)

[Ne-Ba]   Newman, M.E.J., Barkema, G.: Monte Carlo Methods in Statistical Physics. Clarendon Press (1999)

[Pa-Fi] Pastur, L., Figotin, A.: On the theory of disordered spin systems. Theor. Math. Phys., **32**, 615–623 (1978)

[Pa-Sh] Pastur, L., Shcherbina, M.: Absence of self-averaging of the order parameter in the Sherrington–Kirkpatrick model. J. Stat. Phys., **62**, 1–19 (1991)

[Pat] Pathria, R.K.: Statistical Mechanics. Butterworth-Heinemann (1996)

[Po-Sz] Pólya, G., Szegö, G.: Problems and Theorems in Analysis, I: Series, Integral Calculus, Theory of Functions. Springer Verlag (1972)

[Ru1] Ruelle, D.: Statistical Mechanics: Rigorous Results. W.A. Benjamin (1969)

[Ru2] Ruelle, D.: Thermodynamic Formalism: The Mathematical Structures of Classical Equilibrium Statistical Mechanics. Addison-Wesley Publishing (1978)

[Sh-Ti] Shcherbina, M., Tirozzi, B.: The free energy of a class of Hopfield models. J. Statist. Phys., **72**, 113–125 (1993)

[Shi] Shiryaev, A.N.: Probability. Springer-Verlag (1995)

[Sim] Simon, B.: The Statistical Mechanics of Lattice Gases. Princeton University Press (1993)

[Sin] Sinai, Ya.: Theory of Phase Transitions: Rigorous Results. Pergamon Press (1982)

[Ta1] Talagrand, M.: The generalized Parisi formula. C.R. Math. Acad. Sci. Paris, **337**, 111–114 (2003)

[Ta2] Talagrand, M.: Spin Glasses: A Challenge for Mathematicians: Cavity and Mean Field Models. Springer-Verlag (2003)

[Th] Thompson, C.J.: Mathematical Statistical Mechanics. Princeton University Press (1979)

[TKS] Toda, M., Kubo, R., Saito, N.: Statistical Physics I: Equilibrium Statistical Mechanics. Springer-Verlag (1992)

[Vu] Vuillermot, P.: Thermodynamics of quenched random spin systems, and application to the problem of phase transitions in magnetic (spin) glasses. J. Phys A: Math. Gen., **10**, 1319–1333 (1977)

[Wi] Wilson, K.G.: The renormalization group and critical phenomena. Rev. Modern Phys., **55**, 583–600 (1983)

[ZJ] Zinn-Justin, J.: Quantum Field Theory and Critical Phenomena. Oxford University Press (1993)

On Read's Proof that $B(\ell_1)$ Is Not Amenable

G. Pisier

Department of Mathematics, Université Paris VI, Paris, France and Texas A&M
University, College Station, TX 77843-3368, USA
gip@ccr.jussieu.fr, pisier@math.tamu.edu

Summary. In this seminar report we present Read's proof that $B(\ell_1)$ is not an
amenable Banach algebra. Our proof is slightly simpler (our use of expanders instead
of Read's use of random hypergraphs seems to clarify the picture), but the ideas can
all be traced back in Read's original proof.

A Banach algebra B is called amenable if every bounded derivation
$\delta\colon B \to Y$ into a dual bi-module Y is inner. (Recall that δ is inner if there
is y in Y such that $\delta(b) = b \cdot y - y \cdot b$). This notion is due to B.E. Johnson [Jo1], and goes back to the early 70's. Surprisingly, many basic questions
about it remain unanswered. For instance, very little is known on the case
when $B = B(X)$ with X a Banach space. It has been known since the late
70's (Connes, Wassermann) that $B(\ell_2)$ was not amenable, and this was essentially the only known case until recently when Charles Read proved that
$B(\ell_1)$ was not amenable. In the same paper Read also manages to show that
$\bigoplus_{n\geq 1} B(\ell_p^n)$ is not amenable if $1 < p \neq 2 < \infty$, but his proof does not yield
the case $p = 2$ which can be obtained by C^*-algebra theory. Quite surprisingly,
however, the following remains open:

Question. Is $B(\ell_p)$ amenable for $1 < p \neq 2 < \infty$?

Actually, it is not known whether there is *any* infinite dimensional Banach
space X at all for which $B(X)$ is amenable, but it is conceivable that some
suitable example of Gowers-Maurey type (yet to be constructed!) may have
so few bounded operators that $B(X)$ might be amenable.

Let B be a unital Banach algebra. B.E. Johnson proved that B is amenable
iff B satisfies the following property:

(**AD**) There is a constant C such that for any $\varepsilon > 0$, and any finite subset
$u_1, \ldots, u_n$ in B there is T in $B \otimes B$, say $T = \sum_{i=1}^{N} a_i \otimes b_i$ $(a_i, b_i \in B)$ with
the following properties:

$$\|T\|_\wedge \leq C \tag{1}$$

$$\sum a_i b_i = 1 \tag{2}$$

$$\forall\, k = 1, \ldots, n, \quad \|u_k \cdot T - T \cdot u_k\|_\wedge < \varepsilon . \tag{3}$$

Here $\|\ \|_\wedge$ denotes the projective tensor norm and we have used the notation:

$$u \cdot T = \sum u a_k \otimes b_k, \quad T \cdot u = \sum a_k \otimes b_k u$$

for any u in B. Johnson's classical proof that $(AD) \Leftrightarrow$ amenable is not difficult. To get a feel for it we show the easier direction $(AD) \Rightarrow$ amenable. Assume that B satisfies (AD) as above and let $\delta\colon\ B \to Y$ be a bounded derivation. By (3) we have for any $k = 1, \ldots, n$

$$\left\| \sum_{i=1}^N u_k a_i \otimes b_i - a_i \otimes b_i u_k \right\|_\wedge \leq \varepsilon\ ;$$

hence by duality

$$\left\| \sum_i \delta(u_k a_i) b_i - \delta(a_i) b_i u_k \right\| \leq \varepsilon \|\delta\|.$$

Now since $\delta(u_k a_i) = u_k \delta(a_i) + \delta(u_k) a_i$ and $\sum a_i b_i = 1$, if we set $y = \sum \delta(a_i) b_i$ we find

$$\| u_k y - y u_k + \delta(u_k) \| \leq \varepsilon \|\delta\|$$

or equivalently, if $\delta_y(b) = yb - by$

$$\| \delta(u_k) - \delta_y(u_k) \| \leq \varepsilon \|\delta\|.$$

Thus we are able to produce a net of y's with

$$\|y\| = \left\| \sum \delta(a_i) b_i \right\| \leq \|\delta\| C$$

such that $\delta_y - \delta$ tends to zero pointwise. Since Y is a *dual* Banach bi-module, we conclude that $\delta = \delta_{y'}$ where y' is any weak* cluster point of the y's in the weak* topology of y (note $\|y'\| \leq \|\delta\| C$) so that B is amenable. $\qquad\square$

We now turn to the proof that $B(\ell_1)$ is not amenable.

We start by some preliminary observations. We will denote as usual by $\{b(s,t) \mid s, t \in \mathbb{N}\}$ the matrix of an element b in $B(\ell_1)$. Let $T = \sum_{i=1}^N a_i \otimes b_i \in B(\ell_1) \otimes B(\ell_1)$. We denote

$$\widehat{T}(s, \theta) = \sum_{i=1}^N a_i(s, \theta) b_i(\theta, s).$$

Observation 1. $\sup_s \sum_\theta |\widehat{T}(s, \theta)| \leq \|T\|_\wedge.$

Observation 2. $\|(|\widehat{T}(s,\theta)|^{1/2})\|_{B(\ell_2)} \le \sqrt{N}(\|T\|_\wedge)^{1/2}.$

The first point is obvious. For the second one, note that for each $i = 1, \ldots, N$ we have (by Cauchy–Schwarz) for any x, y in the unit sphere of ℓ_2

$$\left| \sum_{s,\theta} |a_i(s,\theta)|^{1/2} |b_i(\theta,s)|^{1/2} x(s) y(\theta) \right|$$

$$\le \left(\sum_{s,\theta} |a_i(s,\theta)| |y(\theta)|^2 \right)^{1/2} \left(\sum_{s,\theta} |b_i(\theta,s)| |x(s)|^2 \right)^{1/2};$$

hence

$$\left\| \left(|a_i(s,\theta)|^{1/2} |b_i(\theta,s)|^{1/2} \right) \right\|_{B(\ell_2)} \le \left(\|a_i\| \|b_i\| \right)^{1/2}$$

so that the second point follows by another application of Cauchy–Schwarz.

We will use the so-called "expanders" or "expanding graphs". We only use rather basic facts from the theory of these graphs (see [Lu], [Ma1], [Ma2], [Sa] for more information). It will be convenient for our exposition to use Cayley graphs associated to systems of generators in finite groups. We need some notation for this. Let G be a finite group and let $S \subset G$ be a finite generating subset. For any g in G, we denote by $\lambda(g) \colon \ell_2(G) \to \ell_2(G)$ the unitary operator of left translation by g. We also let

$$\ell_2^0 = \left\{ f \in \ell_2(G) \mid \sum_{t \in G} f(t) = 0 \right\}.$$

In other words, $\ell_2^0(G)$ is the orthogonal of the constant functions. We then denote

$$\varepsilon(S,G) = \left\| |S|^{-1} \sum_{g \in S} \lambda(g)_{|\ell_2^0} \right\|_{B(\ell_2^0)}.$$

Equivalently, $\varepsilon(S,G)$ is the smallest $\varepsilon > 0$ such that for any function $f \colon G \to \mathbb{C}$ we have

$$\left(\sum_s \left| \frac{1}{|S|} \sum_{g \in S} f(gs) - \frac{1}{|G|} \sum_{t \in G} f(t) \right|^2 \right)^{1/2} \le \varepsilon \left(\sum_{t \in G} |f(t)|^2 \right)^{1/2}. \qquad (4)$$

The following result due to Lubotzky, Phillip and Sarnak, and independently to Margulis, (see [Lu], [Ma1], [Ma2], [Sa]) is much more than what we need:

Theorem. *Let $n = p+1$ with p prime ≥ 3. Let $\varepsilon(n) = 2n^{-1}\sqrt{n-1}$ (note $\varepsilon(n) \le 2/\sqrt{n}$). Then there is a sequence of finite groups $\{G_m \mid m \ge 1\}$ and generating subsets $S_m \subset G_m$ with cardinality $|S_m| = n$ for each m, such that*

$$\varepsilon(S_m, G_m) \le \varepsilon(n)$$

for all m and also

$$|G_m| \to \infty \quad when \quad m \to \infty.$$

Remark. All that we use in the sequel is that we have this for infinitely many n's with numbers $\varepsilon(n)$ which tend to zero when $n \to \infty$, so more elementary versions of "expanding graphs" (see [Lu]) are sufficient.

Read's Theorem. $B = B(\ell_1)$ *fails* (AD).

Proof. Assume that B satisfies (AD) as stated above. Let J be the disjoint union of a sequence $\{G_m \mid m \geq 1\}$ as in the preceding statement. We view ℓ_1 as $\ell_1(J)$. We will use an enumeration $S_m = \{g_1(m), \ldots, g_n(m)\}$ of the generators of G_m.

For each $k = 1, \ldots, n$, we can define the permutation $\pi_k \colon J \to J$ which coincides on G_m with left translation by $g_k(m)$. More precisely, if $j \in J$ let m be such that $j \in G_m$; then we set

$$\pi_k(j) = g_k(m)j.$$

Clearly, π_k is a permutation of J (leaving invariant each G_m) which defines an isometric isomorphism $u_k \colon \ell_1(J) \to \ell_1(J)$ simply by

$$u_k(e_j) = e_{\pi_k(j)},$$

where $\{e_j \mid j \in J\}$ is the canonical basis of $\ell_1(J)$. Let T be as in the above definition of (AD). By (2) we have

$$\forall s \quad \sum_\theta \widehat{T}(s, \theta) = 1.$$

Hence using (1) and Observation 1, we find

$$\forall s \quad 1 \leq \sum_\theta |\widehat{T}(s, \theta)| \leq C. \tag{5}$$

By (3) we have

$$\|u_k^{-1} \cdot T \cdot u_k - T\|_\wedge \leq \varepsilon$$

and it is easy to check that if we let $T_k = u_k^{-1} \cdot T \cdot u_k$ then

$$\widehat{T_k}(s, \theta) = \widehat{T}(\pi_k(s), \theta).$$

Therefore by Observation 1 again we have

$$\sup_s \sum_\theta |\widehat{T}(\pi_k s, \theta) - \widehat{T}(s, \theta)| \leq \varepsilon. \tag{6}$$

Now we set $x(s, \theta) = |\widehat{T}(s, \theta)|^{1/2}$. By (5) we have

$$\forall s \qquad 1 \leq \sum_\theta |x(s,\theta)|^2 \leq C. \tag{5}'$$

Also using the elementary, numerical inequality $|\sqrt{|a|} - \sqrt{|b|}| \leq \sqrt{|a-b|}$, we deduce from (6)

$$\sup_s \left(\sum_\theta |x(\pi_k s, \theta) - x(s,\theta)|^2 \right)^{1/2} \leq \sqrt{\varepsilon}. \tag{6}'$$

In particular, for any m and any s in G_m, we have

$$\left(\sum_\theta |x(g_k(m)s, \theta) - x(s,\theta)|^2 \right)^{1/2} \leq \sqrt{\varepsilon}.$$

Let $F_m: G_m \to \ell_2$ be the function defined by $F_m(s) = \sum_\theta x(s,\theta) e_\theta$. The preceding inequality means that for any k and s we have

$$\left\| F_m(g_k(m)s) - F_m(s) \right\| \leq \sqrt{\varepsilon}.$$

By convexity this implies

$$\left\| \frac{1}{n} \sum_{k=1}^n F_m(g_k(m)s) - F_m(s) \right\| \leq \sqrt{\varepsilon},$$

and after summing over G_m we find

$$\left(\sum_{s \in G_m} \left\| \frac{1}{n} \sum_{k=1}^n F_m(g_k(m)s) - F_m(s) \right\|^2 \right)^{1/2} \leq \sqrt{\varepsilon} \, |G_m|^{1/2}.$$

On the other hand, (4) trivially extends to the ℓ_2-valued case and yields

$$\left(\sum_{s \in G_m} \left\| \frac{1}{n} \sum_{k=1}^n F_m(g_k(m)s) - \frac{1}{|G_m|} \sum_{t \in G_m} F_m(t) \right\|^2 \right)^{1/2}$$
$$\leq \varepsilon(G_m) \left(\sum_{s \in G_m} \| F_m(s) \|^2 \right)^{1/2}.$$

Thus we obtain by the triangle inequality

$$\left(\sum_{s \in G_m} \| F_m(s) \|^2 \right)^{1/2} \leq I + II + III$$

where

$$I = \sqrt{\varepsilon} |G_m|^{1/2}$$

$$II = \varepsilon(G_m)\left(\sum_{s \in G_m} \|F_m(s)\|^2 \right)^{1/2}$$

and

$$III = \left(\sum_{s \in G_m} \left\| \frac{1}{|G_m|} \sum_{t \in G_m} F_m(t) \right\|^2 \right)^{1/2}.$$

But by (5)', we have on one hand

$$|G_m|^{1/2} \leq \left(\sum_{s \in G_m} \|F_m(s)\|^2 \right)^{1/2}$$

and on the other hand

$$II \leq \varepsilon(G_m)\sqrt{C} \, |G_m|^{1/2}.$$

Finally, to estimate III note that

$$III = |G_m|^{-1/2} \left(\sum_{\theta} \left| \sum_{t \in G_m} |\widehat{T}(t,\theta)|^{1/2} \right|^2 \right)^{1/2};$$

hence by Observation 2, we have

$$III \leq \sqrt{N} \, \sqrt{C}.$$

Thus we finally obtain

$$|G_m|^{1/2} \leq \left(\sqrt{\varepsilon} + \varepsilon(G_m)\sqrt{C} \right)|G_m|^{1/2} + \sqrt{N} \, \sqrt{C};$$

hence

$$1 \leq \sqrt{\varepsilon} + \varepsilon(n)\sqrt{C} + |G_m|^{-1/2}\sqrt{N} \, \sqrt{C},$$

and if $\varepsilon < 1$ we can choose n so large that $\sqrt{\varepsilon} + \varepsilon(n)\sqrt{C} < 1$ so that letting $m \to \infty$ we obtain a contradiction.   □

We will say that an operator $T \colon \ell_p \to \ell_p$ is "regular" if the matrix $\{|T(s,t)|\}$ is bounded on ℓ_p. We denote by $B_r(\ell_p)$ the unital Banach algebra of all regular operators on ℓ_p, equipped with the following norm

$$\|T\|_{B_r(\ell_p)} = \|(|T(s,t)|)\|_{B(\ell_p)}.$$

This is a special case of a more general notion, see [P] and the references there for more background. Consider a_i, b_i in $B_r(\ell_p)$. As before, let $\widehat{T}(s,\theta) = \sum_1^N a_i(s,\theta)b_i(\theta,s)$. Note $(b_i(\theta,s))$ is the matrix of an element in the unit ball of $B_r(\ell_p)$, so Hölder's inequality gives us Observation 1 again. Moreover, since $B_r(\ell_2) = (B_r(\ell_p), B_r(\ell_{p'}))^{1/2}$ (see [P]) we have for each i

$$\left\| \left(|a_i(s,\theta)|^{1/2} |b_i(\theta,s)|^{1/2} \right)_{s,\theta} \right\|_{B_r(\ell_2)} \leq \left(\|a_i\|_{B_r(\ell_p)} \|b_i\|_{B_r(\ell_p)} \right)^{1/2},$$

from which Observation 2 follows immediately. Thus we obtain by the same proof as for Read's theorem:

Proposition. *For each $1 \leq p < \infty$, $B_r(\ell_p)$ is not amenable.*

Remark. If $p = 1$, $B_r(\ell_1) = B(\ell_1)$ isometrically.

Remark. Let (a_{ij}) be a $k \times k$ matrix. Let $\alpha = \min\{1/p, 1/p'\}$. It is easy to check that

$$\left\| (|a_{ij}|) \right\|_{B(\ell_p^k)} \leq k^\alpha \| (a_{ij}) \|_{B(\ell_p^k)}.$$

Since $\alpha < 1/2$ when $p \neq 2$, the preceding argument proves (as noted by Read [R]) that the Banach algebra $\oplus_k B(\ell_p^k)$ is not amenable when $p \neq 2$ (we obtain only $III \leq \sqrt{N} \sqrt{C} |G_m|^\alpha$ and if $\alpha < 1/2$, this is enough to conclude). Curiously, the case $p = 2$ (known by other methods) seemed to be out of reach by this kind of proof, but Narutaka Ozawa (private communication) found a way to somehow make it fit into a similar frame.

References

[Jo1]   Johnson, B.E.: Approximate diagonals and cohomology of certain annihilator Banach algebras. Amer. J. Math., **94**, 685–698 (1972)

[Lu]   Lubotzky, A.: Discrete groups, expanding graphs and invariant measures. Progress in Math., 125. Birkhauser (1994)

[Ma1]   Margulis, G.A.: Explicit group-theoretic constructions of combinatorial schemes and their applications in the construction of expanders and concentrators (Russian). Problemy Peredachi Informatsii, **24**, no. 1, 51–60 (1988)

[Ma2]   Margulis, G.A.: Discrete Subgroups of Semisimple Lie Groups. Springer-Verlag, Berlin (1991)

[P]   Pisier, G.: Complex interpolation and regular operators between Banach lattices. Archiv der Mat. (Basel), **62**, 261–269 (1994)

[R]   Read, C.: Relative amenability and the non-amenability of $B(\ell_1)$. Preprint 2001. To appear

[Sa]   Sarnak, P.: Some Applications of Modular Forms. Cambridge University Press, Cambridge (1990)

An Isomorphic Version of the Busemann–Petty Problem for Gaussian Measure

A. Zvavitch

Department of Mathematics, University of Missouri, Columbia, MO 65211, USA
zvavitch@math.missouri.edu

Summary. In this paper we provide upper and lower bounds for the Gaussian Measure of a hyperplane section of a convex symmetric body. We use those estimates to give a partial answer to an isomorphic version of the Gaussian Busemann–Petty problem.

1 Introduction

The standard Gaussian measure on $\mathbb{R}^n$ is given by its density:

$$\gamma_n(K) = \frac{1}{(\sqrt{2\pi})^n} \int_K e^{-\frac{|x|^2}{2}} dx,$$

where $|x|^2 = \sum_{i=1}^n |x_i|^2$.

Consider two convex symmetric bodies (convex symmetric sets with nonempty interior) $K, L \subset \mathbb{R}^n$ such that

$$\gamma_{n-1}(K \cap \xi^\perp) \leq \gamma_{n-1}(L \cap \xi^\perp), \quad \forall \xi \in S^{n-1}, \tag{1}$$

where $K \cap \xi^\perp$ denotes the section of K by the hyperplane orthogonal to ξ. Does it follow that

$$\gamma_n(K) \leq \gamma_n(L)?$$

This is a Gaussian analog of the Busemann–Petty problem (see [K] for details about the BP problem). It was shown in [Z], that the answer to the above question is affirmative if $n \leq 4$ and it is negative if $n \geq 5$.

When $n \geq 5$, it is natural to ask an isomorphic version of the Gaussian Busemann–Petty problem: does there exist an absolute constant c so that if K and L are convex symmetric bodies satisfying (1) then $\gamma_n(K) \leq c\gamma_n(L)$?

In this paper we give a partial answer to this question:

Theorem. *For any $n \in \mathbb{N}$ and convex symmetric bodies $K, L \subset \mathbb{R}^n$, if*

$$\gamma_{n-1}(K \cap \xi^\perp) \leq \gamma_{n-1}(L \cap \xi^\perp), \quad \forall \xi \in S^{n-1},$$

then

$$\gamma_n(K) \le c\left(\gamma_n(L)\right)^{\frac{n-1}{n}}, \tag{2}$$

where c is an absolute constant.

This theorem will follow from two results proved below: the first one is that for any convex symmetric body K

$$\gamma_n(K) \le \gamma_{n-1}(K \cap \xi^{\perp}), \quad \forall \xi \in S^{n-1},$$

and the second one is that for any convex symmetric body L there exists a direction $\xi \in S^{n-1}$, such that

$$\gamma_{n-1}(L \cap \xi^{\perp}) \le c\left(\gamma_n(L)\right)^{\frac{n-1}{n}}. \tag{3}$$

It would be interesting to try to remove the power $(n-1)/n$ on the right hand side of (2), since it will then solve the same problem for the Lebesgue measure (the so-called slicing problem, see [MP1] for more details). We must note that (3) is sharp (consider a sequence of Euclidean balls with radii going to 0), so the theorem cannot be improved using the method described below.

Acknowledgement. The author is grateful to Alexander Koldobsky, Mark Rudelson and the anonymous referee for many suggestions that improved the paper.

2 Proof of the Theorem

It was proved in [BS], that if $K \subset \mathbb{R}^n$ is a convex closed set with $\gamma_n(K) \ge \frac{1}{2}$, then $\gamma_{n-1}(K \cap \xi^{\perp}) \ge \frac{1}{2}$ for all $\xi \in S^{n-1}$. The next simple lemma shows that this result can be improved in the case of K being symmetric.

Lemma 1. *For any convex symmetric body $K \in \mathbb{R}^n$:*

$$\gamma_n(K) \le \gamma_{n-1}(K \cap \xi^{\perp}), \quad \forall \xi \in S^{n-1}.$$

Proof. By rotation invariance of the Gaussian measure, it is enough to consider the case $\xi = (0, \ldots, 0, 1)$. Put

$$K(t) = K \cap \left((0, \cdots, 0, 1)^{\perp} + t(0, \cdots, 0, 1)\right)$$

and

$$PK(t) = K(t)|\mathbb{R}^{n-1} = \text{ orthogonal projection of } K(t) \text{ to } \mathbb{R}^{n-1}.$$

Note that K is a convex body, so for $0 < \lambda < 1$

$$\lambda K(a) + (1-\lambda)K(b) \subset K\left(\lambda a + (1-\lambda)b\right),$$

thus

$$\lambda PK(a) + (1 - \lambda)PK(b) \subset PK(\lambda a + (1 - \lambda)b).$$

In particular,

$$\frac{1}{2}PK(t) + \frac{1}{2}PK(-t) \subset PK(0) = K(0). \tag{4}$$

Next use that γ_{n-1} is a log-concave measure ([Bo], [E] or [Li]):

$$\log \gamma_{n-1}\big(\lambda A + (1 - \lambda)B\big) \geq \lambda \log \gamma_{n-1}(A) + (1 - \lambda) \log \gamma_{n-1}(B).$$

The latter inequality together with (4) gives

$$\begin{aligned}
\log \gamma_{n-1}\big(K(0)\big) &\geq \log \gamma_{n-1}\Big(\frac{1}{2}PK(t) + \frac{1}{2}PK(-t)\Big) \\
&\geq \frac{1}{2} \log \gamma_{n-1}\big(PK(t)\big) + \frac{1}{2} \log \gamma_{n-1}\big(PK(-t)\big) \\
&= \log \gamma_{n-1}\big(PK(t)\big).
\end{aligned}$$

So $\gamma_{n-1}(K(0)) \geq \gamma_{n-1}(PK(t))$.

Now we are ready to finish the proof:

$$\begin{aligned}
\gamma_n(K) &= \frac{1}{(\sqrt{2\pi})^n} \int_K e^{-\frac{|x|^2}{2}} dx \\
&= \frac{1}{(\sqrt{2\pi})^n} \int_{-\infty}^{\infty} e^{-\frac{t^2}{2}} \int_{K(t)} e^{-\frac{\sum_{i=1}^{n-1} x_i^2}{2}} dx dt \\
&= \frac{1}{(\sqrt{2\pi})^n} \int_{-\infty}^{\infty} e^{-\frac{t^2}{2}} \int_{PK(t)} e^{-\frac{\sum_{i=1}^{n-1} x_i^2}{2}} dx dt \\
&= \frac{1}{\sqrt{2\pi}} \int_{-\infty}^{\infty} e^{-\frac{t^2}{2}} \gamma_{n-1}\big(PK(t)\big) dt \\
&\leq \gamma_{n-1}\big(K(0)\big) \frac{1}{\sqrt{2\pi}} \int_{-\infty}^{\infty} e^{-\frac{t^2}{2}} dt = \gamma_{n-1}\big(K(0)\big). \tag{5}
\end{aligned}$$

$\square$

Remark 1. It is clear that (5) actually shows:

$$\gamma_n(K) \leq \gamma_{n-1}(K \cap \xi^{\perp}) \gamma_1(K|\xi), \quad \forall \xi \in S^{n-1}, \tag{6}$$

where $K|\xi$ is the orthogonal projection of K onto ξ.

Moreover it was pointed out to us by the referee, that (6) can be generalized to

$$\gamma_n(K) \leq \gamma_{n-k}(K \cap E^{\perp}) \gamma_k(K|E), \tag{7}$$

for any k-dimensional subspace E of $\mathbb{R}^n$. Actually (7) is true for a general convex body K whose Gaussian-barycenter is 0: $\int_K x d\gamma_n(x) = 0$. This can be shown by straightforward generalization of a result of Milman and Pajor (see [MP2], Remark 2 on page 319 and Lemma 1 on page 316).

Our next goal is to find an upper bound for the Gaussian measure of a hyperplane section. First we will consider the case where K is a regular Euclidean ball of radius $t > 0$.

Lemma 2. *For any $t > 0$ and $n \in \mathbb{N}$,*

$$c_1\big(\gamma_n(tB_2^n)\big)^{\frac{n-1}{n}} \leq \gamma_{n-1}(tB_2^{n-1}) \leq c_2\big(\gamma_n(tB_2^n)\big)^{\frac{n-1}{n}}, \tag{8}$$

where $tB_2^n = \{x \in \mathbb{R}^n : |x| \leq t\}$ and c_1, c_2 are absolute constants.

Proof. First note that for any $t \geq 0$

$$\int_0^t r^{n-2} e^{-\frac{r^2}{2}}\, dr \leq \frac{n^{\frac{n-1}{n}}}{n-1} \left(\int_0^t r^{n-1} e^{-\frac{r^2}{2}}\, dr \right)^{\frac{n-1}{n}}. \tag{9}$$

To prove this inequality consider a function

$$f(t) = \int_0^t r^{n-2} e^{-\frac{r^2}{2}}\, dr - \frac{n^{\frac{n-1}{n}}}{n-1} \left(\int_0^t r^{n-1} e^{-\frac{r^2}{2}}\, dr \right)^{\frac{n-1}{n}},$$

then

$$f'(t) = t^{n-2} e^{-\frac{t^2}{2}} - n^{-\frac{1}{n}} t^{n-1} e^{-\frac{t^2}{2}} \left(\int_0^t r^{n-1} e^{-\frac{r^2}{2}}\, dr \right)^{-\frac{1}{n}}$$

$$= t^{n-2} e^{-\frac{t^2}{2}} \left(1 - n^{-\frac{1}{n}} t \Big(\int_0^t r^{n-1} e^{-\frac{r^2}{2}}\, dr \Big)^{-\frac{1}{n}} \right)$$

$$< t^{n-2} e^{-\frac{t^2}{2}} \left(1 - n^{-\frac{1}{n}} t \Big(\int_0^t r^{n-1}\, dr \Big)^{-\frac{1}{n}} \right) = 0.$$

Thus $f(t)$ is a decreasing function and $f(t) \leq 0$ for $t > 0$, which proves the inequality (9).

Passing to polar coordinates and applying (9) we get

$$\gamma_{n-1}(tB_2^{n-1}) = \frac{1}{(\sqrt{2\pi})^{n-1}} \int_{S^{n-2}} \int_0^t r^{n-2} e^{-\frac{r^2}{2}}\, dr d\theta \tag{10}$$

$$= \frac{|S^{n-2}|}{(\sqrt{2\pi})^{n-1}} \int_0^t r^{n-2} e^{-\frac{r^2}{2}}\, dr$$

$$\leq \frac{|S^{n-2}| n^{\frac{n-1}{n}}}{(\sqrt{2\pi})^{n-1}(n-1)} \left(\int_0^t r^{n-1} e^{-\frac{r^2}{2}}\, dr \right)^{\frac{n-1}{n}}$$

$$= \frac{|S^{n-2}| n^{\frac{n-1}{n}}}{|S^{n-1}|^{\frac{n-1}{n}}(n-1)} \big(\gamma_n(tB_2^n)\big)^{\frac{n-1}{n}}.$$

Finally, from $|S^{n-1}| = 2\pi^{n/2}/\Gamma(\frac{n}{2})$ and the Stirling asymptotic formula for the Gamma-function we get

$$\gamma_{n-1}(tB_2^{n-1}) \le c_2\big(\gamma_n(tB_2^n)\big)^{\frac{n-1}{n}}.$$

To prove the lower bound for $\gamma_{n-1}(tB_2^{n-1})$, we first note that if $t \ge 2\sqrt{n}$ then $\gamma_{n-1}(tB_2^{n-1}) > c$ (see [Li] or [Ba], p. 236), and thus

$$\gamma_{n-1}(tB_2^{n-1}) > c\big(\gamma_n(tB_2^n)\big)^{\frac{n-1}{n}}.$$

If $t < 2\sqrt{n}$ it is easy to show that

$$\left(\int_0^t r^{n-1} e^{-\frac{r^2}{2}} dr\right)^{\frac{n-1}{n}} < C \int_0^t r^{n-2} e^{-\frac{r^2}{2}} dr. \tag{11}$$

Indeed

$$\left(\int_0^t r^{n-1} e^{-\frac{r^2}{2}} dr\right)^{-\frac{1}{n}} \le \left(\int_0^t r^{n-1} e^{-2n} dr\right)^{-\frac{1}{n}} \le Ct^{-1}$$

and

$$\left(\int_0^t r^{n-1} e^{-\frac{r^2}{2}} dr\right)^{\frac{n-1}{n}} < C \int_0^t r^{n-1} t^{-1} e^{-\frac{r^2}{2}} dr < C \int_0^t r^{n-2} e^{-\frac{r^2}{2}} dr.$$

Finally, the lemma follows from (11) and computations similar to (10).   □

To prove the next lemma we first recall the definition of an intersection body. Let K and M be origin symmetric star bodies in $\mathbb{R}^n$. Following Lutwak [Lu], we say that K is the *intersection body of* M if the radius of K in every direction is equal to the volume of the central hyperplane section of M perpendicular to this direction:

$$\|\xi\|_K^{-1} = \text{Vol}_{n-1}(M \cap \xi^\perp), \quad \forall \xi \in S^{n-1}.$$

A more general class of *intersection bodies* is defined as the closure in the radial metric of the class of intersection bodies of star bodies. It is clear that tB_2^n is an intersection body (see [Lu], [K] for more results and examples of intersection bodies). It was proved in [Z] (Theorems 3, 4), that if K is an intersection body then the answer to the Gaussian Busemann–Petty problem is affirmative for K and any convex symmetric body L, whose central sections have greater Gaussian measure than the corresponding sections of K.

Lemma 3. *For any convex symmetric body $K \subset \mathbb{R}^n$, there exists a direction $\xi \in S^{n-1}$ such that*

$$\gamma_{n-1}(K \cap \xi^\perp) \le c\big(\gamma_n(K)\big)^{\frac{n-1}{n}},$$

where c is an absolute constant.

Proof. Assume that for any $\xi \in S^{n-1}$

$$\gamma_{n-1}(K \cap \xi^{\perp}) > c_2 \big(\gamma_n(K)\big)^{\frac{n-1}{n}},$$

where c_2 is the absolute constant from Lemma 1. Consider $t > 0$ so that

$$\gamma_{n-1}(tB_2^{n-1}) = c_2 \big(\gamma_n(K)\big)^{\frac{n-1}{n}}. \tag{12}$$

Then

$$\gamma_{n-1}(K \cap \xi^{\perp}) > \gamma_{n-1}(tB_2^n \cap \xi^{\perp}), \ \ \forall \xi \in S^{n-1}. \tag{13}$$

But tB_2^n is an intersection body, therefore (13) implies

$$\gamma_n(K) > \gamma_n(tB_2^n).$$

The latter inequality together with (12) gives:

$$\big(c_2^{-1} \gamma_{n-1}(tB_2^{n-1})\big)^{\frac{n}{n-1}} > \gamma_n(tB^n)$$

or

$$\gamma_{n-1}(tB_2^{n-1}) > c_2 \big(\gamma_n(tB_2^n)\big)^{\frac{n-1}{n}},$$

which contradicts Lemma 2. □

Remark 2. Using a similar method, one can show that for any intersection body $K \subset \mathbb{R}^n$, there exists a direction $\xi \in S^{n-1}$ such that

$$c_1 \big(\gamma_n(K)\big)^{\frac{n-1}{n}} < \gamma_{n-1}(K \cap \xi^{\perp}),$$

where c_1 is the absolute constant from Lemma 1. Indeed, assuming the latter inequality not to be true, we get

$$\gamma_{n-1}(K \cap \xi^{\perp}) < \gamma_{n-1}(tB_2^n \cap \xi^{\perp}), \ \ \forall \xi \in S^{n-1},$$

where t is such that $\gamma_{n-1}(tB_2^{n-1}) = c_1 \big(\gamma_n(K)\big)^{\frac{n-1}{n}}$. Again the Gaussian Busemann–Petty problem has an affirmative answer in this case, thus

$$c_1 \big(\gamma_n(tB_2^n)\big)^{\frac{n-1}{n}} > \gamma_{n-1}(tB_2^{n-1}),$$

which again contradicts Lemma 2.

Proof of the Theorem. From Lemma 3 we get that there is $\xi \in S^{n-1}$ such that

$$\gamma_{n-1}(L \cap \xi^{\perp}) \le c \big(\gamma_n(L)\big)^{\frac{n-1}{n}},$$

but then, from Lemma 1:

$$\gamma_n(K) \le \gamma_{n-1}(K \cap \xi^{\perp}).$$

Using

$$\gamma_{n-1}(K \cap \xi^{\perp}) \le \gamma_{n-1}(L \cap \xi^{\perp}), \ \ \forall \xi \in S^{n-1},$$

we get

$$\gamma_n(K) \le c \big(\gamma_n(L)\big)^{\frac{n-1}{n}}. \quad □$$

References

[BS]   Banaszczyk, W., Szarek, S.J.: Lattice coverings and Gaussian measures of n-dimensional convex bodies. Discrete Comput. Geom., **17**, no. 3, 283–286 (1997)

[Ba]   Barvinok, A.: A course in convexity. Graduate Studies in Mathematics, **54**. Amer. Math. Soc., Providence, RI (2002)

[Bo]   Borell, C.: The Brunn–Minkowski inequality in Gauss space. Invent. Math., **30**, no. 2, 207–216 (1975)

[E]   Ehrhard, A.: Symetrisation dans l'espace de Gauss [Symmetrization in Gaussian spaces] (in French). Math. Scand., **53**, no. 2, 281–301 (1983)

[K]   Koldobsky, A.: Sections of star bodies and the Fourier transform. Harmonic Analysis at Mount Holyoke (South Hadley, MA, 2001), 225–247, Contemp. Math., **320**. Amer. Math. Soc., Providence, RI (2003)

[Li]   Lifshits, M.A.: Gaussian random functions. Mathematics and its Applications, **322**. Kluwer Academic Publishers, Dordrecht (1995)

[Lu]   Lutwak, E.: Intersection bodies and dual mixed volumes. Adv. Math., **71**, 232–261 (1988)

[MP1] Milman, V.D., Pajor, A.: Isotropic position and inertia ellipsoids and zonoids of the unit ball of a normed n-dimensional space. Geometric Aspects of Functional Analysis (1987–88), 64–104, Lecture Notes in Math., **1376**. Springer, Berlin (1989)

[MP2] Milman, V.D., Pajor, A.: Entropy and asymptotic geometry of non-symmetric convex bodies. Adv. Math., **152**, no. 2, 314–335 (2000)

[Z]   Zvavitch, A.: Gaussian measure of sections of convex bodies. Adv. Math., to appear

Israel GAFA Seminar

(2002–2004)

Friday, November 8, 2002

1. *B. Klartag* (Tel Aviv): Isomorphic Steiner symmetrization (joint with V. Milman)
2. *S. Artstein* (Tel Aviv): Entropy increases at every step (joint work with K. Ball, F. Barthe and A. Naor)

Friday, November 29, 2002

1. *D. Leviatan* (Tel Aviv): Introduction to 'best bases selection for approximation in L_p'
2. *M. Sodin* (Tel Aviv): How to control smooth functions? (joint work with F. Nazarov and A. Volberg)

Friday, March 14, 2003

1. *B. Klartag* (Tel Aviv): A reduction of the slicing problem to finite volume ratio bodies (joint with J. Bourgain and V. Milman)
2. *G. Schechtman* (Weizmann Institute): An integral version of Kashin's theorem regarding Euclidean orthogonal splittings of L_1^{2n}

Friday, March 28, 2003

1. *S. Artstein* (Tel Aviv): Recent progress on the duality of entropy numbers conjecture (joint work with V. Milman and S. Szarek)
2. *Y. Lyubich* (Technion, Haifa): Almost Euclidean subspaces of real l_p^n with p an integer

Friday, May 23, 2003

1. *P. Shvartsman* (Technion, Haifa): Barycentric selectors and a Steiner-type point of a convex body in a Banach space
2. *V. Pestov* (University of Ottawa): Ramsey theory and topological dynamics of infinite-dimensional groups

Friday, November 14, 2003

1. *B. Klartag* (Tel Aviv): Spherical harmonics and geometric symmetrization
2. *S. Alesker* (Tel Aviv): Valuations, non-commutative determinants, and pluripotential theory

Friday, December 5, 2003

1. *S. Artstein* (Tel Aviv): Convex separation and the duality of entropy numbers conjecture (joint work with V. Milman, S. Szarek and N. Tomczak-Jaegermann)
2. *A. Giannopoulos* (Crete): Isotropic convex bodies with small diameter (works of A. Giannopoulos and G. Paouris)

Friday, December 19, 2003

1. *M. Sodin* (Tel Aviv): Random complex zeroes (joint work with B. Tsirelson)
2. *S. Mendelson* (Canberra): Empirical minimization from a geometric viewpoint

Friday, January 9, 2004

1. *S. Alesker* (Tel Aviv): What is a continuous valuation?
2. *B. Klartag* (Tel Aviv): An isomorphic version of the slicing problem

Friday, March 12, 2004

1. *S. Artstein* (Tel Aviv): A notion of duality for log-concave measures and a functional version of Santalo inequality (joint work with B. Klartag and V. Milman)
2. *S. Szarek* (Cleveland): The Knaster problem, Grothendieck's theorem and related questions (joint work with B. Kashin)

PIMS*

Thematic Programme on Asymptotic Geometric Analysis at the University of British Columbia** (Summer 2002)

(Organizers: N. Ghoussoub, R. McCann, V. Milman, G. Schechtman, N. Tomczak-Jaegermann)

Conference on Convexity and Asymptotic Theory of Normed Spaces

(Organizers: E. Lutwak and A. Pajor)

Monday, July 1

1. *E. Werner* (Case Western Reserve University): p-affine surface areas and random polytopes
2. *C. Schütt* (Case Western Reserve University): Surface bodies
3. *M. Reitzner* (University of Freiburg): Random polytopes and the Efron–Stein jackknife inequality
4. *K. Leichtweiss*: (University of Stuttgart): Minkowski addition in the hyperbolic plane

Tuesday, July 2

1. *P. Gruber* (Technische Universität): Optimal quantization
2. *D. Hug* (Mathematisches Institut): On the limit shape of the zero cell of a Poisson hyperplane tessellation
3. *W. Banaszczyk* (University of Lodz): Arrangements of signs and rearrangements of vectors in $\mathbb{R}^n$
4. *K. Böröczky* (Renyi Institute of Mathematics): The stability of the Rogers–Shepard and the Zhang projection inequalities

* Pacific Institute for the Mathematical Sciences
** Co-sponsored by the National Science and Engineering Research Council of Canada, with additional support provided by the National Science Foundation, Microsoft, and the Canada Research Chairs Program.

5. *S. Mendelson* (Australian National University): Geometric parameters in learning theory
6. *F. Barthe* (Université de Marne la Vallée): Extremal properties of sections or projections of B_p^n
7. *A. Iosevich* (University of Missouri): Combinatorial approach to orthogonal exponentials
8. *B. Klartag* (Tel Aviv University): Isomorphic Steiner symmetrization
9. *R. Vitale* (University of Connecticut): Two Gaussian notes
10. *A. Zvavitch* (University of Missouri): Volumes of projections of convex bodies via Fourier transform

Wednesday, July 3

1. *R. Schneider* (University of Freiburg): Mixture of convex bodies
2. *S. Reisner* (University of Haifa): An application of convex geometry to approximation theory: approximation by ridge functions
3. *A. Colesanti* (University of Florence): Geometric properties of capacity of convex bodies
4. *W. Weil* (University of Karlsruhe): Directed projection functions
5. *E. Lutwak* (Polytechnic University): L_p-curvature
6. *A. Stancu* (Polytechnic University): An application of curvature flows to the L_0-Minkowski problem
7. *M. Rudelson* (University of Missouri): Asymptotic geometry of non-symmetric convex bodies

Thursday, July 4

1. *K. Ball* (University College London): Convolution inequalities in convex geometry
2. *D. Yang* (Polytechnic University): L_p affine Sobolev–Zhang inequalities
3. *P. Gronchi* (Consiglio Nazionale delle Ricerche): Continuous movements and volume product
4. *M. Fradelizi* (Université de Marne la Vallée): The extreme points of subsets of s-concave probabilities and a geometric localization theorem
5. *V. Milman* (Tel Aviv University): Are randomizing properties of any two convex bodies similar?
6. *M. Ludwig* (Technische Universität Wien): Affinely associated ellipsoids and matrix valued valuations
7. *M. Ghomi* (University of South Carolina): A survey of some recent convexity results and problems in classical differential geometry
8. *E. Grinberg* (Hebrew University of Jerusalem): Inversion of the radon transform via Gärding–Gindikin fractional integrals
9. *B. Rubin* (Hebrew University of Jerusalem): Analytic families associated to radon transforms in integral geometry

Friday, July 5

1. *G. Schechtman* (Weizmann Institute): Non linear type and Pisier's inequality
2. *F. Gao* (University of Idaho): Small ball probability via covering numbers
3. *R. Schneider* (University of Freiburg): A spherical analogue of the Blaschke–Santalo inequality
4. *D. Cordero-Erausquin* (Université de Marne la Vallée): Santaló's inequality on C^n by complex interpolation
5. *A. Giannopoulos* (University of Crete): Ψ_2-estimates for linear functionals and the slicing problem
6. *H. König* (Universität Kiel): Some examples with extremal projection constants
7. *P. Mankiewicz* (Polish Academy of Sciences): On the geometry of proportional quotients of l_1^m
8. *N. Tomczak-Jaegermann* (University of Alberta): Saturation phenomena for normed spaces

Concentration Period on Measure Transportation and Geometric Inequalities

(Organizer: R. McCann)

Monday, July 8

1. *K. Ball* (University College London): Entropy growth for sums of IID random variables
2. *D. Cordero-Erausquin* (Université de Marne-la-Vallée): A new approach to sharp Sobolev and Gaggliardo–Nirenberg inequalities
3. *B. Nazaret* (ENS– Lyon): A new approach to sharp Sobolev and Gagliardo–Nirenberg inequalities
4. *M. Ledoux* (Universite de Toulouse): Measure concentration, transportation cost, and functional inequalities
5. *M. Angeles Hernandez Cifre* (Universidad de Murcia): Complete systems of inequalities
6. *J. Bastero* (Universidad de Zaragoza): Reverse dual isoperimetric inequalities
7. *M. Romance* (Universidad de Zaragoza): Reverse dual isoperimetric inequalities

Tuesday, July 9

1. *F. Barthe* (Université de Marne-la-Vallée): Optimal measure transportation

2. *S. Artstein* (Tel Aviv University): Entropy increases at every step
3. *A. Naor* (Microsoft Corporation): Entropy jumps in the presence of a spectral gap

Wednesday, July 10

1. *V. Milman* (Tel Aviv University): Geometric inequalities of hyperbolic type
2. *Q. Xia* (Rice University): Optimal paths related to transport problems
3. *R. Wagner* (Tel Aviv University): Restriction to a subspace rarely changes measure
4. *M. Hartzoulaki* (University of Crete): On the volume ratio of two convex bodies

Thursday, July 11

1. *R. McCann* (University of Toronto): Nonlinear diffusion to self-similarity: spreading versus shape via gradient flow
2. *J. Denzler* (University of Tennessee): Fast diffusion to self-similarity: complete spectrum, long-time asymptotics and numerology
3. *J.A. Carrillo* (Universidad de Granada): Asymptotic behaviour of fast diffusion equations
4. *C. Borell* (Chalmers University): On risk aversion and optimal terminal wealth
5. *V. Vu* (University of California at San Diego): Concentration of non-Lipschitz functions and combinatorial applications

Friday, July 12

1. *R. Speicher* (Queen's University): Free probability and free diffusion
2. *G. Blower* (Lancaster University): Almost sure weak convergence and concentration for the circular ensembles of Dyson
3. *Y. Brenier* (CNRS): Density and current interpolation

Conference on Phenomena of Large Dimensions

(Organizers: M. Krivelevich, L. Lovasz, V. Milman and L. Pastur)

Monday, July 15

1. *Y. Brenier* (CNRS): On optimal transportation theory
2. *F. Barthe* (Université de Marne la Vallée): Transportation versus rearrangement

3. *E. Gluskin* (Tel Aviv University): On the sections of product spaces and related topics
4. *C. McDiarmid* (Oxford University): Concentration and random permutations
5. *V. Vu* (University of California at San Diego): Divide and conquer martingales and thin waring bases
6. *P. Tetali* (Georgia Tech): The subgaussian constant and concentration inequalities

Tuesday, July 16

1. *V. Milman* (Tel Aviv University): Some phenomena of large dimension in convex geometric analysis
2. *N. Tomczak-Jaegermann* (University of Alberta): Families of random sections of convex bodies
3. *K. Oleszkiewicz* (Warsaw University): On a non-symmetric version of the Khinchine–Kahane inequality
4. *L. Pastur* (Université Pierre & Marie Curie): Some large dimension problems of mathematical physics
5. *J. Han Kim* (Microsoft Research): The Poisson cloning model for random graphs with applications to k-core problems, random 2-SAT, and random digraphs
6. *N. Wormald* (University of Melbourne): The two-choice paradigm and altered thresholds
7. *S. Vempala* (MIT): A spectral algorithm for learning a mixture of isotropic distributions

Wednesday, July 17

1. *N. Berger* (University of California, Berkeley): Phase transition for the biased random walk on percolation clusters
2. *G. Slade* (University of British Columbia): The percolation phase transition on the n-cube
3. *J. Chayes* (Microsoft Research): Graphical models of the Internet and the Web
4. *C. Borgs* (Microsoft Research): Phase transition in the random partitioning problem
5. *M. Csornyei* (University College London): Structure of null sets and related problems of geometric measure theory
6. *M.-C. Chang* (University of California, Riverside): Recent results in combinatorial number theory

Thursday, July 18

1. *N. Alon* (Tel Aviv University): (n, d, λ)-graphs in extremal combinatorics
2. *B. Sudakov* (Princeton University): On the Ramsey- and Turan-type problems
3. *B. Reed* (McGill University): Crayola and dice: graph colouring via the probabilistic method
4. *M. Simonovits* (Hungarian Academy of Science): Introduction to the Szemeredi regularity lemma
5. *Y. Kohyakawa* (University of San Paulo): The regularity lemma for sparse graphs
6. *A. Rucinski* (Adam Mickiewicz University): Ramsey properties of random structures
7. *R. Vershynin* (University of Alberta): Entropy and volume estimates

Friday, July 19

1. *G. Kalai* (Hebrew University): Results and problems around Borsuk's conjecture
2. *I. Barany* (University College London): Sylvester's question, convex bodies, limit shape
3. *A. Giannopoulos* (University of Crete): Random sections and random rotations of high dimensional convex bodies
4. *G. Schechtman* (Weizmann Institute): l_p^n, $1 < p < 2$, well embed in l_1^{an}, for any $a > 1$
5. *M. Rudelson* (University of Missouri): Distances between sections of convex bodies
6. *A. Naor* (Microsoft Corporation): Metric Ramsey-type phenomena
7. *A. Barvinok* (University of Michigan): How to compute a norm?

Monday, July 22

1. *S. Szarek* (Université Paris VI): On pseudorandom matrices
2. *R. Latala* (Warsaw University): Some estimates of norms of random matrices (non iid case)
3. *A. Soshnikov* (University of California at Davis): On the largest eigenvalue of a random subgraph of the hypercube
4. *J. Bourgain* (Institute for Advanced Study): New results on Green's functions and spectra for discrete Schroedinger operators
5. *M. Sodin* (Tel Aviv University): Zeroes of random analytic functions
6. *I. Laba* (University of British Columbia): Tiling problems and spectral sets

Tuesday, July 23

1. *A. Brieden* (TU München): Computing maximal independent sets in graphs by means of polytopal approximations of l_p-spheres
2. *A. Wigderson* (Institute for Advanced Studies): Expander graphs - where combinatorics and algebra compete and cooperate
3. *M. Krivelevich* (Tel Aviv University): Algorithmic applications of graph eigenvalues and related parameters
4. *L. Lovasz* (Microsoft Research): Discrete analytic functions and global information from local observation
5. *R. Kannan* (Yale University): Random submatrices of a given matrix
6. *S. Safra* (Tel Aviv University): Probabilistically checkable proofs (PCP) and hardness of approximation
7. *V. Klee* (University of Washington): Complexity questions involving largest or smallest simplices

Conference on Non-commutative Phenomena and Random Matrices

(Organizers: G. Pisier and S. Szarek)

Tuesday, August 6

1. *E. Effros* (UCLA): Operator spaces as 'quantized' Banach spaces
2. *H. Rosenthal* (Austin): Can non-commutative L^p spaces be renormed to be stable?
3. *E. Ricard* (Paris 6): Hilbertian operator spaces with few completely bounded maps
4. *D. Blecher* (Houston): Noncommutative M-structure and the interplay of algebra and norm for operator algebras
5. *N. Randrianantoanina* (Miami University, Ohio): Weak type inequalities for non-commutative martingales
6. *A. Gamburd* (Stanford): Random matrices and magic squares

Wednesday, August 7

1. *M.B. Ruskai* (Lowell): The role of maximal L_p bounds in quantum information theory
2. *C. Le Merdy* (Besançon): Holomorphic functional calculus and square functions on non-commutative L_p-spaces
3. *F. Lehner* (Graz): A good formula for noncommutative cumulants
4. *M. Junge* (Urbana): The central limit procedure for noncommuting random variables and applications

5. *K. Jung* (Berkeley): Free entropy dimension and hyperfinite von Neumann algebras
6. *M. Kaneda* (Houston): The ideal envelope of an operator algebra
7. *A. Nou* (Besançon): The non injectivity property for algebras of deformed Gaussians

Thursday, August 8

1. *A. Soshnikov* (University of California, Davis): Determinantal random point fields
2. *Z. Jin Ruan* (Urbana): On real operator spaces
3. *A. Nica* (Waterloo): 2-point functions for multi-matrix models, and non-crossing partitions in an annulus
4. *B. Russo* (University of California, Irvine): On contractively complemented subspaces of C^*-algebras
5. *F. Sukochev* (Adelaide): Perturbation and differentiation formulae in non-commutative L_p spaces
6. *G. Aubrun* (ENS Paris): A small deviation inequality for the largest eigenvalue of a random matrix
7. *M. Meckes* (Case Western Reserve): Concentration of the norm of a random matrix

Friday, August 9

1. *N. Ozawa* (Tokyo): $B(H) \otimes B(H)$ fails the WEP
2. *B. Collins* (ENS Paris): Itzykson–Zuber integrals and free probability
3. *R. Speicher* (Queen's University, Kingston): Maximization of free entropy
4. *Q. Xu* (Besançon): On the maximality of subdiagonal algebras

Conference on Banach Spaces

(Organizers: W.B. Johnson and E. Odell)

Monday, August 12

1. *Y. Benyamini* (Technion): An introduction to the uniform classification of Banach spaces
2. *J. Lindenstrauss* (Hebrew University of Jerusalem): On Frechet differentiability of Lipschitz functions, part I
3. *J. Lindenstrauss* (Hebrew University of Jerusalem): On Frechet differentiability of Lipschitz functions, part II
4. *B. Randrianantoanina* (Miami University): The structure of level sets of Lipschitz quotients

5. *D. Alspach* (Oklahoma State University): Embedding $(p, 2)$-partition and weight spaces into L_p
6. *V. Troitsky* (University of Texas, Austin): Semi-transitive operator algebras
7. *T. Kuchurenko* (University of Missouri): Weak topologies and properties that are fulfilled almost everywhere

Tuesday, August 13

1. *V. Zizler* (University of Alberta): Sigma shrinking Markushevich bases and Corson compacts
2. *T. Schlumprecht* (Texas A & M University): How many operators do there exist on a Banach space?
3. *A. Zsak* (Texas A & M University): Banach spaces with few operators
4. *S. Dilworth* (University of South Carolina): Bases for which the greedy algorithm works
5. *J. Duda* (Charles University): On the set of points that have no nearest point in a closed set

Wednesday, August 14

1. *T. Figiel* (Polish Academy of Sciences): Selecting unconditional basic sequences
2. *V. Farmaki* (Athens University): Baire-1 functions and spreading models
3. *M. Grzech* (Polish Academy of Sciences): Enflo and Rosenthal were right: there do not exist unconditional bases in L_p spaces for nonseparable measures (joint work with R. Frankiewicz and R. Komorowski)
4. *M. Ostrovskii* (Catholic University of America): Minimal volume sufficient enlargements of unit balls of finite dimensional normed spaces
5. *A. Koldobsky* (University of Missouri): Fourier analytic tools in the study of sections and projections of convex bodies
6. *M. Hoffman* (University of Missouri): The Banach envelope of Paley-Wiener type spaces E_p for $0 < p < 1$

Thursday, August 15

1. *L. Tzafriri* (Hebrew University of Jerusalem): Lambda$_p$ sets for some orthogonal systems
2. *G. Androulakis* (University of South Carolina): The method of minimal vectors
3. *A. Martinez* (Universidad de Oviedo): Some aspects of semigroups of operators concerning local theory in Banach spaces
4. *A. Litvak* (University of Alberta): Randomized isomorphic Dvoretzky theorem (joint work with P. Mankiewicz and N. Tomczak-Jaegermann)

5. *R. Anisca* (University of Alberta): Local unconditional structure in subspaces of $l_2(X)$

6. *H. Rosenthal* (University of Texas): The asymptotic structure space of a Banach space

7. *N. Dow* (Oxford University): Smooth renormings of asymptotic-c_0 spaces

Banach Spaces and Convex Geometric Analysis

(April, 2003)

(Organizers: H. Konig and J. Lindenstrauss; Sponsored by University of Kiel and the Landau Center of the Hebrew University, Jerusalem)

Monday, April 7

1. *G. Schechtman* (Weizmann Institute): Integral Kashin splittings
2. *S. Artstein* (Tel Aviv University): Recent progress on the duality of entropy numbers conjecture
3. *A. Koldobsky* (University of Missouri): The Busemann-Petty problem via spherical harmonics
4. *B. Klartag* (Tel Aviv University): A reduction of the slicing problem to finite volume ratio bodies
5. *J. Bastero* (Universidad de Zaragoza): 1-position and isotropy
6. *J.M. Wills* (Mathematisches Institut): On a sphere packing inequality
7. *S. Reisner* (University of Haifa): On the best approximation by ridge functions in the uniform norm

Tuesday, April 8

1. *R. Schneider* (University of Freiburg): Stability versions of inequalities, applied to asymptotic shapes of random cells
2. *D. Hug* (Mathematisches Institut): Large cells in Poisson–Delaunay tessellations
3. *S. Szarek* (Université Paris VI): The Knaster problem and the geometry of high-dimensional cubes
4. *A. Giannopoulos* (University of Crete): The central limit property of convex bodies
5. *B. Kashin* (Steklov Mathematical Institute): Lower estimates for n-term approximation and complexity of functional classes
6. *M. Rudelson* (University of Missouri): Random coordinate projections
7. *C. Michels* (Oldenburg University): Gelfand numbers of finite-dimensional identity operators

Wednesday, April 9

1. *V.D. Milman* (Tel Aviv University): Ellipsoids and isomorphic ellipsoids in asymptotic theory: some facts and questions
2. *M. Ludwig* (Technische Universität Wien): Projection bodies and valuations
3. *N. Tomczak-Jaegermann* (University of Alberta): Saturating constructions for normed spaces
4. *A. Litvak* (University of Alberta): When random proportional subspaces are also random quotients

Thursday, April 10

1. *M. Reitzner* (University of Freiburg): The combinatorial structure of random polytopes
2. *E. Werner* (Case Western Reserve University): Surface bodies and p-affine surface area
3. *A. Naor* (Microsoft Corporation): Recent progress in finite metric space theory
4. *F. Barthe* (Université de Marne la Vallée): Sobolev inequalities for probability measures on the line
5. *D. Cordero-Erausquin* (Université de Marne la Vallée): Entropic inequalities by mass transport
6. *R. Latala* (Warsaw University): Moment estimates for multidimensional chaoses generated by positive random variables with logarithmically concave tails
7. *R. Vershynin* (University of Alberta): Sections of convex bodies via combinatorics
8. *T. Kühn* (Mathematisches Institut): Compact embeddings of Besov spaces in Lorentz–Zygmund spaces

Friday, April 11

1. *K.M. Ball* (University College London): Information decrease along semigroups
2. *M. Meyer* (Université de Marne la Vallée): Positions of maximal volume
3. *A. Defant* (Oldenburg University): Unconditionality in spaces of m-homogeneous polynomials
4. *P. Mankiewicz* (Polish Academy of Sciences): Volumetric invariants and operators on random families of Banach spaces
5. *A. Khrabrov* (St. Petersburg State University):Modified Banach–Mazur distance

Paris GAFA Seminar

(Summer 2003)

(Organizer: V. Milman; supported by University Paris VI and IHES)

July 1, 2003

D. *Cordero* (Marne-la-Vallee): Prekopa–Leindler inequalities for potentials on Riemannian manifolds

July 8, 2003

C. *Richter* (Paris): New counterexamples to Knaster's conjecture

July 15, 2003

B. *Klartag* (Tel Aviv): A geometric inequality and a low M estimate

July 21, 2003

F. *Barthe* (Marne-la-Vallee): Hardy and Sobolev inequalities

July 29, 2003

B. *Klartag* (Tel Aviv): Rate of convergence of geometric symmetrization

July 30, 2003

A. *Koldobsky* (Columbia, Missouri): Cube slicing

August 12, 2003

B. *Klartag* (Tel Aviv): Fast multiplication of large numbers (after a survey of results by Knuth)

August 26, 2003

B. *Klartag* (Tel Aviv): An observation related to John ellipsoid

September 2, 2003

O. *Guédon* (Paris): Perturbation of log-concave probabilities

GAFA Session

Joint Meeting of the New Zealand Mathematical Society and Israel Mathematical Union
(Wellington, February 2004)

(Organizers: V. Milman and V. Pestov; supported in part by the Foreign Ministry of Israel and by the Royal Society of New Zealand through the Marsden Grant)

Monday, February 9

1. *B. Klartag* (Tel Aviv University): An isomorphic version of the slicing problem
2. *V. Milman* (Tel Aviv University): Symmetrizations and isotropicity
3. *S. Mendelson* (Australian National University): Lipschitz embeddings of function classes
4. *N. Tomczak-Jaegermann* (University of Alberta): Geometric inequalities for a class of exponential measures
5. *A. Litvak* (University of Alberta): Random matrices and geometry of random polytopes

Tuesday, February 10

1. *G. Martin* (University of Auckland): Holomorphic motions
2. *V. Pestov* (University of Ottawa): Oscillation stability in topological groups
3. *M. Harmer* (University of Auckland): Spectral properties of the triangle groups
4. *D. Yost* (University of Ballarat): Reducible convex sets
5. *D. Gauld* (University of Auckland): Function spaces and metrisability of manifolds
6. *T. Rangiwhetu* (Victoria University of Wellington): Concentration of measure in Banach spaces and criteria for amenability

Thursday, February 12

1. *E. Khmaladze* (Victoria University of Wellington): Local Poisson processes in the neighbourhood of convex body

2. *M. Hegland* (Australian National University): On the sparse grid approximation in radial basis function spaces – an application of the concentration function

3. *B. Pavlov* (University of Auckland): Modelling of quantum networks

4. *A. Stojmirovic* (Victoria University of Wellington): Quasi-metric spaces with measure

5. *C. Atkin* (Victoria University of Wellington): Weakly compact Banach lie groups

Printing and Binding: Strauss GmbH, Mörlenbach